An Introduction to Modern Econometrics Using Stata

An Introduction to Modern Econometrics Using Stata

CHRISTOPHER F. BAUM
Department of Economics
Boston College

A Stata Press Publication
StataCorp LP
College Station, Texas

Stata Press, 4905 Lakeway Drive, College Station, Texas 77845

Typeset in $\LaTeX\,2_\varepsilon$
Printed in the United States of America
10 9 8 7 6 5 4

ISBN-10: 1-59718-013-0
ISBN-13: 978-1-59718-013-9

I have incurred many intellectual debts during the creation of this book. Bill Gould, David Drukker, and Vince Wiggins of StataCorp have been enthusiastic about the need for such a book in the economics and finance community. David Drukker, Vince Wiggins, Gabe Waggoner, and Brian Poi have provided invaluable editorial comments throughout the process. My coauthors of various Stata routines—Nicholas J. Cox, Mark Schaffer, Steven Stillman, and Vince Wiggins—have contributed a great deal, as have many other members of the Stata user community via their own routines, advice, and Statalist inquiries. Dr. Chuck Chakraborty has been helpful in identifying topics of interest to the consulting community. Dr. Petia Petrova provided a thoughtful review of portions of the manuscript.

At Boston College, I must thank Nadezhda Karamcheva for able research assistance in constructing example datasets, as well as colleagues in Research Services for many useful conversations on using statistical software. I am deeply thankful to Academic Vice President John Neuhauser and Dean Joseph Quinn of the College of Arts and Sciences for their acceptance of this project as a worthy use of my fall 2004 sabbatical term.

I have adapted some materials in the book from course notes for undergraduate and graduate-level econometrics. I thank many generations of Boston College students who have pressed me to improve the clarity of those notes and helped me to understand the aspects of theoretical and applied econometrics that are the most difficult to master.

Last but by no means least, I am most grateful to my wife Paula Arnold for graciously coping with a grumpy author day after day and offering encouragement (and occasionally grammar tips) throughout the creative process.

Christopher F. Baum

Oak Square School
Brighton, Massachusetts
July 2006

Contents

Illustrations

Table

Figures

Preface

This book is a concise guide for applied researchers in economics and finance to learn basic econometrics and use Stata with examples using typical datasets analyzed in economics. Readers should be familiar with applied statistics at the level of a simple linear regression (ordinary least squares, or OLS) model and its algebraic representation, equivalent to the level of an undergraduate statistics/econometrics course sequence.[1] The book also uses some multivariate calculus (partial derivatives) and linear algebra.

I presume that the reader is familiar with Stata's windowed interface and with the basics of data input, data transformation, and descriptive statistics. Readers should consult the appropriate *Getting Started with Stata* manual if review is needed. Meanwhile, readers already comfortable interacting with Stata should feel free to skip to chapter 4, where the discussion of econometrics begins in earnest.

In any research project, a great deal of the effort is involved with the preparation of the data specified as part of an econometric model. While the primary focus of the book is placed upon applied econometric practice, we must consider the considerable challenges that many researchers face in moving from their original data sources to the form needed in an econometric model—or even that needed to provide appropriate tabulations and graphs for the project. Accordingly, Chapter 2 focuses on the details of data management and several tools available in Stata to ensure that the appropriate transformations are accomplished accurately and efficiently. If you are familiar with these aspects of Stata usage, you should feel free to skim this material, perhaps returning to it to refresh your understanding of Stata usage. Likewise, Chapter 3 is devoted to a discussion of the organization of economic and financial data, and the Stata commands needed to reorganize data among the several forms of organization (cross section, time series, pooled, panel/longitudinal, etc.) If you are eager to begin with the econometrics of linear regression, skim this chapter, noting its content for future reference.

Chapter 4 begins the econometric content of the book and presents the most widely used tool for econometric analysis: the multiple linear regression model applied to continuous variables. The chapter also discusses how to interpret and present regression estimates and discusses the logic of hypothesis tests and linear and nonlinear restrictions. The last section of the chapter considers residuals, predicted values, and marginal effects.

Applying the regression model depends on some assumptions that real datasets often violate. Chapter 5 discusses how the crucial zero-conditional-mean assumption of the errors may be violated in the presence of specification error. The chapter also

1. Two excellent texts at this level are Wooldridge (2006) and Stock and Watson (2006).

discusses statistical and graphical techniques for detecting specification error. Chapter 6 discusses other assumptions that may be violated, such as the assumption of independent and identically distributed (i.i.d.) errors, and presents the generalized linear regression model. It also explains how to diagnose and correct the two most important departures from i.i.d., heteroskedasticity and serial correlation.

Chapter 7 discusses using indicator variables or dummy variables in the linear regression models containing both quantitative and qualitative factors, models with interaction effects, and models of structural change.

Many regression models in applied economics violate the zero-conditional-mean assumption of the errors because they simultaneously determine the response variable and one or more regressors or because of measurement error in the regressors. No matter the cause, OLS techniques will no longer generate unbiased and consistent estimates, so you must use instrumental-variables (IV) techniques instead. Chapter 8 presents the IV estimator and its generalized method-of-moments counterpart along with tests for determining the need for IV techniques.

Chapter 9 applies models to panel or longitudinal data that have both cross-sectional and time-series dimensions. Extensions of the regression model allow you to take advantage of the rich information in panel data, accounting for the heterogeneity in both panel unit and time dimensions.

Many econometric applications model categorical and limited dependent variables: a binary outcome, such as a purchase decision, or a constrained response such as the amount spent, which combines the decision whether to purchase with the decision of how much to spend, conditional on purchasing. Because linear regression techniques are generally not appropriate for modeling these outcomes, chapter 10 presents several limited-dependent-variable estimators available in Stata.

The appendices discuss techniques for importing external data into Stata and explain basic Stata programming. Although you can use Stata without doing any programming, learning how to program in Stata can help you save a lot of time and effort. You should also learn to generate reproducible results by using do-files that you can document, archive, and rerun. Following Stata's guidelines will make your do-files shorter and easier to maintain and modify.

Notation and typography

I designed this book for you to learn by doing, so I expect you to read this book while sitting at a computer so that you can try the commands in the book to replicate my results. You can then generalize the commands to suit your own needs.

Generally, I use the typewriter font `command` to refer to Stata commands, syntax, and variables. A "dot" prompt followed by a command indicates that you can type what is displayed after the dot (in context) to replicate the results in the book.

I follow some conventions in my mathematical notation to clarify what I mean:

- Matrices are bold, capital letters, such as $\mathbf{X}$.
- Vectors are bold, lowercase letters, such as $\mathbf{x}$.
- Scalars are lowercase letters in standard font, such as x.
- Data vectors ($\mathbf{x}_i$) are $1 \times k$; think of them as being rows from the data matrix.
- Coefficient vectors ($\boldsymbol{\beta}$) are $k \times 1$ column vectors.

The ubiquitous use of N instead of n to denote the sample size forced me to make an exception to convention and let N be the sample size. Similarly, T denotes the number of time-series observations, M is the number of clusters, and L is the maximum lag length. I also follow the universal convention that the Ljung–Box statistic is denoted by Q and similarly denote the difference-in-Sargan test by C.

To simplify the notation, I do not use different fonts to distinguish random variables from their realizations. When one models the dependent variable y, y is the random variable. Observations on y are realizations of this random variable, and I refer to the ith observations on y as y_i and all the observations as $\mathbf{y}$. Similarly, regressors $\mathbf{x}$ are random variables in the population, and I denote the ith observation on this vector of random variables as $\mathbf{x}_i$, which is the ith row of the data matrix $\mathbf{X}$.

This text complements but does not replace the material in the Stata manuals, so I often refer to the Stata manuals by using [R], [P], etc. For example, [R] **xi** refers to the *Stata Base Reference Manual* entry for `xi`, and [P] **syntax** refers to the entry for `syntax` in the *Stata Programming Reference Manual*.

1 Introduction

This book focuses on the tools needed to carry out applied econometric research in economics and finance. These include both the theoretical foundations of econometrics and a solid understanding of how to use those econometric tools in the research process. That understanding is motivated in this book through an integration of theory with practice, using Stata on research datasets to illustrate how those data may be organized, transformed, and used in empirical estimation. My experience in working with students of econometrics and doctoral candidates using econometric tools in their research has been that you learn to *use* econometrics only by *doing* econometrics with realistic datasets. Thankfully, a growing number of introductory econometrics textbooks[1] follow this approach and focus on the theoretical aspects that are likely to be encountered in empirical work. This book is meant to complement those textbooks and provide hands-on experience with a broad set of econometric tools using Stata.

The rest of this chapter presents my "top 11" list of Stata's distinctive features: aspects of Stata's design and capabilities that make the program an excellent tool for applied econometric research. Sections 1.2 and 1.3 provide essential information for those who want to execute the examples used in the text. Many of those examples use user-written Stata commands that must be installed in your copy of Stata. A convenience program described in that section, `itmeus`, will make doing so a painless task.

1.1 An overview of Stata's distinctive features

Stata is a powerful tool for researchers in applied economics. Stata can help you analyze research easily and efficiently—no matter what kind of data you are working with—whether time-series, panel, or cross-sectional data. Stata gives you the tools you need to organize and manage your data and then to obtain and analyze statistical results.

For many users, Stata is a statistical package with menus that allow users to read data, generate new variables, compute statistical analyses, and draw graphs. To others, Stata is a command line–driven package, commonly executed from a do-file of stored commands that will perform all the steps above without intervention. Some consider Stata to be a programming language for developing ado-files that define programs or new Stata commands that extend Stata by adding data-management, statistics, or graphics capabilities.

1. E.g., Wooldridge (2006) and Stock and Watson (2006).

Understanding some of Stata's distinctive features will help you use Stata more effectively and efficiently. You will be able to avoid typing (or copying and pasting) repetitive commands and constantly reinventing the wheel. Learning to write computationally efficient do-files (say, one that runs in 10 seconds rather than in 2 minutes) is helpful, but more importantly you need to be able to write do-files that can be easily understood and modified. This book will save you time by teaching you to generate comprehensible and extensible do-files that you can rerun with one command.

Consider several of Stata's distinctive features, which I discuss in more detail later:

You can easily learn Stata commands, even if you do not know the syntax. Stata has a dialog for almost every official command, and when you execute a command with a dialog, the Review window displays the command syntax, just as if you had typed it. Although you can submit a command without closing the dialog, you will often want to execute several commands in succession (for instance, generating a new variable and then summarizing its values). Even if you are using Stata dialogs, you can reissue, modify, and resubmit commands by using the Review and Command windows. You can save the contents of the Review window to a file or copy them into the Do-file Editor window so that you can modify and resubmit them. To use these options, control-click or right-click on the Review window.

You can use Stata's Do-file Editor to save time developing your analysis. Once you are familiar with common commands, you will find it easier to place them in a do-file and execute that file rather than entering them interactively (using dialogs or the Command window). Using your mouse, you can select any subset of the commands appearing in the Do-file Editor and execute only those commands. That ability makes it easy to test whether these commands will perform the desired analysis. If your do-file does the entire analysis, it provides a straightforward, reproducible, and documented record of your research strategy (especially if you add comments to describe what is being done, by whom, on what date, etc.).

A simple command performs all computations for all the desired observations. Stata differs from several other statistical packages in its approach to variables. When you read in a Stata dataset, Stata puts in memory a matrix with rows corresponding to the observations and columns representing the variables. You can see this matrix by clicking the Data Viewer or Data Editor icon on Stata's toolbar. Most Stata commands do not require you to explicitly specify observations. Unlike with other statistical package languages, few circumstances in Stata require you to refer to the specific observation, and Stata will run much faster if you avoid doing so. When you must explicitly refer to the prior observation's value—for instance, when you are generating a lagged value in time-series data—*always* use Stata's time-series operators, such as `L.x` for the lagged value of `x` or `D.x` for the first difference.

Looping over variables saves time and effort. One of Stata's most valuable features is the ability to repeat steps (data transformations, estimation, or creating graphics) over several variables. The relevant commands are documented in [P] **forvalues**, [P] **foreach**, and [P] **macro**; see the online help (e.g., `help forvalues`) and appendix B for more details. Using these commands can help you produce a do-file that will loop over variables rather than issuing a separate command for each one; you can easily modify your file later if you need a different list of variables; see chapter 2.

Stata's by-groups reduce the need for programming. Stata lets you define *by-groups* from one or more categorical (integer valued) variables, so you can do sophisticated data transformations with short, simple commands; see chapter 2.

Stata has many statistical features that make it uniquely powerful. Stata can calculate robust and cluster–robust estimates of the variance–covariance matrix of the estimator for nearly all the estimation commands.[2] The `mfx` command estimates marginal effects after estimation. `test`, `testnl`, `lincom`, and `nlcom` provide Wald tests of linear and nonlinear restrictions and confidence intervals for linear and nonlinear functions of the estimated parameters.

You can avoid problems by keeping Stata up to date. If you have an Internet connection, Stata's [R] **update** facility periodically updates Stata's executable and ado-files, free of charge. Most updates contain bug fixes and enhancements to existing commands (and sometimes brand-new commands). To find available updates, use the command `update query` and follow its recommendations. Many problems identified by Stata users have already been addressed by updates, so you should always update your Stata executable and ado-files before reporting any apparent error in the program. Be sure to update your copy of Stata when you reinstall the program on a new computer or hard disk since the installation CD contains the original code (i.e., version 9.0 without updates versus version 9.2 with updates, which is available at this writing).

Stata is infinitely extensible. You can create your own commands that are indistinguishable from official Stata commands. You can add a new command to Stata, whether you or someone else developed it, by writing an ado-file and help file. Any properly constructed ado-files on the `adopath` will define new commands with those names, so Stata's capabilities are open ended (see [P] **sysdir**). Since most Stata commands are written in the do-file language, they are available for viewing and modification, and they demonstrate good programming practice.

Stata's user community provides a wealth of useful additions to Stata. StataCorp's development strategy gives users the same development tools used by the company's own professional programmers. This practice has encouraged a vibrant user community of Stata developers who freely share their contributions. Although any Stata developers may set up their own `net from` sites, most user-written programs are available from the Statistical Software Components (SSC) archive that I maintain at Boston College, which you can access by using Stata's `ssc` command; see [R] **ssc**. You can use a web browser to search the SSC archive, but you should use the `ssc` command to download any of its contents to ensure that the files are handled properly and installed in the appropriate directory. Typing `ssc whatsnew` lists recent additions and updates in the SSC archive. Typing `adoupdate` updates the packages you have installed from the SSC archive, the *Stata Journal*, or individual users' sites as needed.

2. Don't worry if you do not know what these are; I discuss them in detail in the text.

Stata is cross-platform compatible. Unlike many statistical packages, Stata's feature set does not differ across the platforms (Windows, Macintosh, Linux, and Unix) on which it runs. The Stata documentation is not platform specific (with the exception of the *Getting Started with Stata* manuals). A do-file that runs on one platform will run on another (as long as each system has enough memory). This compatibility allows you to move binary data files easily among platforms: that is, all Stata `.dta` files have the same binary data format, so any machine running the same version of Stata can read and write to those files. Stata can also read a data file stored on a web server with the command `use http://...` regardless of platform.

Stata can be fun. Although empirical research is serious business, you need only follow a few threads in Statalist[3] discussions to learn that many users greatly enjoy using Stata and participating in the Stata user community. Although learning to use Stata effectively—like learning to speak a foreign language—is hard work, learning to solve data-management and statistical analysis problems is rewarding. Who knows? Someday your colleagues may turn to you, asking for help with Stata.

1.2 Installing the necessary software

This book uses Stata to illustrate many aspects of applied econometric research. As mentioned, Stata's capabilities are not limited to the commands of official Stata documented in the manuals and in online help but include a wealth of commands documented in the *Stata Journal*, *Stata Technical Bulletin*, and the SSC archive.[4] Those commands will not be available in your copy of Stata unless you have installed them. Because the book uses several of those user-written commands to illustrate the full set of tools available to the Stata user, I have provided a utility command, `itmeus`, that will install all the unofficial commands used in the book's examples. To install that command, you must be connected to the Internet and type

```
ssc install itmeus
```

which will retrieve the command from the SSC archive. When the `ssc` command succeeds, you may type

```
help itmeus
```

as you would with any Stata command, or just

```
itmeus
```

to start the download procedure. All necessary commands will be installed in your copy of Stata. Any example in the book (see the next section to obtain the do-files and datasets used to produce the examples) may then be executed.

3. See http://www.stata.com/statalist/.
4. Type `help ssc` for information on the SSC ("Boston College") archive.

Newer versions of the user-written commands that you install today may become available. The official Stata command `adoupdate`, which you may give at any time, will check to see whether newer versions of these user-written commands are available. Just as the command `update query` will determine whether your Stata executable and official ado-files are up to date, `adoupdate` will perform the same check for user-written commands installed in your copy of Stata.

1.3 Installing the support materials

Except for some small expository datasets, all the data I use in this book are freely available for you to download from the Stata Press web site, http://www.stata-press.com. In fact, when I introduce new datasets, I merely load them into Stata the same way that you would. For example,

```
. use http://www.stata-press.com/data/imeus/tablef7-1.dta, clear
```

Try it.

To download the datasets and do-files for this book, type

```
. net from http://www.stata-press.com/data/imeus/
. net describe imeus
. net get imeus-dta
. net get imeus-do
```

The materials will be downloaded to your current working directory. I suggest that you create a new directory and copy the materials there.

2 Working with economic and financial data in Stata

Economic research always involves several data-management tasks, such as data input, validation, and transformation, which are crucial for drawing valid conclusions from statistical analysis of the data. These tasks often take more time than the statistical analyses themselves, so learning to use Stata efficiently can help you perform these tasks and produce a well-documented Stata dataset supporting your research project.

The first section of this chapter discusses the basics of working with data in Stata. Section 2 discusses common data transformations. The third section discusses the types of data commonly used in microeconomic analysis: cross-sectional, time-series, pooled cross-section time-series, and panel (longitudinal) data. Section 4 discusses using do-files to create reproducible research and perform automated data-validation tasks.

2.1 The basics

To effectively manage data with Stata, you will need to understand some of Stata's basic features. A small Stata dataset will illustrate.

2.1.1 The use command

Open an existing Stata data (`.dta`) file with the `use` command. You can specify just the name of the dataset, such as `use census2c`, or give the complete path to the dataset, such as

```
. use "/Users/baum/doc/SFAME/stbook.5725/dof/census2c.dta"
```

depending on your operating system. With the `use` command, you can also open a file on a web server, such as

```
. use http://www.stata-press.com/data/r9/census2
```

In either of these formats, the quotation marks are required if there are spaces in the directory or filenames. If you are using Stata's menus, specify a filename by selecting File ▷ Open.... You can obtain the full file path from the Review window and store it in a do-file.

Let's `use` a Stata data file and `list` its contents:

```
. use http://www.stata-press.com/data/imeus/census2c, clear
(1980 Census data for NE and NC states)
. list, sep(0)
```

	state	region	pop	popurb	medage	marr	divr
1.	Connecticut	NE	3107.6	2449.8	32.00	26.0	13.5
2.	Illinois	N Cntrl	11426.5	9518.0	29.90	109.8	51.0
3.	Indiana	N Cntrl	5490.2	3525.3	29.20	57.9	40.0
4.	Iowa	N Cntrl	2913.8	1708.2	30.00	27.5	11.9
5.	Kansas	N Cntrl	2363.7	1575.9	30.10	24.8	13.4
6.	Maine	NE	1124.7	534.1	30.40	12.0	6.2
7.	Massachusetts	NE	5737.0	4808.3	31.20	46.3	17.9
8.	Michigan	N Cntrl	9262.1	6551.6	28.80	86.9	45.0
9.	Minnesota	N Cntrl	4076.0	2725.2	29.20	37.6	15.4
10.	Missouri	N Cntrl	4916.7	3349.6	30.90	54.6	27.6
11.	Nebraska	N Cntrl	1569.8	987.9	29.70	14.2	6.4
12.	New Hampshire	NE	920.6	480.3	30.10	9.3	5.3
13.	New Jersey	NE	7364.8	6557.4	32.20	55.8	27.8
14.	New York	NE	17558.1	14858.1	31.90	144.5	62.0
15.	N. Dakota	N Cntrl	652.7	318.3	28.30	6.1	2.1
16.	Ohio	N Cntrl	10797.6	7918.3	29.90	99.8	58.8
17.	Pennsylvania	NE	11863.9	8220.9	32.10	93.7	34.9
18.	Rhode Island	NE	947.2	824.0	31.80	7.5	3.6
19.	S. Dakota	N Cntrl	690.8	320.8	28.90	8.8	2.8
20.	Vermont	NE	511.5	172.7	29.40	5.2	2.6
21.	Wisconsin	N Cntrl	4705.8	3020.7	29.40	41.1	17.5

The contents of this dataset, `census2c`, are arranged in tabular format, similar to a spreadsheet. The rows of the table are the *observations*, or cases, and the columns are the *variables*. There are 21 rows, each corresponding to a U.S. state in the Northeast or North Central regions, and seven columns, or variables: `state`, `region`, `pop`, `popurb`, `medage`, `marr`, and `divr`. Variable names must be unique and follow certain rules of syntax. For instance, they cannot contain spaces or hyphens (`-`) or nonalphabetic or nonnumeric characters, and they must start with a letter.[1] Stata is case sensitive, so `STATE`, `State`, and `state` are three different variables to Stata. Stata recommends that you use *lowercase* names for all variables.

2.1.2 Variable types

Unlike some statistical packages, Stata supports a full range of *variable types*. Many of the data econometricians use are *integer* values. They are often very small integers, such as $\{0,1\}$ for indicator (dummy) variables or values restricted to a single-digit range. To save memory and disk space, Stata lets you define variables as integers, reals, or strings, as needed. There are three integer data types: `byte` for one- or two-digit signed integers, `int` for integers up to $\pm 32{,}740$, and `long` for integers up to ± 2.14 billion. Two real data

1. A variable name may start with an underscore (_), but doing so is not a good idea since many Stata programs create temporary variables with names beginning with an underscore.

types are available for decimal values: `float` and `double`. Variables stored as `floats` have seven digits of precision; `double` variables have 15 digits of precision. Numeric variables are stored as `floats`, unless you specify otherwise. For more details, see `data types`.

String variables may optionally be declared as having a specific length, from `str1` to `str244` characters. If you store a string longer than the specified length, Stata automatically increases the storage size of the variable to accommodate that string, up to a maximum of 244 characters.

Typing the `describe` command displays the contents of a dataset, including the data type of each variable. For example,

```
. describe
Contains data from census2c.dta
  obs:            21                          1980 Census data for NE and NC
                                                states
 vars:             7                          9 Jun 2006 14:50
 size:         1,134 (99.9% of memory free)
-------------------------------------------------------------------------------
              storage  display     value
variable name   type   format      label      variable label
-------------------------------------------------------------------------------
state           str13  %-13s                  State
region          byte   %-8.0g      cenreg     Census region
pop             double %8.1f                  1980 Population, '000
popurb          double %8.1f                  1980 Urban population, '000
medage          float  %9.2f                  Median age, years
marr            double %8.1f                  Marriages, '000
divr            double %8.1f                  Divorces, '000
-------------------------------------------------------------------------------
Sorted by:
```

Stata indicates that the dataset contains 21 observations (`obs`) and 7 variables (`vars`). The variable `state` is a `str13`, so no state names in the dataset exceed 13 characters. `pop`, `popurb`, and `divr` are stored as `doubles`, not as integers, because they are expressed in thousands of residents and therefore contain fractional parts if any of their values are not multiples of 1,000. `medage` is stored as a `float`, whereas `region` is stored as a `byte`, although it appears to have values of `NE` and `N Cntrl`; however, these are not the true contents of `region` but rather its *value label*, as described below.

2.1.3 _n and _N

The observations in the dataset are numbered 1, 2, ..., 21 in the list above, so you can refer to an observation by number. Or you can use `_N` to refer to the highest observation number—the total number of observations—and `_n` to refer to the current observation number, although these notations can vary over subgroups in the data; see section 2.2.8. The observation numbers will change if a `sort` command (see section 2.1.5) changes the order of the dataset in memory.

2.1.4 generate and replace

Stata's basic commands for data transformation are `generate` and `replace`, which work similarly, but with some important differences. `generate` creates a *new* variable with a name not currently in use. `replace` modifies an *existing* variable, and unlike other Stata commands, `replace` may not be abbreviated.

To illustrate `generate`, let's create a new variable in our dataset that measures the fraction of each state's population living in urban areas in 1980. We need only specify the appropriate formula and Stata will automatically apply it to every observation specified by the `generate` command according to the rules of algebra. For instance, if the formula would result in a division by zero for a given state, the result for that state would be flagged as missing. We generate the fraction, `urbanized`, and use the `summarize` command to display its descriptive statistics:

```
. generate urbanized = popurb/pop

. summarize urbanized

    Variable |       Obs        Mean    Std. Dev.       Min        Max
-------------+--------------------------------------------------------
   urbanized |        21    .6667691    .1500842   .3377319   .8903645
```

The average state in this part of the United States is 66.7% urbanized, with that fraction ranging from 34% to 89%.

If the `urbanized` variable had already existed, but we wanted to express it as a percentage rather than a decimal fraction, we would use `replace`:

```
. replace urbanized = 100*urbanized
(21 real changes made)

. summarize urbanized

    Variable |       Obs        Mean    Std. Dev.       Min        Max
-------------+--------------------------------------------------------
   urbanized |        21    66.67691    15.00843   33.77319   89.03645
```

`replace` reports the number of changes it made—all 21 observations.

You should write the data transformations as a simple, succinct set of commands that you can easily modify as needed. There are usually several ways to create the same variable with `generate` and `replace`, but stick with the simplest and clearest form of these statements.

2.1.5 sort and gsort

The `sort` command puts observations in the dataset into a certain order; see [D] **sort**. If your `sort` command includes a single variable name, Stata sorts the data in ascending order based on that variable, whether numeric or string. If you indicate a *varlist* of two or more variables, the data are sorted as follows: typing `sort` orders the observations by the first variable, and then, for observations with equal values for the first variable,

Stata orders those observations by the second variable, and so on. After the `sort`, the dataset will be marked as sorted by those variables, but you must save the dataset to disk if you want to keep the new sort order.

You can use the `gsort` command (see [D] **gsort**) to sort data in descending or ascending order, such as when you have quiz scores or patients' blood pressure readings. A minus sign (-) preceding the variable indicates a descending-order sort on that variable, whereas a plus sign (+) indicates an ascending-order sort. For instance, to sort the states by region and, within region, by population from largest to smallest, type

```
. gsort region -pop
. list region state pop, sepby(region)

     +----------------------------------+
     | region   state               pop |
     |----------------------------------|
  1. | NE       New York        17558.1 |
  2. | NE       Pennsylvania    11863.9 |
  3. | NE       New Jersey       7364.8 |
  4. | NE       Massachusetts    5737.0 |
  5. | NE       Connecticut      3107.6 |
  6. | NE       Maine            1124.7 |
  7. | NE       Rhode Island      947.2 |
  8. | NE       New Hampshire     920.6 |
  9. | NE       Vermont           511.5 |
     |----------------------------------|
 10. | N Cntrl  Illinois        11426.5 |
 11. | N Cntrl  Ohio            10797.6 |
 12. | N Cntrl  Michigan         9262.1 |
 13. | N Cntrl  Indiana          5490.2 |
 14. | N Cntrl  Missouri         4916.7 |
 15. | N Cntrl  Wisconsin        4705.8 |
 16. | N Cntrl  Minnesota        4076.0 |
 17. | N Cntrl  Iowa             2913.8 |
 18. | N Cntrl  Kansas           2363.7 |
 19. | N Cntrl  Nebraska         1569.8 |
 20. | N Cntrl  S. Dakota         690.8 |
 21. | N Cntrl  N. Dakota         652.7 |
     +----------------------------------+
```

2.1.6 if exp and in range

Stata commands operate on all observations in memory by default. Almost all Stata commands accept `if` *exp* and `in` *range* clauses, which restrict the command to a subset of the observations.

To list the five U.S. states with the smallest populations, type the `sort` command, followed by `list` with an `in` *range* qualifier specifying the first and last observations to be listed. The range `1/5` refers to the first 5 observations, and the range `-5/l` refers to the last 5 observations (fifth from last, fourth from last, ..., last). To illustrate,

```
. sort pop
. list state region pop in 1/5

     +-------------------------------------+
     | state           region        pop   |
     |-------------------------------------|
  1. | Vermont         NE          511.5   |
  2. | N. Dakota       N Cntrl     652.7   |
  3. | S. Dakota       N Cntrl     690.8   |
  4. | New Hampshire   NE          920.6   |
  5. | Rhode Island    NE          947.2   |
     +-------------------------------------+

. list state region pop in -5/l

     +--------------------------------------+
     | state          region          pop   |
     |--------------------------------------|
 17. | Michigan       N Cntrl      9262.1   |
 18. | Ohio           N Cntrl     10797.6   |
 19. | Illinois       N Cntrl     11426.5   |
 20. | Pennsylvania   NE          11863.9   |
 21. | New York       NE          17558.1   |
     +--------------------------------------+
```

These two lists give us the five smallest and five largest states, but the latter table is in ascending rather than descending order. Since the `sort` command performs only ascending-order sorts, to list the largest states in decreasing order, type

```
. gsort -pop
. list state region pop in 1/5

     +--------------------------------------+
     | state          region          pop   |
     |--------------------------------------|
  1. | New York       NE          17558.1   |
  2. | Pennsylvania   NE          11863.9   |
  3. | Illinois       N Cntrl     11426.5   |
  4. | Ohio           N Cntrl     10797.6   |
  5. | Michigan       N Cntrl      9262.1   |
     +--------------------------------------+
```

To restrict an operation to observations that meet some logical condition, use the `if` *exp* qualifier. For example, to generate a new `medage` variable, `medagel`, defined only for states with populations of more than 5 million, we could specify

```
. generate medagel = medage if pop > 5000
(13 missing values generated)

. sort state

. list state region pop medagel, sep(0)

     +-------------------------------------------------+
     | state           region         pop    medagel |
     |-------------------------------------------------|
  1. | Connecticut     NE          3107.6          . |
  2. | Illinois        N Cntrl    11426.5       29.9 |
  3. | Indiana         N Cntrl     5490.2       29.2 |
  4. | Iowa            N Cntrl     2913.8          . |
  5. | Kansas          N Cntrl     2363.7          . |
  6. | Maine           NE          1124.7          . |
  7. | Massachusetts   NE          5737.0       31.2 |
  8. | Michigan        N Cntrl     9262.1       28.8 |
  9. | Minnesota       N Cntrl     4076.0          . |
 10. | Missouri        N Cntrl     4916.7          . |
 11. | N. Dakota       N Cntrl      652.7          . |
 12. | Nebraska        N Cntrl     1569.8          . |
 13. | New Hampshire   NE           920.6          . |
 14. | New Jersey      NE          7364.8       32.2 |
 15. | New York        NE         17558.1       31.9 |
 16. | Ohio            N Cntrl    10797.6       29.9 |
 17. | Pennsylvania    NE         11863.9       32.1 |
 18. | Rhode Island    NE           947.2          . |
 19. | S. Dakota       N Cntrl      690.8          . |
 20. | Vermont         NE           511.5          . |
 21. | Wisconsin       N Cntrl     4705.8          . |
     +-------------------------------------------------+
```

`medagel` is defined for the states that meet this condition and set to *missing* for all other states (the value `.` is Stata's missing-value indicator). When you use an `if` *exp* clause with `generate`, observations not meeting the logical condition are set to missing.

To calculate summary statistics for `medage` for the larger states, we could either `summarize` the new variable, `medagel`, or apply an `if` *exp* to the original variable:

```
. summarize medagel

    Variable |       Obs        Mean    Std. Dev.       Min        Max
-------------+--------------------------------------------------------
     medagel |         8       30.65    1.363818       28.8       32.2

. summarize medage if pop > 5000

    Variable |       Obs        Mean    Std. Dev.       Min        Max
-------------+--------------------------------------------------------
      medage |         8       30.65    1.363818       28.8       32.2
```

Either method will produce the same statistics for the eight states that meet the logical condition.

2.1.7 Using if exp with indicator variables

Many empirical research projects in economics require an *indicator variable*, which takes on values {0,1} to indicate whether a particular condition is satisfied. These variables are commonly known as *dummy variables* or *Boolean variables*. To create indicator (dummy) variables, use a *Boolean condition*, which evaluates to true or false for each observation. You also need to use the `if` *exp* qualifier. Using our dataset, you could generate indicator variables for small and large states with

```
. generate smallpop = 0

. replace smallpop = 1 if pop <= 5000
(13 real changes made)

. generate largepop = 0

. replace largepop = 1 if pop > 5000
(8 real changes made)

. list state pop smallpop largepop, sep(0)
```

	state	pop	smallpop	largepop
1.	Connecticut	3107.6	1	0
2.	Illinois	11426.5	0	1
3.	Indiana	5490.2	0	1
4.	Iowa	2913.8	1	0
5.	Kansas	2363.7	1	0
6.	Maine	1124.7	1	0
7.	Massachusetts	5737.0	0	1
8.	Michigan	9262.1	0	1
9.	Minnesota	4076.0	1	0
10.	Missouri	4916.7	1	0
11.	N. Dakota	652.7	1	0
12.	Nebraska	1569.8	1	0
13.	New Hampshire	920.6	1	0
14.	New Jersey	7364.8	0	1
15.	New York	17558.1	0	1
16.	Ohio	10797.6	0	1
17.	Pennsylvania	11863.9	0	1
18.	Rhode Island	947.2	1	0
19.	S. Dakota	690.8	1	0
20.	Vermont	511.5	1	0
21.	Wisconsin	4705.8	1	0

You need to use both **generate** and **replace** to define both the 0 and 1 values. Typing `generate smallpop = 1 if pop <= 5000` would set the variable `smallpop` to missing, not zero, for all observations that did not meet the `if` *exp*. Using a Boolean condition is easier:

```
. generate smallpop = (pop <= 5000)

. generate largepop = (pop > 5000)
```

But if you use this approach, any values of `pop` that are missing ([U] **12.2.1 Missing values**) will be coded as 1 in the variable `largepop` and 0 in the variable `smallpop` since all Stata's missing-value codes are represented as the largest positive number. To resolve this problem, add an `if` *exp*: `if pop < .` statement to the **generate** statements so that measurable values are less than the missing value:

```
. generate smallpop = (pop <= 5000) if pop < .
. generate largepop = (pop > 5000) if pop < .
```

Even if you think your data do not contain missing values, you should account for any missing values by using `if` *exp* qualifiers.

Make sure that you use the `if` *exp* in the example outside the Boolean expression: placing the `if` *exp* inside the Boolean expression (e.g., `(pop > 5000 & pop < .)`) would assign a value of 0 to `largepop` for missing values of `pop`. Properly used, the `if` *exp* qualifier will cause any missing values of `pop` to be correctly reflected in `largepop`.[2]

2.1.8 Using if exp versus by varlist: with statistical commands

You can also use the `if` *exp* qualifier to perform a statistical analysis on a subset of the data. You can `summarize` the data for each `region`, where `NE` is coded as region 1 and `N Cntrl` is coded as region 2, by using an `if` *exp*:

```
. summarize medage marr divr if region==1

    Variable |       Obs        Mean    Std. Dev.       Min        Max
-------------+--------------------------------------------------------
      medage |         9    31.23333    1.023474       29.4       32.2
        marr |         9    44.47922    47.56717      5.226    144.518
        divr |         9    19.30433    19.57721      2.623     61.972

. summarize medage marr divr if region==2

    Variable |       Obs        Mean    Std. Dev.       Min        Max
-------------+--------------------------------------------------------
      medage |        12      29.525    .7008113       28.3       30.9
        marr |        12    47.43642    35.29558      6.094    109.823
        divr |        12    24.33583      19.684      2.142     58.809
```

If your data have discrete categories, you can use Stata's `by` *varlist*`:` prefix instead of the `if` *exp* qualifier.

If you use `by` *varlist*`:` with one or more categorical variables, the command is repeated automatically for each value of the `by` *varlist*`:`, no matter how many subsets are expressed by the `by` *varlist*`:`. However, `by` *varlist*`:` can execute only one command.

To illustrate how to use `by` *varlist*`:`, let's generate the same summary statistics for the two census regions:

2. An even simpler approach would be to type `generate largepop = 1 - smallpop`. If you properly define `smallpop` to handle missing values, the algebra of the `generate` statement will ensure that they are handled in `largepop`, since any function of missing data produces missing data.

```
. by region, sort: summarize medage marr divr

-------------------------------------------------------------------------------
-> region = NE
    Variable |       Obs        Mean    Std. Dev.       Min        Max
-------------+--------------------------------------------------------
      medage |         9    31.23333    1.023474       29.4       32.2
        marr |         9    44.47922    47.56717      5.226    144.518
        divr |         9    19.30433    19.57721      2.623     61.972

-------------------------------------------------------------------------------
-> region = N Cntrl
    Variable |       Obs        Mean    Std. Dev.       Min        Max
-------------+--------------------------------------------------------
      medage |        12      29.525    .7008113       28.3       30.9
        marr |        12    47.43642    35.29558      6.094    109.823
        divr |        12    24.33583      19.684      2.142     58.809
```

Here we needed to `sort` by `region` with the `by` *varlist*`:` prefix. The statistics indicate that Northeasterners are slightly older than those in North Central states, although the means do not appear to be statistically distinguishable.

Do not confuse the `by` *varlist*`:` prefix with the `by()` option available on some Stata commands. For instance, we could produce the summary statistics for `medage` by using the `tabstat` command, which also generates statistics for the entire sample:

```
. tabstat medage, by(region) statistics(N mean sd min max)

Summary for variables: medage
     by categories of: region (Census region)

  region |         N      mean        sd       min       max
---------+--------------------------------------------------
      NE |         9  31.23333  1.023474      29.4      32.2
 N Cntrl |        12    29.525  .7008113      28.3      30.9
---------+--------------------------------------------------
   Total |        21  30.25714  1.199821      28.3      32.2
------------------------------------------------------------
```

Using `by()` as an option modifies the command, telling Stata that we want to compute a table with summary statistics for each `region`. On the other hand, the `by` *varlist*`:` prefix used above repeats the entire command for each value of the by-group.

The `by` *varlist*`:` prefix may include more than one variable, so all combinations of the variables are evaluated, and the command is executed for each combination. Say that we combine `smallpop` and `largepop` into one categorical variable, `popsize`, which equals 1 for small states and 2 for large states. Then we can compute summary statistics for small and large states in each `region`:

```
. generate popsize = smallpop + 2*largepop
. by region popsize, sort: summarize medage marr divr

-> region = NE, popsize = 1
    Variable |       Obs        Mean    Std. Dev.       Min        Max
-------------+--------------------------------------------------------
      medage |         5       30.74    1.121606       29.4         32
        marr |         5      12.011    8.233035      5.226     26.048
        divr |         5      6.2352    4.287408      2.623     13.488

-> region = NE, popsize = 2
    Variable |       Obs        Mean    Std. Dev.       Min        Max
-------------+--------------------------------------------------------
      medage |         4       31.85    .4509245       31.2       32.2
        marr |         4     85.0645    44.61079     46.273    144.518
        divr |         4    35.64075    18.89519     17.873     61.972

-> region = N Cntrl, popsize = 1
    Variable |       Obs        Mean    Std. Dev.       Min        Max
-------------+--------------------------------------------------------
      medage |         8     29.5625    .7998885       28.3       30.9
        marr |         8    26.85387    16.95087      6.094     54.625
        divr |         8    12.14637    8.448779      2.142     27.595

-> region = N Cntrl, popsize = 2
    Variable |       Obs        Mean    Std. Dev.       Min        Max
-------------+--------------------------------------------------------
      medage |         4       29.45    .5446711       28.8       29.9
        marr |         4     88.6015    22.54513     57.853    109.823
        divr |         4    48.71475    8.091091     40.006     58.809
```

The youngest population is found in large North Central states. Remember that large states have `popsize` = 2. We will see below how to better present the results.

2.1.9 Labels and notes

Stata makes it easy to provide labels for the dataset, for each variable, and for each value of a categorical variable, which will help readers understand the data. To label the dataset, use the `label` command:

```
. label data "1980 US Census data with population size indicators"
```

The new label overwrites any previous dataset label.

Say that we want to define labels for the `urbanized`, `smallpop`, `largepop`, and `popsize` variables:

```
. label variable urbanized "Population in urban areas, %"
. label variable smallpop "States with <= 5 million pop, 1980"
. label variable largepop "States with > 5 million pop, 1980"
. label variable popsize "Population size code"
```

```
. describe pop smallpop largepop popsize urbanized

              storage  display     value
variable name   type   format      label      variable label
-------------------------------------------------------------------------------
pop             double %8.1f                  1980 Population, '000
smallpop        float  %9.0g                  States with <= 5 million pop,
                                                1980
largepop        float  %9.0g                  States with > 5 million pop,
                                                1980
popsize         float  %9.0g                  Population size code
urbanized       float  %9.0g                  Population in urban areas, %
```

Now if we give this dataset to another researcher, the researcher will know how we defined `smallpop` and `largepop`.

Last, consider value labels, such as the one associated with the `region` variable:

```
. describe region

              storage  display     value
variable name   type   format      label      variable label
-------------------------------------------------------------------------------
region          byte   %-8.0g      cenreg     Census region
```

`region` is a `byte` (integer) variable with the *variable label* `Census region` and the *value label* `cenreg`. Unlike other statistical packages, Stata's value labels are not specific to a particular variable. Once you define a label, you can assign it to any number of variables that share the same coding scheme. Let's examine the `cenreg` value label:

```
. label list cenreg
cenreg:
           1 NE
           2 N Cntrl
           3 South
           4 West
```

`cenreg` contains codes for four Census regions, only two of which are represented in our dataset.

Because `popsize` is also an integer code, we should document its categories with a value label:

```
. label define popsize 1 "<= 5 million" 2 "> 5 million"
. label values popsize popsize
```

We can confirm that the value label was added to `popsize` by typing the following:

```
. describe popsize

              storage  display     value
variable name   type   format      label      variable label
-------------------------------------------------------------------------------
popsize         float  %12.0g      popsize    Population size code
```

To view the mean for each of the values of `popsize`, type

```
. by popsize, sort: summarize medage

-------------------------------------------------------------------------------
-> popsize = <= 5 million
    Variable |       Obs        Mean    Std. Dev.       Min        Max
-------------+--------------------------------------------------------
      medage |        13    30.01538    1.071483       28.3         32

-------------------------------------------------------------------------------
-> popsize = > 5 million
    Variable |       Obs        Mean    Std. Dev.       Min        Max
-------------+--------------------------------------------------------
      medage |         8       30.65    1.363818       28.8       32.2
```

The smaller states have slightly younger populations.

You can use the `notes` command to add notes to a dataset and individual variables (think of sticky notes, real or electronic):

```
. notes: Subset of Census data, prepared on TS for Chapter 2
. notes medagel: median age for large states only
. notes popsize: variable separating states by population size
. notes popsize: value label popsize defined for this variable
. describe
Contains data from census2c.dta
  obs:            21                          1980 US Census data with
                                                population size indicators
 vars:            12                          9 Jun 2006 14:50
 size:         1,554 (99.9% of memory free)   (_dta has notes)
-------------------------------------------------------------------------------
              storage  display     value
variable name   type   format      label      variable label
-------------------------------------------------------------------------------
state           str13  %-13s                  State
region          byte   %-8.0g      cenreg     Census region
pop             double %8.1f                  1980 Population, '000
popurb          double %8.1f                  1980 Urban population, '000
medage          float  %9.2f                  Median age, years
marr            double %8.1f                  Marriages, '000
divr            double %8.1f                  Divorces, '000
urbanized       float  %9.0g                  Population in urban areas, %
medagel         float  %9.0g                *
smallpop        float  %9.0g                  States with <= 5 million pop,
                                                1980
largepop        float  %9.0g                  States with > 5 million pop,
                                                1980
popsize         float  %12.0g      popsize  * Population size code
                                            * indicated variables have notes
-------------------------------------------------------------------------------
Sorted by:  popsize
     Note:  dataset has changed since last saved
```

```
. notes
_dta:
  1.  Subset of Census data, prepared on 9 Jun 2006 14:50 for Chapter 2
medagel:
  1.  median age for large states only
popsize:
  1.  variable separating states by population size
  2.  value label popsize defined for this variable
```

The string `TS` in the first note is automatically replaced with a time stamp.

2.1.10 The varlist

Many Stata commands accept a *varlist*, a list of one or more variables to be used. A *varlist* may contain the variable names or a wildcard (`*`), such as `*pop` in the *varlist*, meaning any variable name ending in "pop". In the `census2c` dataset, `*pop` will refer to `pop`, `smallpop`, and `largepop`.

A *varlist* may also contain a hyphenated list, such as `cat1-cat4`, which refers to all variables in the dataset between `cat1` and `cat4`, inclusive, in the same order as in the dataset. The order of the variables is that provided by `describe`, or that shown in the Variables window. You can modify the order by using the `order` command.

2.1.11 drop and keep

To discard variables you have created, you can use the `drop` command with a *varlist*. If you have many variables and want to keep only some of them, use the `keep` command with a *varlist* specifying which variables you want to keep. `drop` and `keep` follow the syntax

`drop` *varlist*

`keep` *varlist*

To remove or retain observations, use the syntax

`drop if` *exp*

`drop in` *range*

`keep if` *exp*

`keep in` *range*

With our Census dataset, we could use either `drop if largepop` or `keep if smallpop` to leave only the smaller states' observations.

2.1.12 rename and renvars

To rename a variable, you could generate a new variable equivalent to the old variable and drop the old variable, but a cleaner solution is to use `rename`. Using the syntax

`rename` *old_varname new_varname*

you can rename a variable. To change several variables' prefixes (e.g., `income80`, `income81` to `inc80`, `inc81`), use `renpfix`:

```
. renpfix income inc
```

where `income` is the common prefix of the original variables. For a more general solution to renaming variables, see the `renvars` command of Cox and Weesie (2005).

2.1.13 The save command

To save a dataset for later use, use the `save` *filename* command. We could save the `census2c` dataset to a different file with `save census2d`. To save it with the original name, we would use the `replace` option, which like the `replace` command must be spelled out. However, if you `save, replace`, you cannot restore the contents of the original dataset. Saving to a new filename is generally a better idea.

You can save Stata data to a text file by using the `outsheet` command. Despite its name, this command does not write a spreadsheet file. It writes an ASCII text file that can be read by a spreadsheet or any program that can read tab-delimited or comma-delimited files. Unless you need to transfer the data to another statistical package, you should just `save` the data. If you save a file in a format other than Stata's binary format (a `.dta` file), you may lose some information from your dataset, including labels, notes, value labels, and formats. If you need to move the data to another package, consider using Stat/Transfer (see appendix A), a third-party application available from StataCorp.

The `use` command reads files into Stata much faster than any of Stata's data input commands (`insheet`, `infile`, `infix`) since it does not have to convert text to binary format. Once you bring the data file into Stata, `save` it as a `.dta` file and work with that file henceforth.

2.1.14 insheet and infile

The examples above used an existing Stata data file, `census2c.dta`. But researchers often need to bring data into Stata from an external source: a spreadsheet, a web page, a dataset stored as a text file, or a dataset from another statistics package. Stata provides many options for inputting external data. You could use the `input` command or the Data Editor, but Stata has specialized commands for large datasets.

How do you determine which data input command to use? If your data are in *delimited* format—separated by commas or tab characters—the **insheet** command is usually your best bet. If the data are not delimited, use **infile** or the more specialized **infix** command. These commands work best if data are run together, as they are in many surveys: that is, successive variables' values are in adjacent columns, and you must use the survey's codebook to separate them. Finally, if the data have the format of another statistical package, you may want to use Stat/Transfer to translate the data to Stata format `.dta`. See appendix A for more information about data input.

2.2 Common data transformations

This section discusses data transformations, specifically, the `cond()` and `recode()` functions, missing-data handling, conversion between string and numeric forms, and date handling. The section highlights some useful functions for **generate** and discusses Stata's extended generate command, **egen**. The last two subsections describe by-groups and looping over variables with **forvalues** and **foreach**.

2.2.1 The cond() function

You can code a result variable as x_T when the logical expression C is true and x_F when it is false by using the `cond(`C`,` x_T`,` x_F`)` function.

To separate states having a ratio of marriages to divorces (the net marriage rate) greater than and less than 2, you could define `netmarr2x` as having values 1 and 2 and attach value labels. You could then use the variable in **tabstat**:

```
. generate netmarr2x = cond(marr/divr > 2, 1, 2)
. label define netmarr2xc 1 "marr > 2 divr" 2 "marr <= 2 divr"
. label values netmarr2x netmarr2xc
. tabstat pop medage, by(netmarr2x)
Summary statistics: mean
  by categories of: netmarr2x
```

netmarr2x	pop	medage
marr > 2 divr	5792.196	30.38333
marr <= 2 divr	4277.178	30.08889
Total	5142.903	30.25714

States with a high net marriage rate are larger and have slightly older populations.

You can nest the `cond()` function: that is, you can use other `cond()` functions as its second and third arguments. But using this syntax can be unwieldy, so you might want to use multiple commands.

2.2.2 Recoding discrete and continuous variables

You can use Stata to create a new variable based on the coding of an existing discrete variable. You could write many similar transformation statements such as

```
. replace newcode = 5 if oldcode == 2
. replace newcode = 8 if oldcode == 3
. replace newcode = 12 if oldcode == 5 | oldcode == 6 | oldcode == 7
```

where the vertical bar (`|`) is Stata's "or" operator. But performing transformations this way is inefficient, and copying and pasting to construct these statements will probably lead to typing errors.

Using Stata's `recode` command usually produces more efficient and readable code.[3] For instance,

```
. recode oldcode (2 = 5) (3 = 8) (5/7 = 12), generate(newcode)
```

will perform the above transformation. The equal sign is an assignment operator (old-value(s) → newvalue). Unlike in the line-by-line approach above using `replace`, you can apply `recode` to an entire *varlist*. This approach is handy when a questionnaire-based dataset contains several similar questions with the same coding; you can use the `prefix()` option to define the variable name stub. You can account for missing-data codes, map all unspecified values to one outcome, and specify value labels for the values of the new variables. In fact, you can use `recode`, modifying the existing variables rather than creating new ones, but you should avoid doing this in case any further modifications to the mapping arise.

You can use `generate` and Stata's `recode()` function (not to be confused with the `recode` command discussed above) to map a continuous variable to a new categorical variable.

To generate a histogram of states' median age in whole years,[4] you can use `recode()` to define brackets for `medage` as

```
. generate medagebrack = recode(medage,28,29,30,31,32,33)
. tabulate medagebrack
```

medagebrack	Freq.	Percent	Cum.
29	3	14.29	14.29
30	8	38.10	52.38
31	4	19.05	71.43
32	4	19.05	90.48
33	2	9.52	100.00
Total	21	100.00	

3. Sometimes an algebraic expression offers a one-line solution using `generate`.

4. I present various examples of Stata graphics in this text without explaining the full syntax of Stata's graphics language. For an introduction to Stata graphics, please see **help graph intro** and [G] **graph intro**. For an in-depth presentation of Stata's graphics capabilities, please see *A Visual Guide to Stata Graphics* (Mitchell 2004).

The numeric values (which could be decimal values) label the brackets: e.g., states with `medage` up to 28 years are coded as 28, those greater than 28 but less than or equal to 29 are coded as 29, and so on. If we draw a histogram of `medagebrack`, we can specify that the variable is `discrete`, and the bars will correspond to the breakpoints specified in the `recode()` function:[5]

```
. histogram medagebrack, discrete frequency
> lcolor(black) fcolor(gs15) addlabels
> addlabopts(mlabposition(6)) xtitle(Upper limits of median age)
> title(Northeast and North Central States: Median Age)
(start=29, width=1)
```

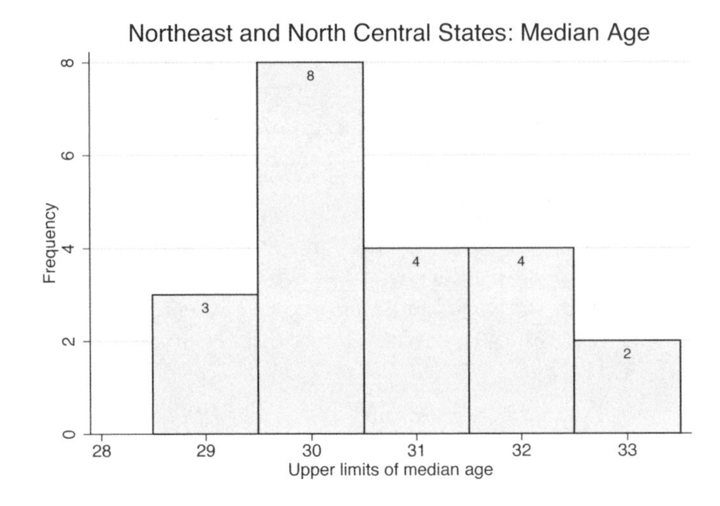

Figure 2.1: `histogram` of values from the `recode()` function

If you do not need specific (or unequally spaced) bracket endpoints, you can use `autocode(`*n*`)`, which generates *n* equally spaced intervals between a minimum and maximum value of the variable; see `help programming functions`.

2.2.3 Handling missing data

Stata has 27 numeric codes representing missing values: `.`, `.a`, `.b`, ..., `.z`. Stata treats missing values as large positive values and sorts them in that order, so that `.` (the system missing-value code; see [U] **12.2.1 Missing values**) is the smallest missing value. Qualifiers such as `if` *x* `< .` can thus exclude all possible missing values.[6]

5. For an excellent presentation of the many styles of graphs supported by Stata, see Mitchell (2004).

6. Stata user code earlier than version 8 often used qualifiers such as `if` *x* `!= .` to rule out missing values. Doing so in versions 8 and later will capture only the `.` missing-data code. If any other missing codes are in the data (for instance, if you used Stat/Transfer to convert an SPSS or SAS dataset to Stata format), you must use `if` *x* `< .` to handle them properly.

By default, Stata omits missing observations from any computation. For `generate` or `replace`, missing values are propagated. Most Stata functions of missing data return missing values. Exceptions include `sum()`, `min()`, and `max`; see [D] **functions** for details. Univariate statistical computations (such as `summarize` computing a mean or standard deviation) consider only nonmissing observations. For multivariate statistical commands, Stata generally practices *casewise deletion* by dropping from the sample observations in which any variable is missing.

Several Stata commands handle missing data in nonstandard ways. For instance, although `correlate` *varlist* uses casewise deletion to exclude any observations containing missing values in any variables of the *varlist* in computing the correlation matrix, the alternative command `pwcorr` computes pairwise correlations using all available data for each pair of variables. The `missing(`*x1*`,`*x2*`,`...`,`*xn*`)` function (see [D] **functions**) returns 1 if any argument is missing and 0 otherwise; that is, it provides the user with a casewise deletion indicator. The `egen` rowwise functions (`rowmax()`, `rowmean()`, `rowmin()`, `rowsd()`, `rowtotal()`) each *ignore* missing values (see section 2.2.7). For example, `rowmean(x1,x2,x3)` computes the mean of three, two, or one of the variables, returning missing only if all three variables' values are missing for that observation. The `egen` functions `rownonmiss()` and `rowmiss()` return, respectively, the number of nonmissing and missing elements in their *varlist*s.

mvdecode and mvencode

Other statistical packages, spreadsheets, or databases may treat missing data differently from how Stata does. Likewise, if the data are to be used in another program that does not use the `.` notation for missing-data codes, you may need to use an alternate representation of Stata's missing data by using the `mvdecode` and `mvencode` commands (see [D] **mvencode**). `mvdecode` allows you to recode numeric values to missing, such as when missing data have been represented as −99, −999, 0.001, and so on. You can use all of Stata's 27 numeric missing-data codes, so you could map −9 to `.a`, −99 to `.b`, and so on. The `mvencode` command provides the inverse function, allowing you to change Stata's missing values to numeric form. Like `mvdecode`, `mvencode` can map each of the 27 numeric missing data codes to a different numeric value.

To transfer missing data values between packages, you may want to use Stat/Transfer (see appendix A). Because this third-party application (distributed by StataCorp) can also transfer variable and value labels between major statistical packages and create subsets of files' contents (e.g., only selected variables are translated into the target format) using Stat/Transfer is well worth the cost for those researchers who often import or export datasets.

2.2.4 String-to-numeric conversion and vice versa

Stata has two types of variables: string and numeric. Often a variable imported from an external source will be misclassified as string rather than numeric. For instance, if the first value read by `insheet` is `NA`, that variable is classified as a string variable. Stata provides several methods for converting string variables to numeric. If the variable has merely been misclassified as string, you use the `real()` function—e.g., `generate hhid = real(household)`, which creates missing values for any observations that cannot be interpreted as numeric. The `destring` command can transform variables in place (with the `replace` option)—although generating a new variable is safer—and may be used with a *varlist* to apply the same transformation to an entire set of variables with one command. If the variable really has string content, and you need a numeric equivalent, you can use the `encode` command.

Do not apply `encode` to a string variable that has purely numeric content (for instance, one that has been misclassified as a string variable) because `encode` will attempt to create a value label for each distinct value of the variable.

Why might you need to `encode` a variable? Consider the `region` identifier in a different U.S. Census dataset, `census2a`:

```
. use http://www.stata-press.com/data/imeus/census2a
(Extracted from http://www.stata-press.com/data/r9/census2.dta)
. describe region
              storage  display     value
variable name   type   format      label      variable label
-------------------------------------------------------------------------------
region          str7   %9s
. tabulate region
     region |      Freq.     Percent        Cum.
------------+-----------------------------------
    N Cntrl |         12       24.00       24.00
         NE |          9       18.00       42.00
      South |         16       32.00       74.00
       West |         13       26.00      100.00
------------+-----------------------------------
      Total |         50      100.00
```

Typing `describe` reveals that `region` is a string variable with a maximum length of 7 characters. You could use this variable in a `tabulate` command to compute the frequencies of U.S. states in each region or (using `tabulate`'s `generate()` option) to create a set of indicator variables, one per region, but you cannot use it in statistical commands. Say that you wanted to retain the readable values of `region` for display but use the variable in statistical commands. You could use the `encode` command: `encode region, generate(cenreg)` to create a new variable `cenreg` (Census region) and a value label (by default, also named `cenreg`) that takes on the values of `region`. The new variable has integer type (`long`), with values assigned from one to the number of distinct values of `region`:

```
. encode region, generate(cenreg)
. describe cenreg
              storage  display     value
variable name   type   format      label      variable label
------------------------------------------------------------------------
cenreg          long   %8.0g       cenreg
. summarize cenreg
    Variable |       Obs        Mean    Std. Dev.       Min        Max
-------------+--------------------------------------------------------
      cenreg |        50         2.6    1.124858          1          4
```

The new variable takes on values 1, 2, 3, or 4. The **tabulate cenreg** command generates the same display as **tabulate region**. However, you can now do statistical analyses by using **cenreg**, such as in a **summarize** command or to define a grouping variable (see section 3.2) for the **tsset** command.

Some string variables may have numeric content but should be stored as strings. For example the U.S. ZIP code, or postal code, is a five-digit integer that may begin with a leading zero. If you need to match household data to, say, Census data, you will want to retain those leading zeros, and you could **encode** a ZIP code variable (assuming that there are not too many of them to create value labels; see **help limits**).

You may also need to generate the string equivalent of a numeric variable. Sometimes it is easier to parse the contents of string variables and extract significant substrings. You can apply such transformations to integer numeric variables through integer division and remainders, but these methods are generally cumbersome and error prone. You can also convert numeric values to string to get around the limits of exact representation of numeric values such as integers with many digits; see Cox (2002b).

We have discussed three methods for string-to-numeric conversion. For each method, the inverse function is available. **string()** lets you use a numeric display format (see [D] **format**), such as a variable with leading zeros used in some ID-number schemes. **tostring** prevents you from losing information while converting variables and can be used with a specific display format. Like **destring**, **tostring** can modify variables specified in a *varlist*.

2.2.5 Handling dates

Stata does not have a special data type for date variables. It understands one date format: *elapsed dates*, representing the number of days since 1 January 1960.[7] Dates before that date (including BCE) are coded as negative integers. Since most of us cannot readily translate 27 July 2005 into the number 16,644, Stata provides several functions for handling dates.

If you have month, day, and year values stored as separate numeric variables **mon**, **day**, and **year**, you can use Stata's **mdy()** function, which takes those three arguments.

7. This system is similar to Excel's handling of dates, with a base date of 1 January 1900 or 1 January 1904, depending on the operating system.

If your year is coded as two digits, you should convert it to a four-digit year before using `mdy()`.

If the date is coded as a string, such as `7/27/2005`, `27jul2005`, `2005-07-27`, or `27.7.2005`, you can use Stata's `date()` function to generate an integer date variable. The `date()` function takes two arguments: the name of a string variable containing the date, and a literal such as `"mdy"`, `"ymd"`, or `"dmy"`, specifying the order of the arguments. The literal allows Stata to tell whether the date `02/03/1996` refers to the third day of February or the second day of March, in European style. You can also use an optional third argument to deal with two-digit years. It is best, though, to generate four-digit years whenever you prepare data for use in Stata. If you are exporting data from Excel, apply a four-digit year format before saving the data in `.csv` or tab-delimited text format.

Once you have defined an integer date variable with `mdy()` or `date()`, use the `format` command to give it a date format for display. `format` *varname* `%td` displays the variable in Stata's default date format, e.g., `27jul2005`; see [U] **12.5.3 Date formats** for other date formats. You can also display only part of the date, e.g., `27 July` or `July 2005`. Formatting the display of Stata variables never changes their content, so you cannot use this method to group observations by month or year. The observations' values remain the underlying integer (elapsed date).

To group observations—for example, by month—you can use one of Stata's many *date functions*. Starting with a daily variable `transdate`, you can use `generate mmyy = mofd(transdate)` to generate a new variable `mmyy`, which is the *elapsed month* in integer form. To put this date in readable form, apply a format: `format mmyy %tm`, where `%tm` is Stata's monthly format. The default format will produce `2005m7` for this date.

Stata supports yearly `%ty`, half-yearly `%th`, quarterly `%tq`, monthly `%tm`, and daily `%td` date types. Stata does not support business-daily data. You can use any of several functions to extract the number of the year, quarter, month, week, day, day of week, or day of year from an elapsed-date variable. Another set of functions, such as `mofd()`, lets you translate elapsed days into other elapsed units of time.

To define *run-together dates*, such as `20050727`, you can either extract the components yourself or use Nicholas Cox's `todate` command, available from `ssc`. This command can also handle other sorts of run-together dates, such as `yyyyww`, where `ww` refers to the week of the year `yyyy`.

Since Stata's dates—whether stored as elapsed dates, or elapsed months, quarters, years, and so on—are integer variables, you can use standard arithmetic to generate elapsed time, measured in whatever time unit you prefer. Stata does not support intraday time measurements, such as the time at which a patient receives a dose of medicine. However, several `egen` functions described in section 2.2.7 have been developed for that purpose.

2.2.6 Some useful functions for generate or replace

Stata provides a useful set of "programming" functions (see [D] **functions** or type `help programming functions`), many of which require no programming. For instance, instead of binning observations by their values relative to breakpoints defined by `recode()` or `autocode()`, you might want to bin the data into equally sized groups: quartiles, quintiles, or deciles. The `group(`*n*`)` function provides this capability, creating n groups of approximately equal size, with the result variable taking on values 1, 2, ..., n. The groups' memberships are defined by the current sort order of the dataset (which you could modify by issuing a `sort` command).

You could replace several `replace` statements with one call to the `inlist()` or `inrange()` function. The former lets you specify a variable and a list of values; it returns 1 for each observation if the variable matches one of the elements of the list and 0 otherwise. You can use the function with either numeric or string variables. For string variables, you can specify up to 10 string values in the list. The `inrange()` function lets you specify a variable and an interval on the real line and returns 1 or 0 to indicate whether the variable's values fall within the interval (which may be open; i.e., one limit may be $\pm\infty$).

Some data transformations involve integer division, that is, truncating the remainder. For instance, four-digit SIC (industry) codes 3211–3299 divided by 100 must all yield 32. You can do this transformation with the `int()` function (defined in `help math functions`). A common task involves extracting one or more digits from an integer code; for instance, the tens and units digits of the codes above can be defined as

```
. gen digit34 = SIC - int(SIC/100)*100
```

or

```
. gen mod34 = mod(SIC,100)
```

where the second construct uses the modulo (`mod()`) function. You could extract the tens digit alone by using

```
. gen digit3 = int((SIC - int(SIC/100)*100)/10)
```

or

```
. gen mod3 = (mod(SIC,100) - mod(SIC,10))/10
```

This method is not useful for long integers such as U.S. Social Security numbers of nine digits or ID codes of 10 or 12 digits. You can use the functions `maxbyte()`, `maxint()`, and `maxlong()` instead. Because 27 values are reserved for missing-value codes, the maximum value that can be stored in a `byte` variable is 100, instead of 127. An `int` variable can hold values up to 32,740, which is insufficient for five-digit U.S. ZIP codes.

The long data type can hold integers up to 2.147 billion (a 10-digit number).[8] Thus all nine-digit numbers can be uniquely represented in Stata, but only a subset of 10-digit numbers will be unique. Integers with more than 10 digits cannot be represented exactly, and because of the mechanics of binary representation and finite-precision arithmetic, representing those values as floats (floating-point numbers) will be problematic if you need exact values. Thus you should store very large integers as string variables if you need their precise values (for example, in matching by patient ID). In other cases, statistical data are reported as very large integers (e.g., the World Bank's *World Development Indicators* database contains gross domestic product in local currency units, rather than millions, billions, or trillions) and should be stored using a double data type since a float data type can retain only about seven digits of precision (per epsfloat()); see Cox (2002b).

One exceedingly useful function for generate is the sum() function, which provides a *running sum* over the specified observations. This function is useful with time-series data in converting a flow variable into a stock variable. If you have an initial capital stock value and a net investment series, the sum() of investment plus the initial capital stock defines the capital stock at each point in time.

2.2.7 The egen command

Whereas the functions available in generate or replace are limited to those listed in [D] **functions** (see also help functions), Stata's egen command provides an open-ended list of capabilities. Just as you can extend Stata's command set by placing other .ado and .hlp files on the adopath, you can invoke egen functions that are defined by ado-files with names starting with _g, stored on the adopath. Many of these functions are part of official Stata (see [D] **egen** and help egen), but your copy of Stata may include other egen functions that you have written or that you have downloaded from the SSC archive ([R] **ssc**) or another Stata user's net site. This section discusses several official Stata functions and several useful additions developed by the Stata user community.

Although egen's syntax is similar to that of generate, there are several differences. Not all egen functions allow a by *varlist*: (see the documentation to determine whether a function is *byable*). Similarly, you cannot use _n and _N explicitly with egen. Since you cannot specify that a variable created with a nonbyable egen function should use the logic of replace, you may need to use a temporary variable as the egen result and then use replace to combine those values over groups.

Official egen functions

To get spreadsheetlike functionality in Stata's data transformations, you will need to understand the rowwise egen functions, which allow you to calculate sums, averages,

8. For each of these data types, negative numbers of similar magnitudes may be stored. For floating-point numbers, see the maxfloat() and maxdouble() functions.

standard deviations, extrema, and counts across several Stata variables. You can also use wildcards. With a list of state-level U.S. Census variables `pop1890`, `pop1900`, ..., `pop2000`, you may use `egen nrCensus = rowmean(pop*)` to compute the average population of each state over those decennial censuses. As discussed in section 2.2.3, the rowwise functions can work with missing values. The mean will be computed for all 50 states, although several were not part of the United States in 1890. You can compute the number of nonmissing elements in the rowwise list with `rownonmiss()`, with `rowmiss()` as the complementary value. Other official rowwise functions include `rowmax()`, `rowmin()`, `rowtotal()`, and `rowsd()` (row standard deviation).

Official `egen` also provides statistical functions for computing a statistic for specified observations of a variable and placing that constant value in each observation of the new variable. Since these functions generally let you use `by` *varlist*`:`, you can use them to compute statistics for each by-group of the data, as discussed in section 2.2.8. Using `by` *varlist*`:` makes it easier to compute statistics for each household for individual-level data or each industry for firm-level data. The `count()`, `mean()`, `min()`, `max()`, and `total()` functions are especially useful.[9]

Other functions in this statistical category include `iqr()` (interquartile range), `kurt()` (kurtosis), `mad()` (median absolute deviation), `mdev()` (mean absolute deviation), `median()`, `mode()`, `pc()` (percent or proportion of total), `pctile()`, `p(`*n*`)` (*n*th percentile), `rank()`, `sd()` (standard deviation), `skew()` (skewness), and `std()` (z-score).

egen functions from the user community

The most comprehensive collection of additional `egen` functions is Nicholas J. Cox's `egenmore` package, available with the `ssc` command.[10] The `egenmore` package contains routines by Cox and others (including me). Some of these routines extend the functionality of official `egen` routines, whereas others provide capabilities lacking in official Stata. Many of the routines require Stata version 8 or later.

For example, extensions have been made to improve the way Stata handles dates. Stata's date variables are stored internally as floating-point values. For a date variable measuring days (rather than weeks, months, quarters, half-years, or years), the integer part records the number of days elapsed since an arbitrary day zero of 1 January 1960. Although you could use the decimal part of a date to represent an elapsed fraction of a day (e.g., 0.25 as 6:00 a.m.), Stata does not support intraday values or time arithmetic. The `egenmore` package contains functions that provide such support. The `dhms()` function creates a date variable with the fractional part reflecting hours, minutes, and seconds, whereas `hms()` computes the number of seconds past midnight for time comparisons (e.g., such as stock-market tick data, which are recorded to the sec-

9. Before Stata 9, `egen total()` was called `egen sum()`, but the name was changed because it was often confused with `generate`'s `sum()` function.

10. The package is labeled `egenmore` since it further extends `egenodd`, which appeared in the *Stata Technical Bulletin* (Cox 1999, 2000). Most of the `egenodd` functions now appear in official Stata, so they will not be discussed here.

ond). The companion function `elap2()` displays an elapsed time between two fractional date variables in days, hours, minutes, and seconds, whereas `elap()` provides a similar function for a number of seconds. Functions `hmm()` and `hmmss()` display fractional days as hours and minutes, or hours, minutes, and seconds.

Several `egenmore` functions work with standard Stata dates, expressed as integer days. The `bom()` and `eom()` functions create date variables corresponding to the first day or last day of a given calendar month. They can be used to generate the offset for any number of months (e.g., the last day of the third month from now). If you use the `work` option, you can specify the first (last) nonweekend day of the month (although this function does not support holidays). You can also use the functions `bomd()` and `eomd()` to find the first (last) day of a month in which their date-variable argument falls, which is useful if you wish to aggregate observations by calendar month.

Several `egenmore` functions extend `egen`'s statistical capabilities. The `corr()` function computes correlations (optionally covariances) between two variables; `gmean()` and `hmean()` compute geometric and harmonic means; `rndint()` computes random integers from a specified uniform distribution; `semean()` computes the standard error of the mean; and `var()` computes the variance. The `filter()` function generalizes `egen`'s `ma()` function, which can produce only two-sided moving averages of an odd number of terms. In contrast, `filter()` can apply any linear filter to data that you have declared to be time-series data by using `tsset`, including panel data, for which the filter is applied separately to each panel (see section 3.4.1). You can use the companion function `ewma()` to apply an exponentially weighted moving average to time-series data.

Useful data-management functions include `rall()`, `rany()`, and `rcount()`. These rowwise functions, working from a *varlist*, evaluate a specified condition and indicate whether all (any) of the variables satisfy the condition or how many variables satisfy the condition. For instance, typing

```
. egen allpos = rall(var1 var2 var3), cond(@ > 0 & @ < .)
. egen anyneg = rany(var1 var2 var3), cond(@ < 0 )
. egen countpos = rcount(dum*), cond(@ > 0 & @ < .)
```

would create `allpos` with a value of 1 for each observation in which all three variables are positive and nonmissing and 0 otherwise; `anyneg` with a value of 1 where any of the three variables were negative and 0 otherwise; and `countpos()` indicating the number of nonmissing dummy variables that are positive. You could use `countpos` to ensure that a set of dummies is mutually exclusive and exhaustive, since it should return 1 for each observation (`countpos` has other uses as well). The `@` symbol is a placeholder, standing for the value of the variable in that observation. You can also apply these functions to string variables.

Another useful data-management function is the `record()` function (the name is meant to evoke "setting a record"). You can use this function to compute the record value, such as the highest wage earned to date by each employee or the lowest stock price encountered to date. If the data contain annual wage rates for several employees over several years,

```
. egen hiwage = record(wage), by(empid) order(year)
```

will compute for each employee (as specified with `by(empid)`) the highest wage earned to date, allowing you to evaluate conditions when wages have fallen because of a job change, etc.[11] Several other `egen` functions are available in the `egenmore` package on the SSC archive.

In summary, `egen` functions handle several common data-management tasks. The open-ended nature of this command implies that new functions often become available, either through ado-file updates to official Stata or through contributions from the user community. The latter will generally be announced on Statalist (with past messages accessible in the Statalist archives), and recent contributions will be highlighted in `ssc whatsnew`.

2.2.8 Computation for by-groups

One of Stata's most useful features is the ability to transform variables or compute statistics over by-groups, which are defined with the `by` *varlist*`:` prefix introduced in section 2.1.8. There, we discussed how to use by-groups in statistical commands. We now discuss how to use them with `generate`, `replace`, and `egen` in data transformations.

If you use a by-group, `_n` and `_N` have alternative meanings (they usually refer to the current observation and last defined observation in the dataset, respectively). Within a by-group, `_n` is the current observation of the group and `_N` is the last observation of the group. Here we `gsort` the state-level data by region and descending order of population. We then use `generate`'s running `sum()` function, `by region:`, to display the total population in each region that lives in the largest, two largest, three largest, ..., states:

```
. use http://www.stata-press.com/data/imeus/census2d, clear
(1980 US Census data with population size indicators)
. gsort region -pop
. by region: generate totpop = sum(pop)
```

11. I am grateful to Nicholas J. Cox for his thorough documentation of `help egenmore`.

```
. list region state pop totpop, sepby(region)

     +---------------------------------------------+
     | region   state              pop     totpop |
     |---------------------------------------------|
  1. | NE       New York       17558.1   17558.07 |
  2. | NE       Pennsylvania   11863.9   29421.97 |
  3. | NE       New Jersey      7364.8   36786.79 |
  4. | NE       Massachusetts   5737.0   42523.83 |
  5. | NE       Connecticut     3107.6    45631.4 |
  6. | NE       Maine           1124.7   46756.06 |
  7. | NE       Rhode Island     947.2   47703.22 |
  8. | NE       New Hampshire    920.6   48623.83 |
  9. | NE       Vermont          511.5   49135.28 |
     |---------------------------------------------|
 10. | N Cntrl  Illinois       11426.5   11426.52 |
 11. | N Cntrl  Ohio           10797.6   22224.15 |
 12. | N Cntrl  Michigan        9262.1   31486.23 |
 13. | N Cntrl  Indiana         5490.2   36976.45 |
 14. | N Cntrl  Missouri        4916.7   41893.14 |
 15. | N Cntrl  Wisconsin       4705.8    46598.9 |
 16. | N Cntrl  Minnesota       4076.0   50674.87 |
 17. | N Cntrl  Iowa            2913.8   53588.68 |
 18. | N Cntrl  Kansas          2363.7   55952.36 |
 19. | N Cntrl  Nebraska        1569.8   57522.18 |
 20. | N Cntrl  S. Dakota        690.8   58212.95 |
 21. | N Cntrl  N. Dakota        652.7   58865.67 |
     +---------------------------------------------+
```

We can use _n and _N in this context. They will be equal in the last observation of each by-group:

```
. by region: list region totpop if _n == _N

-----------------------------------------------------------------------------

-> region = NE

     +-------------------+
     | region     totpop |
     |-------------------|
  9. | NE       49135.28 |
     +-------------------+

-----------------------------------------------------------------------------

-> region = N Cntrl

     +---------------------+
     | region       totpop |
     |---------------------|
 12. | N Cntrl    58865.67 |
     +---------------------+
```

We could have computed the total population by region, stored as a new variable, by using `egen`'s `total()` function. We might instead want to compute states' *average* population by region:

```
. by region: egen meanpop = mean(pop)
. list region state pop meanpop, sepby(region)
```

	region	state	pop	meanpop
1.	NE	New York	17558.1	5459.476
2.	NE	Pennsylvania	11863.9	5459.476
3.	NE	New Jersey	7364.8	5459.476
4.	NE	Massachusetts	5737.0	5459.476
5.	NE	Connecticut	3107.6	5459.476
6.	NE	Maine	1124.7	5459.476
7.	NE	Rhode Island	947.2	5459.476
8.	NE	New Hampshire	920.6	5459.476
9.	NE	Vermont	511.5	5459.476
10.	N Cntrl	Illinois	11426.5	4905.473
11.	N Cntrl	Ohio	10797.6	4905.473
12.	N Cntrl	Michigan	9262.1	4905.473
13.	N Cntrl	Indiana	5490.2	4905.473
14.	N Cntrl	Missouri	4916.7	4905.473
15.	N Cntrl	Wisconsin	4705.8	4905.473
16.	N Cntrl	Minnesota	4076.0	4905.473
17.	N Cntrl	Iowa	2913.8	4905.473
18.	N Cntrl	Kansas	2363.7	4905.473
19.	N Cntrl	Nebraska	1569.8	4905.473
20.	N Cntrl	S. Dakota	690.8	4905.473
21.	N Cntrl	N. Dakota	652.7	4905.473

We could do this same calculation over a by *varlist*: with more than one variable:

```
. by region popsize, sort: egen meanpop2 = mean(pop)
. list region popsize state pop meanpop2, sepby(region)
```

	region	popsize	state	pop	meanpop2
1.	NE	<= 5 million	Rhode Island	947.2	1322.291
2.	NE	<= 5 million	New Hampshire	920.6	1322.291
3.	NE	<= 5 million	Vermont	511.5	1322.291
4.	NE	<= 5 million	Connecticut	3107.6	1322.291
5.	NE	<= 5 million	Maine	1124.7	1322.291
6.	NE	> 5 million	New York	17558.1	10630.96
7.	NE	> 5 million	New Jersey	7364.8	10630.96
8.	NE	> 5 million	Massachusetts	5737.0	10630.96
9.	NE	> 5 million	Pennsylvania	11863.9	10630.96
10.	N Cntrl	<= 5 million	N. Dakota	652.7	2736.153
11.	N Cntrl	<= 5 million	Kansas	2363.7	2736.153
12.	N Cntrl	<= 5 million	Missouri	4916.7	2736.153
13.	N Cntrl	<= 5 million	Minnesota	4076.0	2736.153
14.	N Cntrl	<= 5 million	Iowa	2913.8	2736.153
15.	N Cntrl	<= 5 million	S. Dakota	690.8	2736.153
16.	N Cntrl	<= 5 million	Wisconsin	4705.8	2736.153
17.	N Cntrl	<= 5 million	Nebraska	1569.8	2736.153
18.	N Cntrl	> 5 million	Illinois	11426.5	9244.112
19.	N Cntrl	> 5 million	Indiana	5490.2	9244.112
20.	N Cntrl	> 5 million	Michigan	9262.1	9244.112
21.	N Cntrl	> 5 million	Ohio	10797.6	9244.112

We can now compare each state's population with the average population of states of its size class (large or small) in its region.

Although `egen`'s statistical functions can be handy, creating variables with constant values or constant values over by-groups in a large dataset will consume much of Stata's available memory. If you need the constant values only for a subsequent transformation, such as computing each state population's deviation from average size, and will not use them in later analyses, `drop` those variables at the earliest opportunity. Or consider other Stata commands such as `center`, which transforms a variable into deviation from mean form and works with by-groups.

`egen` can be considerably slower than built-in functions or special-purpose commands. You can use `egen` functions to generate constant values for each element of a by-group: total household income or average industry output, but `egen` is an inefficient tool. Stata's `collapse` command is especially tailored to perform that very function and will generate one summary statistic for each by-group.

2.2.9 Local macros

Using Stata's *local macros* can help you work much more efficiently. In Stata, a local macro is a container that can hold one object—such as a number or variable name—or a set of objects. A local macro may contain any combination of alphanumeric characters and can hold more than 8,000 characters in all versions of Stata. A Stata macro is really an *alias* that has both a name and a value. You can return a macro's value at any time by *dereferencing* its name:

```
. local country US UK DE FR
. local ctycode 111 112 136 134
. display "`country'"
US UK DE FR
. display "`ctycode'"
111 112 136 134
```

The Stata command to define the macro is `local` (see [P] **macro**). A *local macro* is created within a do-file or in an ado-file and ceases to exist when that do-file terminates.

The first `local` command *names* the macro—as `country`—and then defines its *value* to be the list of four two-letter country codes. The next `local` statement does the same for macro `ctycode`. To access the value of the macro, we must *dereference* it. *macroname* refers to the local macro name. To obtain its value, the name of the macro must be preceded by the *left tick* character (`) and followed by the apostrophe ('). To avoid errors, be careful to use the correct punctuation. In the example's `display` statements, we must wrap the dereferenced macro in double quotes since `display` expects a double-quoted string argument or the value of a scalar expression, such as `display log(14)`.

You will need to understand local macros before you work with Stata's looping constructs, as we will now discuss. See section B.1 for more detail on macros.

2.2.10 Looping over variables: forvalues and foreach

In discussing `recode` in section 2.2.2, we stressed the importance of using one command rather than several similar commands to change the values stored in a variable. Likewise, if your dataset contains several variables with similar contents, you would rather loop over those variables than write a line to handle each one. The most powerful loop constructs available in Stata are `forvalues` and `foreach`; see section B.3.

Say that we have a set of variables, gdp1, gdp2, gdp3, and gdp4, containing gross domestic product (GDP) values for four countries. Using [P] **forvalues**, we can take advantage of the similarity of their names to perform [D] **generate** and [R] **summarize** statements:

```
. forvalues i = 1/4 {
  2.         generate double lngdp`i' = log(gdp`i')
  3.         summarize lngdp`i'
  4. }
    Variable |       Obs        Mean    Std. Dev.       Min        Max
-------------+--------------------------------------------------------
      lngdp1 |       400    7.931661      .59451   5.794211   8.768936
    Variable |       Obs        Mean    Std. Dev.       Min        Max
-------------+--------------------------------------------------------
      lngdp2 |       400    7.942132    .5828793   4.892062   8.760156
    Variable |       Obs        Mean    Std. Dev.       Min        Max
-------------+--------------------------------------------------------
      lngdp3 |       400    7.987095     .537941   6.327221   8.736859
    Variable |       Obs        Mean    Std. Dev.       Min        Max
-------------+--------------------------------------------------------
      lngdp4 |       400    7.886774    .5983831   5.665983   8.729272
```

In the `forvalues` command, we define the local macro `i` as the loop index. Following an equal sign is the range of values that `i` will take on as a *numlist*, such as `1/4`, as here, or `10(5)50`, indicating 10 to 50 in steps of 5. The body of the loop contains one or more statements to be executed for each value in the list. Each time through the loop, the local macro contains the subsequent value in the list.

This example shows an important use of `forvalues`: if you loop over variables with names that have an integer component, you do not need a separate statement for each variable. The integer component need not be a suffix; we could loop over variables named cty*N*gdp just as easily.

But variable names may not have a common numeric component or contain numbers that are not consecutive; for example, instead of gdp1, gdp2, gdp3, and gdp4, we might have UKgdp, USgdp, DEgdp, and FRgdp. Here we use `foreach` to handle several different specifications of the variable list. We may perform the same `generate` and `summarize` steps by listing the variables to be manipulated:

```
. foreach c in US UK DE FR  {
  2.          generate double lngdp`c' = log(`c'gdp)
  3.          summarize lngdp`c'
  4. }
    Variable |       Obs        Mean    Std. Dev.       Min        Max
-------------+--------------------------------------------------------
     lngdpUS |       400    7.931661      .59451   5.794211   8.768936
    Variable |       Obs        Mean    Std. Dev.       Min        Max
-------------+--------------------------------------------------------
     lngdpUK |       400    7.942132    .5828793   4.892062   8.760156
    Variable |       Obs        Mean    Std. Dev.       Min        Max
-------------+--------------------------------------------------------
     lngdpDE |       400    7.987095     .537941   6.327221   8.736859
    Variable |       Obs        Mean    Std. Dev.       Min        Max
-------------+--------------------------------------------------------
     lngdpFR |       400    7.886774    .5983831   5.665983   8.729272
```

Like **forvalues**, the block of code is repeated, with the local macro taking on each of the values in the list in turn. We may also place the values in a local macro and use the name of that macro in the **foreach** command. In this syntax, we do not dereference the macro in the **foreach** command.

```
. local country US UK DE FR
. foreach c of local country {
  ...
```

The **foreach** command can have a *varlist*, a *newvarlist* of variables to be created, an explicit list of elements, or a *numlist*. Since a *varlist* may contain wildcards, we could have used `foreach c of varlist *gdp` in the example above.

These examples have used **forvalues** and **foreach** to loop over a set of variables. Sometimes we want to loop over a set of variables and store the result in one variable. Within a loop, you may use only **replace** to accumulate results in one series since trying to **generate** the series twice will fail. But for **replace** to function the first time through the loop, the variable must have been **generate**d previously:

```
. generate double gdptot = 0
. foreach c of varlist *gdp {
  2.          quietly replace gdptot = gdptot + `c'
  3. }
. summarize *gdp gdptot
    Variable |       Obs        Mean    Std. Dev.       Min        Max
-------------+--------------------------------------------------------
       USgdp |       400    3226.703    1532.497    328.393   6431.328
       UKgdp |       400    3242.162    1525.788   133.2281   6375.105
       DEgdp |       400    3328.577    1457.716   559.5993   6228.302
       FRgdp |       400    3093.778    1490.646   288.8719   6181.229
      gdptot |       400    12891.22    3291.412   4294.267   21133.94
```

Here we could have also generated **gdptot** with the **egen** function **rowtotal()**; we used a loop to illustrate the concept. The summand must be initialized to zero outside the loop.

2.2.11 Scalars and matrices

Stata also lets you use *scalars* and *matrices* with analysis commands. Scalars, like local macros, can hold both numeric and string values, but a numeric scalar can hold only one numeric value.[12] Most analysis commands return one or more results as numeric scalars. For instance, `describe` returns the scalars `r(N)` and `r(k)`, corresponding to the number of observations and variables in the dataset. A scalar is also much more useful for storing one numeric result—such as the mean of a variable—than for storing that value in a Stata variable containing `maxobs` copies of the same number. A scalar may be referred to in any subsequent Stata command by its name:

```
. scalar root2 = sqrt(2.0)
. generate double rootGDP = gdp*root2
```

Unlike a macro, a scalar's name gives its value, so it does not have to be dereferenced; see section B.2 for more information about scalars.

Stata's estimation commands create both scalars and Stata matrices: in particular, the matrix `e(b)`, containing the set of estimated parameters, and the matrix `e(V)`, containing the estimated variance–covariance matrix of the estimates (VCE). You can use Stata's `matrix` commands to modify matrices and use their contents in later commands; see section B.4 for more information about Stata's matrices.

2.2.12 Command syntax and return values

Stata's analysis commands follow a regular syntax:

`cmdname` *varlist* [*if*] [*in*] [, *options*]

As discussed in section 2.1.6, most Stata analysis commands let you specify `if` *exp* and `in` *range* clauses. Many analysis commands have options that modify their behavior.

Stata analysis commands can be either e-class commands (estimation commands) or r-class commands (all other analysis commands). The command's class determines whether its saved results are returned in `r()` or `e()`. The r-class commands return those elements in `r()`, which you can view by typing `return list`. Using the census data, we `summarize` the pop variable:

```
. use http://www.stata-press.com/data/imeus/census2c, clear
(1980 Census data for NE and NC states)
. summarize pop
    Variable |       Obs        Mean    Std. Dev.       Min        Max
-------------+--------------------------------------------------------
         pop |        21    5142.903    4675.152    511.456   17558.07
```

12. The length of a string scalar is limited to the length of a string variable (244 characters).

```
. return list
scalars:
                  r(N) =  21
              r(sum_w) =  21
               r(mean) =  5142.902523809524
                r(Var) =  21857049.56321066
                 r(sd) =  4675.152357219031
                r(min) =  511.456
                r(max) =  17558.072
                r(sum) =  108000.953
```

Typing `return list` displays the saved results for `summarize`, which include several scalars, including some not displayed in the output, such as `r(sum_w)`, `r(Var)`, and `r(sum)`. We can access these values and use them in later computations, for instance[13]

```
. display "The standardized mean is `r(mean)'/`r(sd)'"
The standardized mean is 5142.902523809524/4675.152357219031
```

In contrast, if we use an estimation command, such as `mean`, we can display the saved results by using `ereturn list`:

```
. mean pop popurb
Mean estimation                           Number of obs    =        21

             |       Mean   Std. Err.     [95% Conf. Interval]
-------------+------------------------------------------------
         pop |   5142.903   1020.202      3014.799    7271.006
      popurb |   3829.776    840.457      2076.613    5582.938

. ereturn list
scalars:
               e(df_r) =  20
             e(N_over) =  1
                  e(N) =  21
               e(k_eq) =  1
            e(k_eform) =  0
macros:
                e(cmd) : "mean"
              e(title) : "Mean estimation"
         e(estat_cmd) : "estat_vce_only"
            e(varlist) : "pop popurb"
            e(predict) : "_no_predict"
         e(properties) : "b V"
matrices:
                  e(b) :  1 x 2
                  e(V) :  2 x 2
                 e(_N) :  1 x 2
              e(error) :  1 x 2
functions:
             e(sample)
```

13. We may `summarize` any number of variables, but the results of `return list` report only statistics from the last variable in the *varlist*.

```
. matrix list e(b)

e(b)[1,2]
            pop      popurb
y1    5142.9025   3829.7758

. matrix list e(V)

symmetric e(V)[2,2]
              pop      popurb
   pop  1040811.9
popurb   849907.5   706367.96
```

The mean command saves several items in e(), including the matrices e(b) and e(V). Matrix e(b) contains the means of both pop and popurb, whereas e(V) contains the estimated VCE. ereturn list also displays several scalars (such as e(N), the number of observations), macros (such as e(varlist) of the command), matrices, and the function e(sample). Section 4.3.6 discusses how to use estimation results in more detail.

Exercises

1. Using the cigconsump dataset, generate a list of the stateids corresponding to the far western states: Washington, Oregon, California, Nevada, Utah, Idaho, and Arizona. Use this list to keep observations from only those states. Drop the state variable, and create a new string variable, state, that contains the full name of the state. save the dataset as cigconsumpW.
2. Using the cigconsumpW dataset, generate a list of the unique values of stateid and state (with levelsof) as local macros. Use reshape to make a wide-format dataset of the packpc, avgprs, and incpc variables. Using foreach, create a set of tables for each state (labeled by the full state name) listing these three variables by year. Compute correlations of the states' packpc variables.

3 Organizing and handling economic data

This chapter discusses four organizational schemes for economic data: the cross section, the time series, the pooled cross section/time series, and the panel dataset. Section 5 presents some tools for manipulating and summarizing panel data. Sections 6–8 present several Stata commands for combining and transforming datasets: `append`, `merge`, `joinby`, and `reshape`. The last section discusses using do-files to produce reproducible research and automatically validate data.

3.1 Cross-sectional data and identifier variables

A common type of data encountered in applied economics and finance is known as cross-sectional data, which contain measurements on distinct individuals at a given point in time. Those observations (rows in the Data Editor) vary over the units, such as individuals, households, firms, industries, cities, states, or countries. The variables (columns in the Data Editor) are generally measurements taken in a given period, such as household income for 2003, firms' reported profits for the first quarter of 2004, or cities' population in the 2000 Census. However, variables may contain measurements from other periods. For instance, a cross-sectional dataset of cities might contain variables named `pop1970`, `pop1980`, `pop1990`, and `pop2000` containing the cities' populations for those four decennial censuses. But unlike in time-series data, the observations in a cross-sectional dataset are indexed with an i subscript, without reference to t (the time).

In a cross-sectional dataset, the order of the observations in the dataset is arbitrary. We could `sort` the dataset on any of its variables to display or analyze extreme values of that variable without changing the results of statistical analyses, which implies that we can use Stata's `by` *varlist*`:` prefix. As discussed in section 2.1.8, using a `by` *varlist*`:` prefix requires that the data be sorted on the defined by-group, which you can do easily by using the `sort` option of the `by` *varlist*`:` prefix; that is, type `by` *varlist*`, sort:`. Time series, on the other hand, must follow a chronological order to be analyzed meaningfully.

Cross-sectional datasets usually have an identifier variable, such as a survey ID assigned to each individual or household, a firm-level identifier (e.g., a CUSIP code), industry-level identifier (e.g., a two-digit Standard Industrial Classification [SIC]) code, or a state or country identifier (e.g., MA, CT, US, UK, FRA, GER). Often there will be more than one identifier per observation. For instance, a survey might contain both a household ID variable and a state identifier. Since Stata's variables may be declared as either numeric or string data types, practically any identifier available in an external data file may be used to define an identifier variable in Stata.

3.2 Time-series data

Cross-sectional datasets are found most often in applied microeconomic analysis. For example, a dataset might contain the share prices of the Standard and Poor's (S&P) 500 firms at the market close on a given day: a pure cross section. But we might also have a dataset that tracks a particular firm's share price, or that share price and the S&P index, daily for 2000–2003. The latter is a time-series dataset, and each observation would be subscripted by t rather than i. A time series is a sequence of observations on a given characteristic observed at a regular interval, such as x_t, x_{t+1}, $x_{t+\tau}$, with each period having the same length (though not necessarily an equal interval of clock time). The share price on the last trading day of each month may be between 26 and 31 days later than its predecessor (given holidays). For business-daily data, such as stock market prices, Friday is (usually) followed by Monday. But say that you had a list of dates and workers' wage rates, which records the successive jobs held and wages earned at the arbitrary intervals when workers received raises or took a new job. Those data could be placed on a time-series calendar, but they are not time-series data.

Periods in time-series data are identified by a Stata date variable, which can define annual, semiannual, quarterly, monthly, weekly, daily, or generic (undated) time intervals.[1] You can use `tsset` to indicate that this date variable defines the time-series calendar for the dataset. A nongeneric date variable should have one of the date formats (e.g., `%tq` for quarterly or `%td` for daily) so that dates will be reported as calendar units rather than as integers.

A few of Stata's time-series commands cannot handle *gaps*, or missing values, in the sequence of dates. Although an annual, quarterly, monthly, or weekly series might not contain gaps, daily and business-daily series often have gaps for weekends and holidays. You need to define such series in business days; for example, you could define a variable equal to the observation number (`_n`) as the date variable.

You must define a time-series calendar with `tsset` before you can use Stata's time-series commands (and some functions). But even if you do not need a time-series calendar, you should define such a calendar when you transform data so that you can refer to dates in doing statistical analysis: for instance, you may use `generate`, `summarize`, or `regress` with the qualifier `if tin(`*firstdate*`,`*lastdate*`)`. This function—which should be read as "t in"—lets you specify a range of dates, rather than observation numbers using a more cumbersome `in` *range*. The interval from the beginning to a specified date or from a date to the end of the sample may be given as `if tin(,`*lastdate*`)` or `if tin(`*firstdate*`,)`, respectively.

1. Defining the observation number (`_n`) as the date variable is the most often used generic time interval.

3.2.1 Time-series operators

Stata provides *time-series operators*—`L.`, `F.`, `D.`, `S.`—which let you specify *lags*, *leads* (forward values), *differences*, and *seasonal differences*, respectively. The time-series operators make it unnecessary to create a new variable to use a lag, difference, or lead. When combined with a *numlist*, they let you specify a set of these constructs in one expression. Consider the lag operator, `L.`, which when prepended to a variable name refers to the (first-) lagged value of that variable: `L.x`. A number may follow the operator so that `L4.x` would refer to the fourth lag of `x`—but more generally, a *numlist* may be used so that `L(1/4).x` refers to the first through fourth lags of `x` and `L(1/4).(x y z)` defines a list of the first through fourth lagged values of each of the variables `x`, `y`, and `z`. These expressions may be used anywhere that a *varlist* is required.

Like the lag operator, the lead operator `F.` lets you specify future values of one or more variables. The lead operator is unnecessary, since a lead is a negative lag, and an expression such as `L(-4/4).x` will work, labeling the negative lags as leads. The difference operator, `D.`, generates differences of any order. The first difference, `D.x`, is Δx or $x_t - x_{t-1}$. The second difference, `D2.x`, is not $x_t - x_{t-2}$, but rather $\Delta(\Delta x_t)$: that is, $\Delta(x_t - x_{t-1})$ or $x_t - 2x_{t-1} + x_{t-2}$. You can also combine the time-series operators so that `LD.x` is the lag of the first difference of `x` (that is, $x_{t-1} - x_{t-2}$) and refers to the same expression, as does `DL.x`. The seasonal difference `S.` computes the difference between the value in the current period and the period 1 year ago. For quarterly data, `S4.x` generates $x_t - x_{t-4}$, and `S8.x` generates $x_t - x_{t-8}$.

In addition to being easy to use, time-series operators will also never misclassify an observation. You could refer to a lagged value as `x[_n-1]` or a first difference as `x[_n] - x[_n-1]`, but that construction is not only cumbersome but also dangerous. Consider an annual time-series dataset in which the 1981 and 1982 data are followed by the data for 1984, 1985, ..., with the 1983 data not appearing in the dataset (i.e., not recorded as missing values, but physically absent). The observation-number constructs above will misinterpret the lagged value of 1984 to be 1982, and the first difference for 1984 will incorrectly span the 2-year gap. The time-series operators will not make this mistake. Since `tsset` has been used to define `year` as the time-series calendar variable, the lagged value or first difference for 1984 will be properly coded as missing, whether or not the 1983 data are stored as missing in the dataset.[2] Thus you should always use time-series operators when referring to past or future values or computing differences in a time-series dataset.

3.3 Pooled cross-sectional time-series data

Microeconomic data can also be organized into *pooled cross-section time series*, in which every observation has both an i and t subscript.[3] For example, we might have the responses from 3 weeks' presidential popularity polls in which each poll contains 400

2. The time-series operators also provide a similar benefit in panel data, as discussed below.
3. Econometricians often call data with this structure *pseudopanel data*; see Baltagi (2001).

randomly selected respondents. But the randomly sampled individuals who respond to the poll one week will probably not appear in the following poll or in any other poll drawn from a national sample before the election. These data are pooled cross sections over time such that observation j at time 1 has no relation to observation j at time 2 or time 3. We may use `collapse` to compute summary statistics for each cross section over time. For instance, if we have annual data for several random samples of U.S. cities for 1998–2004, we could use

```
. collapse (mean) income (median) medinc=income (sum) population, by(year)
```

which would create a new dataset with 1 observation per year, containing the year, average income, median income, and total population of cities sampled in that year.[4]

Although pooled cross-section/time-series data allow us to examine both the individual and time dimensions of economic behavior, they cannot be used to trace individuals over time. In that sense, they are much less useful than *panel* or *longitudinal* data, which I will now describe.

3.4 Panel data

A common form of data organization in microeconomics, macroeconomics, and finance is a type of pooled cross-sectional time-series data called *panel* or *longitudinal* data. *Panel data* contain measurements on the same individuals over several periods.[5] Perhaps the most celebrated longitudinal study of households is the University of Michigan's *Panel Study of Income Dynamics* (PSID), an annual survey of (originally) 5,000 households carried out since 1968. On the financial side, S&P COMPUSTAT databases of firm-level characteristics are one of the most important sources of panel data for financial research.

In this form of data organization, each individual's observations are identified, allowing you to generate microlevel measures not present in the original data. For example, if we have a pooled cross-sectional time-series dataset gathered from repeated annual surveys of randomly sampled individuals that measure their financial wealth along with demographics, we may calculate only an average net savings rate (the rate at which wealth is being accumulated or decumulated) or an average for subsamples of the data (such as the savings rate of African American respondents or of women under 35). We cannot monitor individual behavior, but if we have panel data on a group of individuals who have responded to annual surveys over the same time span, we can calculate individual savings rates and cohort measures for subsamples.

A panel dataset may be either *balanced* or *unbalanced*. In a balanced panel, each of the units, $i = 1, \ldots, G$, is observed in every period $t = 1, \ldots, T$, resulting in $G \times T$ observations in the dataset. Such a panel is easy to work with because the first T

4. The full syntax of `collapse` is described in section 3.4.

5. Panel data are so called because the first examples of these data were the time series of responses of a panel of experts, such as economic forecasters predicting next year's GDP growth or inflation rate. Panel data are also known as *longitudinal* data since they allow you to longitudinally analyze a cohort of households, workers, firms, stocks, or countries.

observations will correspond to unit 1, the second T to unit 2, and so on. However, economic and financial data are often not available in balanced form because some individuals drop out of a multiyear survey.

Furthermore, if we constrain analysis to a balanced panel, we create *survivorship bias*. For example, the S&P COMPUSTAT database of U.S. firms contains 20 years of annual financial statement data—but only for those firms that have existed for the entire period. The set of firms is thus unrepresentative in omitting startups (even those of age 19) and firms that were taken over during that time. Although the algebra of panel-data transformations and estimators is simplified with a balanced panel, I often prefer to work with an unbalanced panel to avoid such biases and mitigate the loss of sample size that may result from insisting on balanced-panel data.

Stata's tools for panel data make it easy to work with any set of observations, balanced or unbalanced, that can be uniquely identified by i and t. Unlike Stata, many statistical packages and matrix languages require a balanced structure with T observations on each of G units, even if some of them are wholly missing.[6] You can use `tsset` to indicate that the data are panel data. The same command that defines a time-series calendar for a time series may specify the panel variable as well:

`tsset` *panelvar timevar*

The *timevar* must be a date variable, whereas *panelvar* may be any integer variable that uniquely identifies the observations belonging to a given unit. The integer values need not be sequential: that is, we could use three-digit SIC codes `321`, `326`, `331`, and `342` as the *panelvar*. But if the units of the data are identified by a string variable, such as a two-letter state abbreviation, we must `encode` that variable to create a *panelvar* identifier. `tsset` will report the ranges of *panelvar* and *timevar*.

3.4.1 Operating on panel data

Stata contains a thorough set of tools for transforming panel data and estimating the parameters of econometric models that take account of the nature of the data. Any `generate` or `egen` functions that support a `by` *varlist*`:` may be applied to panel data by using the *panelvar* as the *by-variable*. Descriptive data analysis on panel data often involves generating summary statistics that remove one of the dimensions of the data. You may want to compute average tax rates across states for each year or average tax rates over years for each state. You can compute these sets of summary statistics by using the `collapse` command, which produces a dataset of summary statistics over the elements of its `by(`*varlist*`)` option. The command syntax is

6. You can use Stata's `tsfill` command to generate such a structure from an unbalanced panel.

`collapse` *clist* [*if*] [*in*] [*weight*] [, *options*]

where the *clist* is a list of [(*stat*)] *varlist* pairs or a list of [(*stat*)] *target_var*=*varname* pairs. In the first format, the *stat* may be any of the descriptive statistics available with `summarize` (see [R] **summarize**), with some additions: e.g., all 100 percentiles may be computed. If not specified, the default *stat* is `mean`. To compute more than one summary statistic for a given variable, use the second form of the command, where the *target_var* names the new variable to be created. The `by(`*varlist*`)` option specifies that `collapse` generate one result observation for each unique value of the `by(`*varlist*`)`. For more information on the `collapse` syntax, see [D] **collapse**.

The `grunfeld` dataset contains annual firm-level data for 10 U.S. firms over 20 years, as the output from `tsset` shows. We first `summarize` three variables over the entire panel:

```
. use http://www.stata-press.com/data/imeus/grunfeld, clear
. tsset
       panel variable:  company, 1 to 10
        time variable:  year, 1935 to 1954
. summarize mvalue invest kstock
```

Variable	Obs	Mean	Std. Dev.	Min	Max
mvalue	200	1081.681	1314.47	58.12	6241.7
invest	200	145.9583	216.8753	.93	1486.7
kstock	200	276.0172	301.1039	.8	2226.3

After using `preserve` to retain the original data, we use `collapse` by `year` to generate the mean market value, sum of investment expenditures, and mean of firms' capital stock for each year.

```
. preserve
. collapse (mean) mvalue (sum) totinvYr=invest (mean) kstock, by(year)
. graph twoway tsline mvalue totinvYr kstock
```

I plot these time-series data in figure 3.1 to illustrate that the cross-sectional summary statistics trend upward over these two decades.[7]

7. I present Stata graphics in this text without explaining the syntax of Stata's graphics language. For an introduction to Stata graphics, see `help graph intro` and [G] **graph intro**. For an in-depth presentation of Stata's graphics capabilities, see *A Visual Guide to Stata Graphics* (Mitchell 2004).

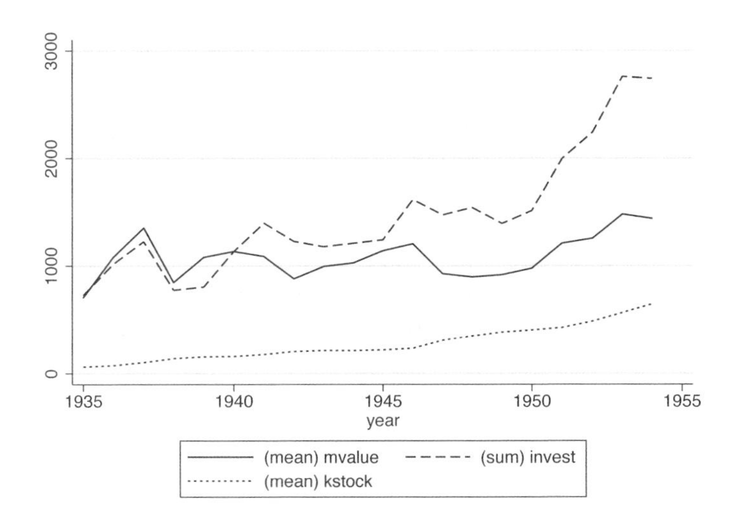

Figure 3.1: Graph of panel data collapsed to time series

When performing data transformations on panel data, you should take advantage of the time-series operators' housekeeping facilities. Consider the dataset above. If the lagged value of `mvalue` is generated with `mvalue[_n-1]`, you must explicitly exclude the first observation of each firm from the computation. Otherwise its lagged value would refer to the last observation of the prior firm for firms 2, ..., 10. In contrast, you can use

```
. generate lagmvalue = L.mvalue
```

without considering the panel nature of the data. Each firm's first observation of `lagmvalue` will be defined as missing.

Stata's commands for panel data are described in [XT] **xt** and [XT] **intro**. Each command's name begins with `xt`. Chapter 9 introduces some estimation techniques for panel data in economic analysis.

3.5 Tools for manipulating panel data

Section 3.4 introduces balanced and unbalanced panel (longitudinal) data and shows how to use the `collapse` command to create a pure time series or a pure cross section from panel data. As described in section 3.4, you should always use `tsset` to set up panel data so that you can use Stata's time-series operators and `xt` commands. Sometimes translating an external date format into Stata's date variables is cumbersome. Say that you import a time series, perhaps with a spreadsheet-formatted date variable, and want to establish a time-series calendar for these data. You must work with the existing date to make it a Stata date variable by using the `date()` or `mdy()` functions, assign a format to the variable (e.g., %ty for annual, %tm for monthly), and then use tsset to define the time-series calendar with that formatted date variable. You can use the tsmktim utility to do all these steps together (Baum and Wiggins 2000). You need only specify the name of a new time-series calendar variable and its start date:

```
. tsmktim datevar, start(1970)
```

A new version of the routine (available from ssc) allows you to generate a date variable for each unit within a panel:

```
. tsmktim datevar, start(1970q3) i(company)
```

Each unit's series must start (but need not end) in the same period.

3.5.1 Unbalanced panels and data screening

Researchers organizing panel data often apply particular conditions to screen data. For instance, unbalanced firm-level panel data may have only one or two annual observations available for particular firms rather than the 20 years available for more-established firms in the sample. Here I discuss several commands that you can use before estimation to manipulate panel data organized in the *long* format (see section 3.6).

You can use the xtdes command to describe the pattern of panel data, particularly to determine whether the panel is balanced or unbalanced.[8] This command leaves scalars in r(): r(N) gives the number of panel units, with r(min) and r(max) giving the minimum and maximum number of observations available per unit. If those two items are equal, the panel is balanced. As discussed earlier, you could use tsfill to "rectangularize" an unbalanced panel by filling in missing observations with missing values. But what if you wanted to remove any panel units with fewer than the maximum number of observations? You could create a new variable counting observations within each unit as a by-group and use that variable to flag units with missing observations:

```
. xtdes
. local maxobs = r(max)
. by company: generate obs = _N
. drop if obs < `maxobs'
```

Often an unbalanced panel is preferable, but we may want to screen out units that have fewer than m observations; if $m = 10$, for instance, drop if obs < 10 would remove units failing to pass that screen.

Using this logic will ensure that each firm has the minimum number of observations, but it cannot guarantee that they are without gaps. Some of Stata's time-series estimation and testing routines do not allow gaps within time series, although the tsset command saves macros r(tmins) and r(tmaxs) to signal the first and last periods of any unit of the panel.

8. It is tempting to imagine that this command is named xtdesc, but it is not.

Nicholas Cox's routine `tsspell` (available from `ssc`) identifies complete runs of data, or "spells", within a time series or within a panel of time series. A gap terminates the spell and starts a new spell following the gap. Thus `obs == 'maxobs'` would correspond to a unit with a spell of `'maxobs'`. The routine is general and may be used to identify spells on the basis of a logical condition (for instance, the sign of a variable such as GDP growth changing from positive to negative or a variable identifying the party in power changing). We will use a simpler aspect of the routine to identify gaps in the time series as shown by the calendar variable.

Consider the missing data in a modified version of the *Stata Longitudinal/Panel Data Reference Manual* `grunfeld` dataset. The original dataset contains a balanced panel of 20 years of annual data on 10 firms. In the modified version, five of those firms lack one or more observations: one firm "starts late", one firm "ends early", and three firms' series have embedded gaps:

```
. use http://www.stata-press.com/data/imeus/grunfeldGaps, clear
. xtdes
 company:  1, 2, ..., 10                                      n =          10
    year:  1935, 1936, ..., 1954                              T =          20
           Delta(year) = 1; (1954-1935)+1 = 20
           (company*year uniquely identifies each observation)
Distribution of T_i:   min      5%     25%       50%       75%     95%     max
                        17      17      18        20        20      20      20
     Freq.  Percent    Cum. |  Pattern
 ---------------------------+----------------------
        5     50.00   50.00 |  11111111111111111111
        1     10.00   60.00 |  ..111111111111111111
        1     10.00   70.00 |  111111111.1111.11111
        1     10.00   80.00 |  111111111111...11111
        1     10.00   90.00 |  1111111111111.111111
        1     10.00  100.00 |  111111111111111111..
 ---------------------------+----------------------
       10    100.00         |  XXXXXXXXXXXXXXXXXXXX
```

We identify these conditions by using `tsspell` with the condition `D.year == 1`.[9] For series with gaps, that condition will fail. The `tsspell` routine creates three new variables, `_spell`, `_seq`, and `_end`, and we are concerned with `_spell`, which numbers the spells in each firm's time series. A firm with one unbroken spell (regardless of starting and ending dates) will have `_spell = 1`. A firm with one gap will have later observations identified by `_spell = 2`, and so on. To remove all firms with embedded gaps, we may apply similar logic to that above to drop firms with more than one reported spell:

```
. tsspell year, cond(D.year == 1)
. egen nspell = max(_spell), by(company)
. drop if nspell > 1
(54 observations deleted)
```

9. This condition would work for any other Stata data frequency, since half-years, quarters, months, weeks, and days are also stored as successive integers.

```
. xtdes
 company:  2, 3, ..., 10                                    n =          7
    year:  1935, 1936, ..., 1954                            T =         20
           Delta(year) = 1; (1954-1935)+1 = 20
           (company*year uniquely identifies each observation)
Distribution of T_i:   min      5%     25%       50%       75%     95%     max
                        18      18      18        20        20      20      20
     Freq.  Percent    Cum. |  Pattern
 ---------------------------+----------------------
        5     71.43   71.43 |  11111111111111111111
        1     14.29   85.71 |  ..111111111111111111
        1     14.29  100.00 |  111111111111111111..
 ---------------------------+----------------------
        7    100.00         |  XXXXXXXXXXXXXXXXXXXX
```

Or if we were willing to retain firms with gaps but did not want to keep any spell shorter than a certain length (say, 5 years) we could use the `_seq` variable, which counts the length of each spell, and include `_spell` in the `egen` command that computes the maximum over each firm and spell within the firm's observations:

```
. use http://www.stata-press.com/data/imeus/grunfeldGaps, clear
. tsspell year, cond(D.year == 1)
. replace _spell = F._spell if _spell == 0
(14 real changes made)
. egen maxspell = max(_seq+1), by(company _spell)
. drop if maxspell < 5
(4 observations deleted)
. xtdes
 company:  1, 2, ..., 10                                    n =         10
    year:  1935, 1936, ..., 1954                            T =         20
           Delta(year) = 1; (1954-1935)+1 = 20
           (company*year uniquely identifies each observation)
Distribution of T_i:   min      5%     25%       50%       75%     95%     max
                        14      14      18        20        20      20      20
     Freq.  Percent    Cum. |  Pattern
 ---------------------------+----------------------
        5     50.00   50.00 |  11111111111111111111
        1     10.00   60.00 |  ..111111111111111111
        1     10.00   70.00 |  111111111......11111
        1     10.00   80.00 |  111111111111...11111
        1     10.00   90.00 |  1111111111111.111111
        1     10.00  100.00 |  111111111111111111..
 ---------------------------+----------------------
       10    100.00         |  XXXXXXXXXXXXXXXXXXXX
```

The resulting dataset includes firms' spells of 5 years or more. The `_spell` variable is recoded from 0 to its following value; by default, `tsspell` places a zero in the first observation of each spell, but we want that value to be the current spell number. Likewise, because the `_seq` variable starts counting from zero, we consider the maximum value of (`_seq + 1`) to evaluate the spell length. Experimenting with `tsspell` will reveal its usefulness in enforcing this type of constraint on the data.

3.5.2 Other transforms of panel data

Some analyses require smoothing the data in each panel; tssmooth (see [TS] **tssmooth**) provides the most widely used smoothers, all of which can be applied to the data in each panel. For example, we might want a weighted moving average of four prior values, with arithmetic weights 0.4(0.1)0.1. That construct can be viewed as a filter applied to a series in the time domain and computed with tssmooth ma:

```
. tssmooth ma wtavg = invest, weights(0.1(0.1)0.4 <0>)
```

The weights are applied to the fourth, third, second, and first lags of invest, respectively, to generate the variable wtavg. The <0> is a placeholder to instruct Stata that the zero-lag term should be given a weight of zero. This command can also be used to impose a two-sided filter with varying weights:

```
. tssmooth ma wtavg  = invest, weights(1 4 <6> 4 1)
```

This command specifies that a two-sided centered moving average be computed, with weights 1/16, 4/16, 6/16, 4/16, and 1/16. You can apply the tssmooth ma command to panel data because the filter is automatically applied separately to each time series within the panel.

Other analyses use functions of the extreme values in each series. For example, the record() egen part of Nicholas Cox's egenmore package (available from ssc) provides one solution. For example,

```
. egen maxtodate = record(wage), by(id) order(year)
. egen hiprice = record(share_price), by(firm) order(quote_date)
```

The first example identifies the highest wage to date in a worker's career, whereas the second identifies the highest price received to date for each firm's shares.

3.5.3 Moving-window summary statistics and correlations

When working with panel data, you often want to calculate summary statistics for subperiods of the time span defined by the panel calendar. For instance, if you have 20 years' data on each of 100 firms, you may want to calculate 5-year averages of their financial ratios. You can calculate these averages with the tabstat (see [R] **tabstat**) command. You need only define the 5-year periods as the elements of a by-group and specify that selector variable as the arguments to tabstat's by() option, while prefixing the command with by firmid:. However, to retrieve the computed statistics, you will need to use the save option, which stores them in several matrices. Or you could use several egen statements to generate these statistics as new variables, using the same by-group strategy.

To compute summary statistics from *overlapping* subsamples, we could define a by-group, but here Stata's by capabilities cannot compute statistics from a sequence of by-groups that are formed by a "moving window" with, for example, 11 months' overlap. The mvsumm routine of Baum and Cox (available from [R] ssc) computes any of the univariate statistics available from summarize, detail and generates a time series containing that statistic over the defined time-series sample. You can specify the window width (the number of periods included in the statistic's computation) as an option, as well as the alignment of the resulting statistic with the original series. This routine is especially handy for financial research, in which some measure of recent performance—the average share price over the last 12 months or the standard deviation (volatility) of the share price over that interval—is needed as a regressor. The mvsumm routine will operate separately on each time series of a panel, as long as a panel calendar has been defined with tsset.

Another way to generate moving-window results is to use Stata's rolling prefix, which can execute any statistical command over moving windows of any design. However, rolling is more cumbersome than mvsumm, since it creates a new dataset containing the results, which then must be merged with the original dataset.

To calculate a moving correlation between two time series for each unit of a panel, you can use Baum and Cox's mvcorr routine (available from ssc). This computation is useful in finance, where computing an optimal hedge ratio involves computing just such a correlation, for instance, between spot and futures prices of a particular commodity. Since the mvcorr routine supports time-series operators, it allows you to compute moving autocorrelations. For example,

```
. mvcorr invest L.invest, win(5) gen(acf) end
```

specifies that the first sample autocorrelation of an investment series be computed from a five-period window, aligned with the last period of the window (via option end), and placed in the new variable acf. Like mvsumm, the mvcorr command operates automatically on each time series of a panel:[10]

```
. use http://www.stata-press.com/data/imeus/grunfeld, clear
. drop if company>4
(120 observations deleted)
. mvcorr invest mvalue, window(5) generate(rho)
. xtline rho, yline(0) yscale(range(-1 1))
> byopts(title(Investment vs. Market Value: Moving Correlations by Firm))
```

Figure 3.2 shows the resulting graph of four firms' investment-market value correlations.

10. For a thorough presentation of the many styles of graphs supported by Stata, see Mitchell (2004).

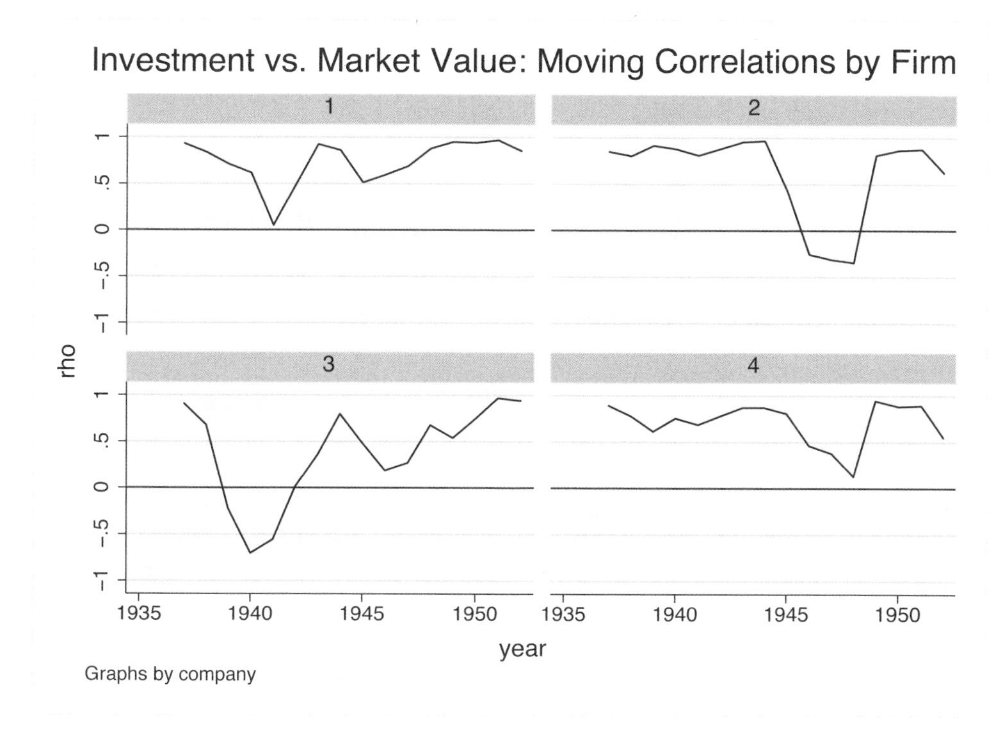

Figure 3.2: Moving-window correlations

3.6 Combining cross-sectional and time-series datasets

Applied economic analysis often involves combining datasets. You may want to pool the data over different cross-sectional units or build a dataset with both cross-sectional and time-series characteristics. In the former case, you may have 200 observations that reflect probable voters' responses to a telephone survey carried out in Philadelphia, 300 observations from that same survey administered in Chicago, and 250 observations from voters in Kansas City. In the latter case, you may have a dataset for each of the six New England states containing annual state disposable personal income and population for 1981–2000. You may want to combine those six datasets into one dataset. How you combine them—over cross sections or over the cross-section and time-series dimensions—depends on the type of analysis you want to do. We may want to work with data in what Stata calls *wide format*, in which measurements on the same variable at different points in time appear as separate variables. For instance, we might have time series of population for each of the New England states, with variables named `CTpop`, `MApop`, `MEpop`, ... In contrast, you may find it easier to work with *long-format* data, in which those same data are *stacked*, so that you have one variable, `pop`, with sets of observations associated with each state. You then must define another variable that identifies the unit (here the state). Stata has commands for each type of combination, as well as the `reshape` command for transforming data between the wide and long formats. We first discuss combining cross-sectional datasets to produce a pooled dataset in the long format.

3.7 Creating long-format datasets with append

If we have the three voter survey datasets for Philadelphia, Chicago, and Kansas City mentioned above and want to combine them into one pooled dataset, we must first ensure that each dataset's variable names are identical. We will use the `append` command, and since `prefBush` and `prefBUSH` are different variables to Stata, we cannot combine files containing those variables and expect them to properly align. We can use `rename` to ensure that variable names match. We will also want to be able to recover the city identifier from the combined dataset, even if it is not present in the individual datasets. We could remember that the first 200 observations come from Philadelphia, the next 300 from Chicago, and so on, but if the dataset is ever sorted into a different order, we will lose this identifier. Thus we should insert a new variable, `city`, into each dataset, which could be either a numeric variable with a value label of the city's name or a string variable that can be `encode`d into numeric form for use in a `by` *varlist*`:`. We can then use `append` to combine them:

```
. use philadelphia, clear
. append using chicago
. append using kcity
. save vote3cities, replace
```

If we have a string variable, `city`, containing the city name, we can use

```
. encode city, gen(citycode)
```

to create a numeric city identifier. We could then use the `citycode` variable in a statistical command or to create a set of city-specific dummy variables with the `tabulate` command's `generate()` option; see [R] **tabulate oneway**. Although our example illustrates how three datasets can be combined, any number of datasets could be combined in this manner.

This dataset is in Stata's long format; for each variable, the measurements for each unit within the panel are stored in separate observations. Since we have combined three pure cross sections, each of arbitrary order, there is no relationship among respondent #1 in Philadelphia, respondent #1 in Chicago, and respondent #1 in Kansas City. But this long format makes it easy to do many computations that would be cumbersome if the data were combined any other way. For instance, you can easily compute whether Philadelphia's mean of `prefBush` is equal to Chicago's or whether all three means of `prefBush` are statistically distinguishable. That would not be the case if the three datasets' variables were recombined horizontally rather than vertically.

3.7.1 Using merge to add aggregate characteristics

The long-format dataset we constructed above is useful if we want to add aggregated information to individual records. For instance, imagine that the voter survey data contain each individual's five-digit ZIP code (zipcode) along with his or her preferences for the presidential candidates, and we want to evaluate whether the voter's income level as proxied by the average income in her ZIP code affects her voting preferences. How may we append the income information to each record? We could use a sequence of replace statements or a complicated nested cond() function. But you can easily create a new dataset containing ZIP codes (in the same five-digit integer form) and average income levels. If these data are acquired from the U.S. Census, you should have records for every Philadelphia (or Chicago, or Kansas City) ZIP code or for the entire state in which each city is located. But you will be using this file merely as a lookup table. We can then sort the dataset by ZIP code and save it as incbyzip.dta. We can combine this information with the original file by using the following commands:

```
. use vote3cities, clear
. sort zipcode
. merge zipcode using incbyzip, nokeep
```

Using merge in this way is known as a *one-to-many* match-merge where the income for each ZIP code is added to each voter record in that ZIP code. The zipcode variable is the *merge key*. Both the *master file* (vote3cities.dta) and the *using file* (incbyzip.dta) must be sorted by the merge key. By default, merge creates a new variable, _merge, which takes on an integer value of 1 if that observation was found only in the master dataset, 2 if it was found only in the using dataset, or 3 if it was found in both datasets. Here we expect tab _merge to reveal that all values equal 3. Each voter's ZIP code should be mapped to a known value in the using file. Although many ZIP codes in the using file may not be associated with any voter in the sample, which would yield a _merge of 2, we specified the nokeep option to drop the unneeded entries in the using file from the merged file. We could then use

```
. assert _merge == 3
. drop _merge
```

to verify that the match was successful. Using merge is much easier than using a long and complicated do-file that uses replace. By merely modifying the using file, we can correct any problems in the one-to-many merge. If we had several ZIP code–specific variables to add to the record, such as average family size, average proportion minority, and average proportion of home ownership, we could handle these variables with one merge command. This technique is useful for working with individual data for persons, households, plants, firms, states, provinces, or countries when we want to combine the microdata with aggregates for ZIP codes, cities, or states at the individual level; industries or sectors at the plant or firm level; regions or the macroeconomy at the state or province level; and global regions or world averages at the country level.

3.7.2 The dangers of many-to-many merges

You can also use the `merge` command to combine datasets by using a *one-to-one* match-merge. For example, we might have two or more datasets whose observations pertained to the same units: e.g., U.S. state population figures from the 1990 and 2000 Censuses.

The many-to-many merge is a potential problem that arises when there are multiple observations in both datasets for some values of the merge key variable(s). Match-merging two datasets that both have more than one value of the merge key variable(s) can cause repeated execution of the same do-file to have a different number of cases in the result dataset without indicating an error. A coding error in one of the files usually causes such a problem. You can use the `duplicates` command to track down such errors. To prevent such problems, specify either the `uniqmaster` or `uniqusing` option in a match-merge. For instance, the ZIP code data should satisfy `uniqusing` in that each ZIP code should be represented only once in the file. In a one-to-one match-merge, such as the state income and population data, you could use the `unique` option, as it implies both `uniqmaster` and `uniqusing` and asserts that the merge key be unique in both datasets.[11]

Beyond `append` and `merge`, Stata has one more command that combines datasets: `joinby` is used less often since its task is more specialized. The command creates a new dataset by forming all possible pairwise combinations of the two datasets, given the merge key. Usually, you will want to use `merge` instead of `joinby`.[12]

3.8 The reshape command

If your dataset is organized in long or wide format, you may need to reorganize it to more easily obtain data transformations, statistics, or graphics. To solve this problem, you can use the `reshape` (see [D] **reshape**) command, which reorganizes a dataset in memory without modifying data files. Some statistical packages do not have a reshape feature, so you would need to write the data to one or more external text files and read it back in. With `reshape`, this extra step is not necessary in Stata, but you must label the data appropriately. You may need to do some experimentation to construct the appropriate command syntax, which is all the more reason for using a do-file, since someday you are likely to come upon a similar application for `reshape`.

Consider a wide-format dataset with variables labeled `pop1970`, `pop1980`, `pop1990`, `pop2000`, and `area`, and with observations identified by each of the six New England state codes.

11. Those familiar with relational database-management systems, such as SQL, will recognize that *uniqueness* means that the merge key is a valid and unique *primary key* for the dataset.

12. Those familiar with relational database-management systems will recognize `joinby` as the SQL *outer join*: a technique to be avoided in most database tasks.

```
. use http://www.stata-press.com/data/imeus/reshapeState, clear
. list
```

	state	pop1970	pop1980	pop1990	pop2000	area
1.	CT	.1369841	.6184582	.4241557	.2648021	.871691
2.	MA	.6432207	.0610638	.8983462	.9477426	.4611429
3.	ME	.5578017	.5552388	.5219247	.2769154	.4216726
4.	NH	.6047949	.8714491	.8414094	.1180158	.8944746
5.	RI	.684176	.2551499	.2110077	.4079702	.0580662
6.	VT	.1086679	.0445188	.5644092	.7219492	.6759487

We want to reshape the dataset into long format so that each state's and year's population value will be recorded in one variable. We specify `reshape long pop` so that the variable to be placed in the long format will be derived from all variables in the dataset whose names start with `pop`. The command works with $x_{i,j}$ data; in Stata, i defines a panel, and j defines an identifier that varies within each panel. Here the state defines the panel, so we specify that `state` is the `i()` variable; here the `j()` option specifies that the suffixes of the `pop` variable be retained as a new variable, `year`:

```
. reshape long pop, i(state) j(year)
(note: j = 1970 1980 1990 2000)
Data                               wide   ->   long
-----------------------------------------------------------------------------
Number of obs.                        6   ->     24
Number of variables                   6   ->      4
j variable (4 values)                     ->   year
xij variables:
              pop1970 pop1980 ... pop2000 ->   pop
-----------------------------------------------------------------------------
```

The 6 observations of the original wide dataset have been expanded to 24, since each state had four population figures in the original form:

(Continued on next page)

```
. list

     +-----------------------------------+
     | state   year        pop      area |
     |-----------------------------------|
  1. |    CT   1970   .1369841   .871691 |
  2. |    CT   1980   .6184582   .871691 |
  3. |    CT   1990   .4241557   .871691 |
  4. |    CT   2000   .2648021   .871691 |
  5. |    MA   1970   .6432207  .4611429 |
     |-----------------------------------|
  6. |    MA   1980   .0610638  .4611429 |
  7. |    MA   1990   .8983462  .4611429 |
  8. |    MA   2000   .9477426  .4611429 |
  9. |    ME   1970   .5578017  .4216726 |
 10. |    ME   1980   .5552388  .4216726 |
     |-----------------------------------|
 11. |    ME   1990   .5219247  .4216726 |
 12. |    ME   2000   .2769154  .4216726 |
 13. |    NH   1970   .6047949  .8944746 |
 14. |    NH   1980   .8714491  .8944746 |
 15. |    NH   1990   .8414094  .8944746 |
     |-----------------------------------|
 16. |    NH   2000   .1180158  .8944746 |
 17. |    RI   1970    .684176  .0580662 |
 18. |    RI   1980   .2551499  .0580662 |
 19. |    RI   1990   .2110077  .0580662 |
 20. |    RI   2000   .4079702  .0580662 |
     |-----------------------------------|
 21. |    VT   1970   .1086679  .6759487 |
 22. |    VT   1980   .0445188  .6759487 |
 23. |    VT   1990   .5644092  .6759487 |
 24. |    VT   2000   .7219492  .6759487 |
     +-----------------------------------+
```

In the wide format, the observations are labeled $i = 1, \ldots,$ _N, and each measurement to be transformed to the long form consists of variables indexed by $j = 1, \ldots, J$. The *varlist* for `reshape long` lists the *base names* or *stubs* of all variables that are in the $x_{i,j}$ form and should be `reshaped` to the long form. Here we have only `pop` with $J = 4$ because there are four decennial census years of population data; this same $x_{i,j}$ format may include several variables. For instance, our dataset might contain additional variables `popM1970`, `popF1970`, ..., `popM2000`, `popF2000` with gender breakdowns of state population. The `reshape long` statement's *varlist* would then read `pop popM popF` because these variables are to be treated analogously.

You must specify `j()`, as the long format requires the j identifier. Here the j dimension is the year, but it could be any characteristic. Instead of state population measures for different years, we could have `popWhite`, `popBlack`, `popHispanic`, and `popAsian`. Then we would use the options `j(race) string`, to specify that the j identifier is the `string` variable `race` (which you could then `encode`). Variables that do not vary over j (`year` or `race`) are not specified in the `reshape` statement. In the example above, the state's `area` is time invariant, so it is automatically replicated for each `year`.

We continue with the long-format dataset that results from the example above but now want the data in wide format. We then use `reshape wide` to specify that the `pop` variable be spread over the values of `j(year)`. The rows of the resulting wide-format dataset are defined by the `i(state)` option:

```
. reshape wide pop, i(state) j(year)
(note: j = 1970 1980 1990 2000)
Data                               long   ->   wide
-----------------------------------------------------------------------------
Number of obs.                       24   ->      6
Number of variables                   4   ->      6
j variable (4 values)              year   ->   (dropped)
xij variables:
                                    pop   ->   pop1970 pop1980 ... pop2000
-----------------------------------------------------------------------------
```

This command is the same as the `reshape long` in the prior example, with `long` replaced by `wide`. The same information is required: you must specify the variables to be widened (here named explicitly, not by stubs), the panel's i variable, and the within-panel identifier (j variable). In creating the wide-format data, the j variable is dropped because its values are now spread over the columns `pop1970`, `pop1980`, `pop1990`, and `pop2000`. To illustrate,

```
. list
```

	state	pop1970	pop1980	pop1990	pop2000	area
1.	CT	.1369841	.6184582	.4241557	.2648021	.871691
2.	MA	.6432207	.0610638	.8983462	.9477426	.4611429
3.	ME	.5578017	.5552388	.5219247	.2769154	.4216726
4.	NH	.6047949	.8714491	.8414094	.1180158	.8944746
5.	RI	.684176	.2551499	.2110077	.4079702	.0580662
6.	VT	.1086679	.0445188	.5644092	.7219492	.6759487

You need to choose appropriate variable names for `reshape`. If our wide dataset contained `pop1970`, `Pop1980`, `popul1990`, and `pop2000census`, you would not be able to specify the common stub labeling the choices. However, say that we have for each state the measures `pop1970`, `pop1970M`, and `pop1970F`. The command

```
. reshape long pop pop@M pop@F, i(state) j(year)
```

uses `@` as a placeholder for the location of the j component of the variable name. Similarly, in our `race` example, if the variables were named `Whitepop`, `Blackpop`, `Hispanicpop`, and `Asianpop`, the command

```
. reshape long @pop, i(state) j(race) string
```

would handle those names. In more difficult cases, where repeatedly using `rename` may be tedious, `renvars` (Cox and Weesie 2001) may be useful.

This discussion has only scratched the surface of `reshape`. See [D] **reshape** for more information, and experiment with the command.

3.8.1 The xpose command

You can use the `xpose` command to make radical changes to the organization of your data. This command turns observations into variables and vice versa. This functionality is common in spreadsheets and matrix languages, but it is rarely useful in Stata because applying `xpose` will usually destroy the contents of string variables. If all variables in the dataset are numeric, this command may be useful. Rather than using `xpose`, consider reading in the raw data with the `byvariable()` option of `infile` (see [D] **infile (free format)**). If you need to transpose the data, they were probably not created sensibly in the first place.

3.9 Using Stata for reproducible research

3.9.1 Using do-files

Stata's command-line syntax makes it easy to document your data transformations, statistical analyses, and graphics creation. For some users, the command line is a nuisance, so they applauded Stata's dialogs when they appeared in version 8. But even Stata's dialogs produce complete commands in the Review window.

Stata does not require you to keep a record of the commands you issued, but most research requires that you be able to reproduce findings. Unless you carefully document every step of the process, you will not be able to reproduce your findings later, which can be disastrous. With Stata, you can document your research by using a do-file.

A do-file contains a sequence of Stata commands and can be invoked by selecting File ▷ Do... from the menu, by double-clicking on its icon, or by issuing the `do` command at the command line. Normally, a do-file will stop if it encounters an error. You can construct a do-file by using the Do-file Editor in Stata or any text editor. A do-file is merely a text file. You can include comments in a do-file, as long as they follow Stata conventions. Including such comments can help you remember what you did in your research. Placing a creation or revision date in a do-file is good practice.

An entire research project can involve hundreds or thousands of Stata commands from start to finish, so including all of them in a massive do-file would be cumbersome. Instead, consider writing a *master* do-file that calls a sequence of do-files to perform each step of the process: data input, data validation and cleaning, data transformation, statistical analyses, and production of graphical and tabular output. Each step might in turn be carried out by several do-files. If you follow this strategy, then it becomes straightforward to redo some step of the analysis in case of error or to rebuild a `.dta` file if it is inadvertently modified.

This strategy of using modular do-files for each step of the research project works well when you need to conduct a parallel analysis on another dataset. Many survey datasets have annual waves. Your processing of the latest wave of the survey will probably follow many of the same steps as last year's. With a well-organized and well-documented set of do-files, you need only copy those files and apply them to the new set of data.

No software package will force you to be a responsible researcher, but Stata makes following good research habits easy so that you can return to a research project and see exactly what you did to generate the final results.

3.9.2 Data validation: assert and duplicates

Before you can effectively manage data, you need to make sure that all the values of the raw data make sense. Are there any apparent coding errors in the data values? Should any values of numeric variables properly be coded as some sort of missing data, as discussed above? As mentioned, to construct an audit trail for your data management, you can create a do-file that reads the raw data, applies several checks to ensure that data values are appropriate, and writes the initial Stata binary data file. This data file should not be modified in later programs or interactive analysis. Each program that uses the file and creates additional variables, subsets, or merges of the data should save the resulting modified file under a new name. Each step in the data validation and transformation process can then be documented and reexecuted as needed. Even if the raw data are provided in Stata binary format from an official source, you should assume that there are coding errors.

You should follow this method from the beginning of the data-management process. On Statalist, users often say such things as, "I did the original data transformations (or merges) in Excel, and now I need to ...." Even if you are more familiar with a spreadsheet syntax than with the Stata commands needed to replicate that syntax, you should use Stata so that you can document and reproduce its operations on the data. Consider two research assistants starting with the same set of 12 spreadsheets. They are instructed to construct one spreadsheet performing some complicated append or merge processes by using copy and paste. What is the probability that the two assistants will produce identical results? Probably less than one.

The proposed solution: export the 12 spreadsheets to text format, and read them into Stata by using a do-file that loops over the `.txt` or `.csv` files and applies the same transformations to each one, performing the appropriate `append` or `merge` operations. That do-file, once properly constructed, will produce a reproducible result. You can easily modify the do-file to perform a similar task, such as handling 12 spreadsheets containing cost elements rather than revenues. You can (and should) add comments to the do-file documenting its purpose, dates of creation/modification, and creator/modifier. You may either place an asterisk (`*`) at the beginning of each comment line, use the block comment syntax (`/*` to begin a comment, `*/` to end it) to add several lines of comments to a do-file, or use two forward slashes (`//`) to add a comment after a command but on the same line. Although you will need to learn Stata's programming features to set up these do-files, doing so is well worth the effort.

First, use describe and summarize to get useful information about the data you have imported (typically by using insheet, infile, or infix).[13] Consider a version of the census2a dataset that has been altered to illustrate data validation:

```
. use http://www.stata-press.com/data/imeus/census2b, clear
(Version of census2a for data validation purposes)
. describe
Contains data from census2b.dta
  obs:            50                          Version of census2a for data
                                                validation purposes
 vars:             5                          23 Sep 2004 15:49
 size:         1,850 (99.9% of memory free)
-------------------------------------------------------------------------------
              storage  display     value
variable name   type   format      label      variable label
-------------------------------------------------------------------------------
state           str14  %14s
region          str7   %9s
pop             float  %9.0g
medage          float  %9.0g
drate           float  %9.0g
-------------------------------------------------------------------------------
Sorted by:
. summarize
    Variable |       Obs        Mean    Std. Dev.       Min        Max
-------------+--------------------------------------------------------
       state |         0
      region |         0
         pop |        49     4392737     4832522         -9   2.37e+07
      medage |        50       35.32    41.25901       24.2        321
       drate |        50       104.3    145.2496         40       1107
```

The log displays the data types of the five variables. The first two are string variables (of maximum length 14 and 7 characters, respectively), whereas other three are float variables. These data types appear to be appropriate to the data.

The descriptive statistics reveal several anomalies for the numeric variables. Population data appear to be missing for one state, which is clearly an error. Furthermore, population takes on a negative value for at least one state, indicating some coding errors. We know that the values of U.S. states' populations in recent decades should be greater than several hundred thousand but no more than about 30 million. A median age of 321 would suggest that Ponce de Leon is alive and well. Likewise, the drate (death rate) variable has a mean of 104 (per 100,000), so a value of 10 times that number suggests a coding error.

Rather than just firing up the Data Editor and visually scanning for the problems sprinkled through this small illustrative dataset, we are interested in data-validation techniques that we can apply to datasets with thousands of observations. We use assert

13. See appendix A for an explanation of these input commands.

to check the validity of these three variables, and in case of failure, we list the offending observations. If all checks are passed, this do-file should run without error:

```
use http://www.stata-press.com/data/imeus/census2b, clear
                                                    // check pop
list if pop < 300000 | pop > 3e7
assert pop < . & pop > 300000 & pop <= 3e7
                                                    // check medage
list if medage <= 20 | medage >= 50
assert  medage > 20  & medage < 50
                                                    // check drate
list if drate < 10   | drate >= 104+145
assert  drate < 10   & drate < 104+145
```

The first `list` command shows that population should not be missing (`< .`, as above), that it should be at least 300,000, and that it should be less than 30 million (3.0×10^7). By reversing the logical conditions in the `list` command, we can assert that all cases have valid values for `pop`.[14] Although the `list` command uses `|`, Stata's "or" operator, the `assert` command uses `&`, Stata's "and" operator, because each condition must be satisfied. Likewise, we `assert` that each state's median age should be between 20 and 50 years. Finally, we assert that the death rate should be at least 10 per 100,000 and less than $\widehat{\mu} + \widehat{\sigma}$ from that variable's descriptive statistics. Let's run the data-validation do-file:

```
. use http://www.stata-press.com/data/imeus/census2b, clear
(Version of census2a for data validation purposes)
. list if pop < 300000 | pop > 3e7
```

	state	region	pop	medage	drate
4.	Arkansas	South	-9	30.6	99
10.	Georgia	South	.	28.7	81
15.	Iowa	N Cntrl	0	30	90

```
. assert pop <. & pop > 300000 & pop <= 3e7
3 contradictions in 50 observations
assertion is false
r(9);

end of do-file
r(9);
```

The do-file fails to run to completion because the first `assert` finds three erroneous values of `pop`. We should now correct these entries and rerun the do-file until it executes without error. This little example could be expanded to a really long do-file that checked each of several hundred variables, and it would exit without error if all assertions are satisfied.

We can use `tabulate` to check the values of string variables in our dataset. In the `census2b` dataset, we will want to use `region` as an identifier variable in later analysis, expecting that each state is classified in one of four U.S. regions.

14. Strictly speaking, we need not apply the `< .` condition, but it is good form to do so since we might not have an upper bound condition.

```
. use http://www.stata-press.com/data/imeus/census2b, clear
(Version of census2a for data validation purposes)

. list state if region==""

       +--------+
       |  state |
       |--------|
    2. | Alaska |
   11. | Hawaii |
       +--------+

. tabulate region

     region |      Freq.     Percent        Cum.
------------+-----------------------------------
    N Cntrl |         12       25.00       25.00
         NE |          9       18.75       43.75
      South |         16       33.33       77.08
       West |         11       22.92      100.00
------------+-----------------------------------
      Total |         48      100.00

. assert r(N) == 50
assertion is false
r(9);

end of do-file
r(9);
```

The tabulation reveals that only 48 states have `region` defined. We can use one of the items left behind by the `tabulate` command: `r(N)`, the total number of observations tabulated.[15] Here the assertion that we should have 50 defined values of `region` fails, and a list of values where the variable equals string missing (the null string) identifies Alaska and Hawaii as the misclassified entries.

Validating data with `tabulate` can also generate cross tabulations. Consider, for instance, a dataset of medical questionnaire respondents in which we construct a two-way table of `gender` and `NCPreg`, the number of completed pregnancies. Not only should the latter variable have a lower bound of zero and a sensible upper bound, its cross tabulation with `gender=="Male"` should yield only zero values.

We can use `duplicates` to check string variables that should take on unique values. This command can handle much more complex cases—in which a combination of variables must be unique (or a so-called *primary key* in database terminology),[16] but we will apply it to the single variable `state`:

15. Section 2.2.12 discusses the `return list` from r-class commands, such as `tabulate`.

16. As an example: U.S. senators' surnames may not be unique, but the combination of surname and state code almost surely will be unique.

```
. use http://www.stata-press.com/data/imeus/census2b, clear
(Version of census2a for data validation purposes)
. duplicates list state
Duplicates in terms of state

  +----------------+
  | obs:     state |
  |----------------|
  |   16    Kansas |
  |   17    Kansas |
  +----------------+

. assert r(sum) == 0
assertion is false
r(9);
end of do-file
r(9);
```

The return item `r(sum)` is set equal to the total number of duplicate observations found (here, 2), so the identification of duplicates implies that you need to correct the dataset. The `duplicates` command could also be applied to numeric variables.

In summary, following sound data-management principles can improve the quality of your data analysis. You should bring the data into Stata as early in the process as possible. Use a well-documented do-file to validate the data, ensuring that variables that should be complete are complete, that unique identifiers are such, and that only sensible values are present in every variable. That do-file should run to completion without error if all data checks are passed. Last, you should not modify the validated and, if necessary, corrected file in later analysis. Subsequent data transformations or merges should create new files rather than overwriting the original contents of the validated file. Following these principles, although time consuming, will ultimately save a good deal of your time and ensure that the data are reproducible and well documented.

Exercises

1. Using the `cigconsumpW` dataset (in long format), merge the `state` variable with that of the *Stata Data Management Reference Manual* dataset `census5` (apply the `uniqusing` option since this dataset is a pure cross section). Compute the averages of `packpc` for subsamples of `median_age` above and below its annual median value (hint: `egen`, `tabstat`). Does smoking appear to be age related?
2. Using the `cigconsumpW` dataset (in long format), compute 4-year, moving-window averages of `packpc` for each state with `mvsumm`. List the `year`, `packpc`, and `mw4packpc` for California. What do you notice about the moving average relative to the series itself?

4 Linear regression

This chapter presents the most widely used tool in applied economics: the linear regression model, which relates a set of continuous variables to a continuous outcome. The explanatory variables in a regression model often include one or more binary or indicator variables; see chapter 7. Likewise, many models seek to explain a binary response variable as a function of a set of factors, which linear regression does not handle well. Chapter 10 discusses several forms of that model, including those in which the response variable is limited but not binary.

4.1 Introduction

This chapter discusses multiple regression in the context of a prototype economic research project. To carry out such a research project, we must

1. lay out a research framework—or economic model—that lets us specify the questions of interest and defines how we will interpret the empirical results;
2. find a dataset containing empirical counterparts to the quantities specified in the economic model;
3. use exploratory data analysis to familiarize ourselves with the data and identify outliers, extreme values, and the like;
4. fit the model and use specification analysis to determine the adequacy of the explanatory factors and their functional form;
5. conduct statistical inference (given satisfactory findings from specification analysis) on the research questions posed by the model; and
6. analyze the findings from hypothesis testing and the success of the model in terms of predictions and marginal effects. On the basis of these findings, we may have to return to one of the earlier stages to reevaluate the dataset and its specification and functional form.

Section 2 reviews the basic regression analysis theory on which regression point and interval estimates are based. Section 3 introduces a prototype economic research project studying the determinants of communities' single-family housing prices and discusses the various components of Stata's results from fitting a regression model of housing prices. Section 4 discusses how to transform Stata's estimation results into publication-quality tables. Section 5 discusses hypothesis testing and estimation subject to constraints on the parameters. Section 6 deals with computing residuals and predicted values. The last section discusses computing marginal effects. In the following chapters, we take up violations of the assumptions on which regression estimates are based.

4.2 Computing linear regression estimates

The linear regression model is the most widely used econometric model and the baseline against which all others are compared. It specifies the conditional mean of a response variable y as a linear function of k independent variables

$$E[y \mid x_1, x_2, \dots, x_k] = \beta_1 x_1 + \beta_2 x_2 + \cdots + \beta_k x_k$$

Given values for the βs, which are fixed parameters, the linear regression model predicts the average value of y in the population for different values of $x_1, x_2, \dots, x_k$.

Suppose that the mean value of single-family home prices in Boston-area communities, conditional on the student–teacher ratios, is given by

$$E[\texttt{price} \mid \texttt{stratio}] = \beta_1 + \beta_2 \, \texttt{stratio}$$

where `price` is the mean value of single-family home prices and `stratio` is the student–teacher ratio. This relationship reflects the hypothesis that the quality of communities' school systems is capitalized into housing prices. Here the population is the set of communities in the Commonwealth of Massachusetts. Each town or city in Massachusetts is generally responsible for its own school system.

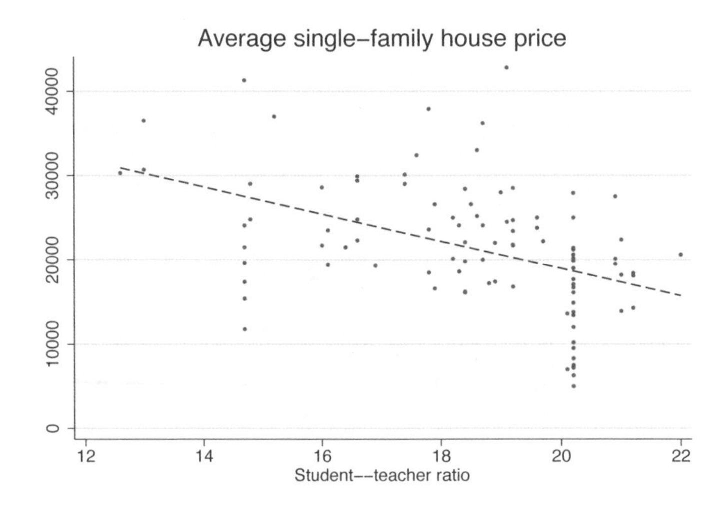

Figure 4.1: Conditional mean of single-family house price

Figure 4.1 shows average single-family housing prices for 100 Boston-area communities, along with the linear fit of housing prices to student–teacher ratios. The conditional mean of `price` for each value of `stratio` is shown by the appropriate point on the line. As theory predicts, the mean house price conditional on the student–teacher ratio is inversely related to that ratio. Communities with more crowded schools are considered less desirable. Of course, this relationship between house price and the student–teacher ratio must be considered ceteris paribus: all other factors that might affect the price of the house are held constant when we evaluate the effect of a measure of community schools' quality on the house price.

In working with economic data, we do not know the population values of $\beta_1, \beta_2, \ldots, \beta_k$. We work with a sample of N observations of data from that population. Using the information in this sample, we must

1. obtain good estimates of the coefficients $(\beta_1, \beta_2, \ldots, \beta_k)$;
2. determine how much our coefficient estimates would change if we were given another sample from the same population;
3. decide whether there is enough evidence to rule out some values for some of the coefficients $(\beta_1, \beta_2, \ldots, \beta_k)$; and
4. use our estimated $(\beta_1, \beta_2, \ldots, \beta_k)$ to interpret the model.

To obtain estimates of the coefficients, some assumptions must be made about the process that generated the data. I discuss those assumptions below and describe what I mean by good estimates. Before performing steps 2–4, I check whether the data support these assumptions by using a process known as *specification analysis*.

If we have a cross-sectional sample from the population, the linear regression model for each observation in the sample has the form

$$y_i = \beta_1 + \beta_2 x_{i,2} + \beta_3 x_{i,3} + \cdots + \beta_k x_{i,k} + u_i$$

for each observation in the sample $i = 1, 2, \ldots, N$. The u process is a stochastic disturbance, representing the net effect of all other unobservable factors that might influence y. The variance of its distribution, σ_u^2, is an unknown population parameter to be estimated along with the β parameters. We assume that $N > k$: to conduct statistical inference, there must be more observations in the sample than parameters to be estimated. In practice, N must be much larger than k.

We can write the linear regression model in matrix form as

$$\mathbf{y} = \mathbf{X}\boldsymbol{\beta} + \mathbf{u} \tag{4.1}$$

where $\mathbf{X}$ is an $N \times k$ matrix of sample values.[1]

This population regression function specifies that a set of k regressors in $\mathbf{X}$ and the stochastic disturbance u are the determinants of the response variable (or regressand)

1. Some textbooks use k in this context to refer to the number of slope parameters rather than the number of columns of $\mathbf{X}$. That will explain the deviations in the formulas given below; where I write k some authors write $(k + 1)$.

y. We usually assume that the model contains a constant term, so x_1 is understood to equal one for each observation.

The key assumption in the linear regression model involves the relationship in the population between the regressors $\mathbf{x}$ and u.[2] We may rewrite (4.1) as

$$u = y - \mathbf{x}\boldsymbol{\beta}$$

We assume that

$$E\left[u \mid \mathbf{x}\right] = 0 \tag{4.2}$$

i.e., that the u process has a zero-conditional mean. This assumption is that the unobserved factors involved in the regression function are not related systematically to the observed factors. This approach to the regression model lets us consider both nonstochastic and stochastic regressors in $\mathbf{X}$ without distinction, as long as they satisfy the assumption of (4.2).[3]

4.2.1 Regression as a method-of-moments estimator

We may use the zero-conditional-mean assumption shown in (4.2) to define a *method-of-moments* estimator of the regression function. Method-of-moments estimators are defined by *moment conditions* that are assumed to hold for the population moments. When we replace the unobservable population moments by their sample counterparts, we derive feasible estimators of the model's parameters. The zero-conditional-mean assumption gives rise to a set of k moment conditions, one for each x. In particular, the zero-conditional-mean assumption implies that each regressor is uncorrelated with u.[4]

$$\begin{aligned} E[\mathbf{x}'u] &= \mathbf{0} \\ E[\mathbf{x}'(y - \mathbf{x}\boldsymbol{\beta})] &= \mathbf{0} \end{aligned} \tag{4.3}$$

Substituting calculated moments from our sample into the expression and replacing the unknown coefficients $\boldsymbol{\beta}$ with estimated values $\widehat{\boldsymbol{\beta}}$ in (4.3) yields the *ordinary least squares* (OLS) estimator

$$\begin{aligned} \mathbf{X}'\mathbf{y} - \mathbf{X}'\mathbf{X}\widehat{\boldsymbol{\beta}} &= \mathbf{0} \\ \widehat{\boldsymbol{\beta}} &= (\mathbf{X}'\mathbf{X})^{-1}\mathbf{X}'\mathbf{y} \end{aligned} \tag{4.4}$$

We may use $\widehat{\boldsymbol{\beta}}$ to calculate the regression residuals:

$$\widehat{\mathbf{u}} = \mathbf{y} - \mathbf{X}\widehat{\boldsymbol{\beta}}$$

2. $\mathbf{x}$ is a vector of random variables and u is scalar random variable. In (4.1), $\mathbf{X}$ is a matrix of realizations of the random vector $\mathbf{x}$, $\mathbf{u}$ and $\mathbf{y}$ are vectors of realizations of the scalar random variables u and y.

3. Chapter 8 discusses how to use the instrumental-variables estimator when the zero-conditional-mean assumption is encountered.

4. The assumption of zero-conditional mean is stronger than that of a zero covariance, because covariance considers only *linear* relationships between the random variables.

Given the solution for the vector $\widehat{\boldsymbol{\beta}}$, the additional parameter of the regression problem σ_u^2—the population variance of the stochastic disturbance—may be estimated as a function of the regression residuals $\widehat{u}_i$

$$s^2 = \frac{\sum_{i=1}^{N} \widehat{u}_i^2}{N-k} = \frac{\widehat{\mathbf{u}}'\widehat{\mathbf{u}}}{N-k} \tag{4.5}$$

where $(N-k)$ are the residual *degrees of freedom* of the regression problem. The positive square root of s^2 is often termed the standard error of regression, or root mean squared error. Stata uses the latter term and displays s as `Root MSE`.

The method of moments is not the only approach for deriving the linear regression estimator of (4.4), which is the well-known formula from which the OLS estimator is derived.[5]

4.2.2 The sampling distribution of regression estimates

The OLS estimator $\widehat{\boldsymbol{\beta}}$ is a vector of random variables because it is a function of the random variable y, which in turn is a function of the stochastic disturbance u. The OLS estimator takes on different values for each sample of N observations drawn from the population. Because we often have only one sample to work with, we may be unsure of the usefulness of the estimates from that sample. The estimates are the realizations of the random vector $\widehat{\boldsymbol{\beta}}$ from the *sampling distribution* of the OLS estimator. To evaluate the precision of a given vector of estimates $\widehat{\boldsymbol{\beta}}$, we use the sampling distribution of the regression estimator.

To learn more about the sampling distribution of the OLS estimator, we must make further assumptions about the distribution of the stochastic disturbance u_i. In classical statistics, the u_i were assumed to be independent draws from the same normal distribution. The modern approach to econometrics drops the normality assumption and simply assumes that the u_i are independent draws from an identical distribution (i.i.d.).[6]

Using the normality assumption, we were able to derive the exact finite-sample distribution of the OLS estimator. In contrast, under the i.i.d. assumption, we must use large-sample theory to derive the sampling distribution of the OLS estimator. Basically, large-sample theory supposes that the sample size N becomes infinitely large. Since no real sample is infinitely large, these methods only approximate the sampling distribution of the OLS estimator in finite samples. With a few hundred observations or more, the large-sample approximation works well, so these methods work well with applied economic datasets.

5. The treatment here is similar to that of Wooldridge (2006). See Stock and Watson (2006) and appendix 4.A for a derivation based on minimizing the squared-prediction errors.

6. Both frameworks also assume that the (constant) variance of the u process is finite. Formally, i.i.d. stands for independently and identically distributed.

Although large-sample theory is more abstract than finite-sample methods, it imposes weaker assumptions on the data-generating process. We will use large-sample theory to define "good" estimators and to evaluate the precision of the estimates produced from a given sample.

In large samples, *consistency* means that as N goes to ∞, the estimates will converge to their respective population parameters. Roughly speaking, if the probability that the estimator produces estimates arbitrarily close to the population values goes to one as the sample size increases to infinity, the estimator is said to be *consistent*.

The sampling distribution of an estimator describes the set of estimates produced when that estimator is applied to repeated samples from the underlying population. You can use the sampling distribution of an estimator to evaluate the precision of a given set of estimates and to statistically test whether the population parameters take on certain values.

Large-sample theory shows that the sampling distribution of the OLS estimator is approximately normal.[7] Specifically, when the u_i are i.i.d. with finite variance σ_u^2, the OLS estimator $\widehat{\boldsymbol{\beta}}$ has a large-sample normal distribution with mean β and variance $\sigma_u^2\mathbf{Q}^{-1}$, where $\mathbf{Q}^{-1}$ is the variance–covariance matrix of X in the population. The variance–covariance of the estimator, $\sigma_u^2\mathbf{Q}^{-1}$, is also referred to as a VCE. Because it is unknown, we need a consistent estimator of the VCE. Although neither σ_u^2 nor $\mathbf{Q}^{-1}$ is actually known, we can use consistent estimators of them to construct a consistent estimator of $\sigma_u^2\mathbf{Q}^{-1}$. Given that s^2 consistently estimates σ_u^2 and $1/N(\mathbf{X}'\mathbf{X})$ consistently estimates $\mathbf{Q}$, $s^2(\mathbf{X}'\mathbf{X})^{-1}$ is a VCE of the OLS estimator.[8]

4.2.3 Efficiency of the regression estimator

Under the assumption of i.i.d. errors, the Gauss–Markov theorem holds. Among linear, unbiased estimators, the OLS estimator has the smallest sampling variance, or the greatest precision.[9] In that sense, it is *best*, so that "ordinary least squares is BLUE" (the *best linear unbiased estimator*) for the parameters of the regression model. If we consider only unbiased estimators that are linear in the parameters, we cannot find a more *efficient* estimator. The property of efficiency refers to the precision of the estimator. If estimator A has a smaller sampling variance than estimator B, estimator A is said to be *relatively efficient*. The Gauss–Markov theorem states that OLS is relatively efficient

7. More precisely, the distribution of the OLS estimator converges to a normal distribution. Although appendix B provides some details, in the text I will simply refer to the "approximate" or "large-sample" normal distribution. See Wooldridge (2006) for an introduction to large-sample theory.

8. At first glance, you might think that the expression for the VCE should be multiplied by $1/N$, but this assumption is incorrect. As discussed in appendix B, because the OLS estimator is consistent, it is converging to the constant vector of population parameters at the rate $1/\sqrt{N}$, implying that the variance of the OLS estimator is going to zero as the sample size gets larger. Large-sample theory compensates for this effect in how it standardizes the estimator. The loss of the $1/N$ term in the estimator of the VCE is a product of this standardization.

9. For a formal presentation of the Gauss–Markov theorem, see any econometrics text, e.g., Wooldridge (2006, 108–109). The OLS estimator is said to be "unbiased" because $E[\widehat{\boldsymbol{\beta}}] = \beta$.

versus all other linear, unbiased estimators of the parameterization model. However, this statement rests upon the hypotheses of an appropriately specified model and an i.i.d. disturbance process with a zero-conditional mean, as specified in (4.2).

4.2.4 Numerical identification of the regression estimates

As in (4.4) above, the solution to the regression problem involves a set of k moment conditions, or equations to be jointly solved for the k parameter estimates $\widehat{\beta}_1, \widehat{\beta}_2, \ldots, \widehat{\beta}_k$. When will these k parameter estimates be uniquely determined, or *numerically identified*? We must have more sample observations than parameters to be estimated, or $N > k$. That condition is not sufficient, though. For the simple "two-variable" regression model $y_i = \beta_1 + \beta_2 x_{i,2} + u_i$, $\text{Var}[x_2]$ must be greater than 0. If there is no variation in x_2, the data do not provide sufficient information to determine estimates of β_1 and β_2.

In multiple regression with many regressors, $\mathbf{X}_{N \times k}$ must be a matrix of full column rank k, which implies two things. First, only one column of $\mathbf{X}$ can take on a constant value, so each of the other regressors must have a positive sample variance. Second, there are no exact linear dependencies among the columns of the matrix $\mathbf{X}$. The assumption that $\mathbf{X}$ is of full column rank is often stated as "$(\mathbf{X}'\mathbf{X})$ is of full rank" or "$(\mathbf{X}'\mathbf{X})$ is nonsingular (or invertible)." If the matrix of regressors $\mathbf{X}$ contains k linearly independent columns, the cross-product matrix $(\mathbf{X}'\mathbf{X})$ will have rank k, its inverse will exist, and the parameters $\beta_1, \ldots, \beta_k$ in (4.4) will be numerically identified.[10] If numerical identification fails, the sample does not contain enough information for us to use the regression estimator on the model as it is specified. That model may be valid as a description of the data-generating process, but the particular sample may lack the necessary information to generate a regressor matrix of full column rank. Then we must either respecify the model or acquire another sample that contains the information needed to uniquely determine the regression estimates.

4.3 Interpreting regression estimates

This section illustrates using regression by an example from a prototype research project and discusses how Stata presents regression estimates. We then discuss how to recover the information displayed in Stata's estimation results for further computations within your program and how to combine this information with other estimates to present them in a table. The last subsection considers problems of numerical identification, or collinearity, that may appear when you are estimating the regression equation.

10. When computing infinite precision, we must be concerned with numerical singularity and a computer program's ability to reliably invert a matrix regardless of whether it is analytically invertible. As we discuss in section 4.3.7, computationally *near-linear dependencies* among the columns of $\mathbf{X}$ should be avoided.

4.3.1 Research project: A study of single-family housing prices

As an illustration, we present regression estimates from a model fitted to 506 Boston-area communities' housing price data, in which the response variable is the logarithm of the median price of a single-family home in each community. The dataset (`hprice2a`) contains an attribute of each community's housing stock that we would expect to influence price: `rooms`, the average number of rooms per house. Our research question relates to the influences on price exerted by several external factors. These factors, measured at the community level, include a measure of air pollution (`lnox`, the log of nitrous oxide in parts per 100m), the distance from the community to employment centers (`ldist`, the log of the weighted distance to five employment centers), and the average student–teacher ratio in local schools (`stratio`). From economic theory, we would expect the average number of rooms to increase the price, ceteris paribus. Each of the external factors is expected to decrease the median housing price in the community. More polluted communities, those less conveniently situated to available jobs, and those with poorly staffed schools should all have less expensive housing, given the forces of supply and demand.

We present the descriptive statistics with `summarize` and then fit a regression equation.

```
. use http://www.stata-press.com/data/imeus/hprice2a, clear
(Housing price data for Boston-area communities)
. summarize price lprice lnox ldist stratio, sep(0)
    Variable |       Obs        Mean    Std. Dev.       Min        Max
-------------+--------------------------------------------------------
       price |       506    22511.51    9208.856       5000      50001
      lprice |       506    9.941057     .409255   8.517193    10.8198
        lnox |       506    1.693091    .2014102   1.348073   2.164472
       ldist |       506    1.188233     .539501   .1222176   2.495682
     stratio |       506    18.45929     2.16582       12.6         22
```

The `regress` command, like other Stata estimation commands, requires us to specify the response variable followed by a *varlist* of the explanatory variables.

```
. regress lprice lnox ldist rooms stratio
      Source |       SS       df       MS              Number of obs =     506
-------------+------------------------------           F(  4,   501) =  175.86
       Model |  49.3987735     4  12.3496934           Prob > F      =  0.0000
    Residual |  35.1834974   501  .070226542           R-squared     =  0.5840
-------------+------------------------------           Adj R-squared =  0.5807
       Total |  84.5822709   505  .167489645           Root MSE      =    .265

------------------------------------------------------------------------------
      lprice |      Coef.   Std. Err.      t    P>|t|     [95% Conf. Interval]
-------------+----------------------------------------------------------------
        lnox |    -.95354   .1167418    -8.17   0.000    -1.182904   -.7241762
       ldist |  -.1343401   .0431032    -3.12   0.002    -.2190255   -.0496548
       rooms |   .2545271   .0185303    13.74   0.000     .2181203    .2909338
     stratio |  -.0524512   .0058971    -8.89   0.000    -.0640373   -.0408651
       _cons |   11.08387   .3181115    34.84   0.000     10.45887    11.70886
------------------------------------------------------------------------------
```

The header of the regression output describes the overall model estimates, whereas the table presents the point estimates, their precision, and their interval estimates.

4.3.2 The ANOVA table: ANOVA F and R-squared

The regression output for this model includes the analysis of variance (ANOVA) table in the upper left, where the two sources of variation are displayed as `Model` and `Residual`. The `SS` are the sums of squares, with the `Residual SS` corresponding to $\widehat{\mathbf{u}}'\widehat{\mathbf{u}}$ and the total `Total SS` to $\widetilde{\mathbf{y}}'\widetilde{\mathbf{y}}$ in (4.6) below. The next column of the table reports the `df`: the degrees of freedom associated with each sum of squares. The degrees of freedom for total `SS` are $(N-1)$ since the total `SS` have been computed by using one sample statistic, $\overline{y}$. The degrees of freedom for the model are $(k-1)$, equal to the number of slopes (or explanatory variables), or one fewer than the number of estimated coefficients due to the constant term. The model `SS` refer to the ability of the four regressors to jointly explain a fraction of the variation of y about its mean (the total `SS`). The residual degrees of freedom are $(N-k)$, indicating that $(N-k)$ residuals may be freely determined and still satisfy the constraint from the first normal equation of least squares that the regression surface passes through the multivariate point of means $(\overline{y}, \overline{x}_2, \ldots, \overline{x}_k)$:

$$\overline{y} = \widehat{\beta}_1 + \widehat{\beta}_2\overline{x}_2 + \widehat{\beta}_3\overline{x}_3 + \cdots + \widehat{\beta}_k\overline{x}_k$$

In the presence of the constant term $\widehat{\beta}_1$, the first normal equation implies that $\overline{\widehat{u}} = \overline{y} - \Sigma_i \overline{\mathbf{x}}_i \widehat{\boldsymbol{\beta}}_i$ must be identically zero.[11] This is not an assumption but is an algebraic implication of the least-squares technique, which guarantees that the sum of least-squares residuals (and their mean) will be very close to zero.[12]

The last column of the ANOVA table reports the `MS`, the mean squares due to regression and error, or the `SS` divided by the `df`. The ratio of the `Model MS` to `Residual MS` is reported as the ANOVA F statistic, with numerator and denominator degrees of freedom equal to the respective `df` values. This ANOVA F statistic is a test of the null hypothesis[13] that the slope coefficients in the model are jointly zero: that is, the null model of $y_i = \mu + u_i$ is as successful in describing y as the regression alternative. The `Prob > F` is the tail probability or p-value of the F statistic. Here we can reject the null hypothesis at any conventional level of significance. Also the `Root MSE` for the regression of 0.265, which is in the units of the response variable y, is small relative to the mean of that variable, 9.94.

The upper-right section of the `regress` output contains several *goodness-of-fit* statistics, which measure the degree to which a fitted model can explain the variation of the response variable y. All else equal, we should prefer a model with a better fit to the data. For the sake of parsimony, we also prefer a simpler model. The mechanics of

11. Recall that the first column of $\mathbf{X} = \iota$, an N-element unit vector.

12. Since computers use finite arithmetic, the sum will differ from zero. A well-written computer program should result in a difference similar to machine precision. For this regression, Stata reports a mean residual of -1.4×10^{-15}, comparable to the `epsdouble()` value of 2.2×10^{-16}, which is the smallest number distinguishable by Stata.

13. I discuss hypothesis testing in detail in section 4.5.

regression imply that a model with a great many regressors can explain y arbitrarily well. Given the least-squares residuals, the most common measure of goodness of fit, regression R^2, may be calculated (given a constant term in the regression function) as

$$R^2 = 1 - \frac{\widehat{\mathbf{u}}'\widehat{\mathbf{u}}}{\widetilde{\mathbf{y}}'\widetilde{\mathbf{y}}} \tag{4.6}$$

where $\widetilde{\mathbf{y}} = y - \overline{y}$: the regressand with its sample mean removed. This calculation emphasizes that the object of regression is not to explain $\mathbf{y}'\mathbf{y}$, the raw sum of squares of the response variable y, which would merely explain why $E[y] \neq 0$—not an interesting question. Rather, the object is to explain the variations in the response variable.

With a constant term in the model, the least-squares approach seeks to explain the largest possible fraction of the sample *variation* of y about its mean (and not the associated *variance*). The null model with which (4.1) is contrasted is $y = \mu + u_i$, where μ is the population mean of y. In estimating a regression, we want to determine whether the information in the regressors $\mathbf{x}$ is useful. Is the conditional expectation $E[y|\mathbf{x}]$ more informative than the unconditional expectation $E[y] = \mu$? The null model above has an $R^2 = 0$, whereas virtually any set of regressors will explain some fraction of the variation of y around $\overline{y}$, the sample estimate of μ. R^2 is that fraction in the unit interval, the proportion of the variation in y about $\overline{y}$ explained by $\mathbf{x}$.

4.3.3 Adjusted R-squared

What about the `Adj R-squared`? The algebra of least squares dictates that adding a $(k+1)$st column to $\mathbf{X}$ will result in a regression estimate with $R^2_{k+1} \geq R^2_k$. R^2 cannot fall with the addition of $\mathbf{x}_{k+1}$ to the regression equation, as long as the observations on the marginal regressor are linearly independent of the previous k columns from a numerical standpoint.[14] Indeed, we know that R^2_N (that is, R^2 calculated from a regression in which there are N linearly independent columns of $\mathbf{X}$ and N observations in the sample) must equal 1.0. As we add regressors to $\mathbf{x}$, R^2 cannot fall and is likely to rise, even when the marginal regressor is irrelevant econometrically.

What if we have a competing model that cannot be expressed as nested within this model, and this model does not nest within the competing model? A nonstatistical approach to this problem, especially where the two models differ widely in their numbers of regressors (or `Model df`), is to consider their $\overline{R}^2$ values, the statistic Stata labels as `Adj R-squared`.[15] The $\overline{R}^2$ considers the explained *variance* of y, rather than the explained *variation*, as does ordinary R^2. That is, rather than merely considering $\widehat{\mathbf{u}}'\widehat{\mathbf{u}}$, the residual sum of squares, $\overline{R}^2$ takes into account the degrees of freedom lost in fitting

14. In this sense, the limitations of finite arithmetic using the binary number system intrude: since 0.100 cannot be exactly expressed in a finite number of digits in the binary system, even a column that should be perfectly collinear with the columns of $\mathbf{X}_k$ may not be so computationally. The researcher should know her data and recognize when a candidate regressor cannot logically add information to an existing regressor matrix, whether or not the resulting regressor matrix is judged to possess full column rank by Stata.

15. A formal statistical approach to the nonnested models problem is presented below in section 4.5.5.

the model and scales $\widehat{\mathbf{u}}'\widehat{\mathbf{u}}$ by $(N-k)$ rather than N.[16] $\overline{R}^2$ can be expressed as a corrected version of R^2 in which the degrees-of-freedom adjustments are made, penalizing a model with more regressors for its loss of parsimony:

$$\overline{R}^2 = 1 - \frac{\widehat{\mathbf{u}}'\widehat{\mathbf{u}}/(N-k)}{\widetilde{\mathbf{y}}'\widetilde{\mathbf{y}}/(N-1)} = 1 - (1-R^2)\frac{N-1}{N-k}$$

If an irrelevant regressor is added to a model, R^2 cannot fall and will probably rise, but $\overline{R}^2$ will rise if the benefit of that regressor (reduced variance of the residuals) exceeds the cost of including it in the model: 1 degree of freedom.[17] Therefore, $\overline{R}^2$ can fall when a more elaborate model is considered, and indeed it is not bounded by zero. Algebraically, $\overline{R}^2$ must be less than R^2 since $(N-1)/(N-k) > 1$ for any $\mathbf{X}$ matrix and cannot be interpreted as the "proportion of variation of y", as can R^2 in the presence of a constant term. Nevertheless, you can use $\overline{R}^2$ to informally compare models with the same response variable but differing specifications. You can also compare the equations' s^2 values (labeled `Root MSE` in Stata's output) in units of the dependent variable to judge nonnested specifications.

Two other measures commonly used to compare competing regression models are the Akaike information criterion (AIC; Akaike [1974]) and Bayesian information criterion (BIC; often referred to as the Schwarz criterion: Schwarz [1978]). These measures also account for both the goodness of fit of the model and its parsimony. Each measure penalizes a larger model for using additional degrees of freedom while rewarding improvements in goodness of fit. The BIC places a higher penalty on using degrees of freedom. You can calculate the AIC and BIC after a regression model with the `estat ic` command. `estat ic` will display the log likelihood of the null model (that with only a constant term), the log likelihood of the fitted model, the model degrees of freedom, and the AIC and BIC values. For the regression above, we would type

```
. estat ic
```

Model	Obs	ll(null)	ll(model)	df	AIC	BIC
.	506	-265.4135	-43.49514	5	96.99028	118.123

Least-squares regression can also be considered a maximum likelihood estimator of the vector $\boldsymbol{\beta}$ and ancillary parameter σ_u^2.[18] The degree to which our fitted model improves upon the null model in explaining the variation of the response variable is measured by the (algebraically) larger magnitude of `ll(model)` than that of `ll(null)`.[19]

16. For comparison you may write (4.6), dividing both numerator and denominator by N.

17. This is not a statistical judgment, as $\overline{R}^2_{k+1}$ can exceed $\overline{R}^2_k$ if the t statistic on the added regressor exceeds 1.0 in absolute value.

18. The maximum likelihood estimator requires the normality assumption. See Johnston and DiNardo (1997).

19. A *likelihood-ratio test* formally compares these two magnitudes under the null hypothesis that the null model is adequate. I discuss likelihood-ratio tests in chapter 10.

4.3.4 The coefficient estimates and beta coefficients

Below the ANOVA table and summary statistics, Stata reports the $\widehat{\boldsymbol{\beta}}$ coefficient estimates, along with their estimated standard errors, t statistics, and the associated p-values labeled `P>|t|`: that is, the tail probability for a two-tailed test on the hypothesis H_0: $\widehat{\beta}_j = 0$.[20] The last two columns display an estimated confidence interval, with limits defined by the current setting of `level`. You can use the `level()` option on `regress` (or other estimation commands) to specify a particular level. After performing the estimation (e.g., with the default 95% level), you can redisplay the regression results with, for instance, `regress, level(90)`. You can change the default `level` (see [R] **level**) for the session or permanently with `set level` # [`, permanently`].

Economic researchers often express regressors or response variables in logarithms.[21] A model in which the response variable is the log of the original series and the regressors are in levels is termed a *log-linear* (or *single-log*) model. The rough approximation that $\log(1 + x) \simeq x$ for reasonably small x is used to interpret the regression coefficients. These coefficients are also the *semielasticities* of y with respect to x, measuring the response of y in percentage terms to a unit change in x. When logarithms are used for both the response variable and regressors, we have the *double-log* model. In this model, the coefficients are themselves *elasticities* of y with respect to each x. The most celebrated example of a double-log model is the Cobb–Douglas production function, $q = al^{\alpha}k^{\beta}e^{\epsilon}$, which we can estimate by linear regression by taking logs of q, l, and k.

In other social science disciplines, linear regression results are often reported as estimated *beta coefficients*. This terminology is somewhat confusing for economists, given their common practice of writing the regression model in terms of βs. The beta coefficient is defined as $\partial y^* / \partial x_j^*$, where the starred quantities are z-transformed or standardized variables; for instance, $y^* = (y_i - \overline{y})/s_y$, where $\overline{y}$ is the sample mean and s_y is the sample standard deviation of the response variable. Thus the beta coefficient for the jth regressor tells us how many standard deviations y would change given a 1–standard deviation change in x_j. This measure is useful in disciplines where many empirical quantities are indices lacking a natural scale. We can then rank regressors by the magnitudes of their beta coefficients because the absolute magnitude of the beta coefficient for x_j indicates the strength of the effect of that variable. For the regression model above, we can merely redisplay the regression by using the `beta` option:

We discuss hypothesis testing in detail in section 4.5.

Economists use natural logs exclusively; references to `log` should be taken as the natural log, or `ln`.

```
. regress, beta

      Source |       SS       df       MS              Number of obs =     506
-------------+------------------------------           F(  4,   501) =  175.86
       Model |  49.3987735     4  12.3496934           Prob > F      =  0.0000
    Residual |  35.1834974   501  .070226542           R-squared     =  0.5840
-------------+------------------------------           Adj R-squared =  0.5807
       Total |  84.5822709   505  .167489645           Root MSE      =    .265

------------------------------------------------------------------------------
      lprice |      Coef.   Std. Err.      t    P>|t|                     Beta
-------------+----------------------------------------------------------------
        lnox |    -.95354   .1167418    -8.17   0.000                -.4692738
       ldist |  -.1343401   .0431032    -3.12   0.002                -.1770941
       rooms |   .2545271   .0185303    13.74   0.000                 .4369626
     stratio |  -.0524512   .0058971    -8.89   0.000                -.2775771
       _cons |   11.08387   .3181115    34.84   0.000                        .
------------------------------------------------------------------------------
```

The output indicates that `lnox` has the largest beta coefficient, in absolute terms, followed by `rooms`. In economic and financial applications, where most regressors have a natural scale, it is more common to compute marginal effects such as elasticities or semielasticities (see section 4.7).

4.3.5 Regression without a constant term

With Stata, you can estimate a regression equation without a constant term by using the `noconstant` option, but I do not recommend doing so. Such a model makes little sense if the mean of the response variable is nonzero and all regressors' coefficients are insignificant.[22] Estimating a constant term in a model that does not have one causes a small loss in the efficiency of the parameter estimates. In contrast, incorrectly omitting a constant term produces inconsistent estimates. The tradeoff should be clear: include a constant term, and let the data indicate whether its estimate can be distinguished from zero.

What if we want to estimate a homogeneous relationship between y and the regressors $\mathbf{x}$, where economic theory posits $y \propto \mathbf{x}$? We can test the hypothesis of proportionality by estimating the relationship with a constant term and testing $H_0: \beta_1 = 0$. If the data reject that hypothesis, we should not fit the model with the constant term removed. Many of the common attributes of a linear regression are altered in a model that truly lacks a constant term. For instance, the least-squares residuals are not constrained to have zero sum or mean, and R^2 measured conventionally will be negative when the null model $y_i = \mu + u_i$ is not only preferable but strictly dominates the model $y_i = \beta_2 x_{i,2} + u_i$. Therefore, unless we have a good reason to fit a model without a constant term, we should retain the constant. An estimated $\widehat{\beta}_1$ not significantly different from zero does not harm the model, and it renders the model's summary statistics comparable to those of other models of the response variable y.

22. If we provide the equivalent of a constant term by including a set of regressors that add up to a constant value for each observation, we should specify the `hascons` option as well as `noconstant`. Using the `hascons` option will alter the `Model SS` and `Total SS`, affecting the ANOVA F and R^2 measures; it does not affect the `Root MSE` or the t statistics for individual coefficients.

The noconstant option might be sensible when the regressor matrix contains a set of variables that sum to a constant value. For instance, if the regressors include a set of portfolio shares or budget shares, we cannot include all those regressors in a model with a constant term because the constant is an exact linear combination of the share variables. Fitting a model with $(k+1)$ regressors, adding one variable to the list of k regressors, implies that at the margin there must be some useful information in regressor $(k+1)$: information that cannot be deduced, in linear terms, from the first k regressors. In the presence of accounting constraints or identities among the variables, one item must fail to satisfy that condition. If that condition is detected, Stata will automatically drop one of the regressors and indicate a coefficient value of (dropped). Then, rather than using the noconstant option, we should drop one of the portfolio or budget shares and include a constant term. The significance of the fitted model will be invariant to the choice of the excluded regressor. We may still want to include a complete set of items that sum to a constant value in a regression model, so we must omit the constant term (with the noconstant option) to prevent Stata from determining that the regressor matrix is rank deficient.

4.3.6 Recovering estimation results

The regress command shares the features of all estimation (e-class) commands. As discussed in section 2.2.12, we can view saved results from regress by typing ereturn list. All Stata estimation commands save an estimated parameter vector as matrix e(b) and the estimated variance–covariance matrix of the parameters as matrix e(V). You can refer to an element of the estimated parameter vector as _b[*varname*] and its associated estimated standard error as _se[*varname*] in later commands. However, the contents of e(), _b[], and _se[] are overwritten when the next e-class command is executed, so that if some of these values are to be retained, they should be copied to local macros, scalars, or matrices.

For example, typing ereturn list for the regression above produced

```
. ereturn list
scalars:
                  e(ll_0) =  -265.4134648194153
                    e(ll) =  -43.4951392092929
                  e(r2_a) =  .5807111444517128
                   e(rss) =  35.18349741237626
                   e(mss) =  49.39877352102588
                  e(rmse) =  .2650029089298266
                    e(r2) =  .5840322442976398
                     e(F) =  175.8550695227946
                  e(df_r) =  501
                  e(df_m) =  4
                     e(N) =  506
```

```
macros:
                 e(title) : "Linear regression"
                e(depvar) : "lprice"
                   e(cmd) : "regress"
            e(properties) : "b V"
               e(predict) : "regres_p"
                 e(model) : "ols"
             e(estat_cmd) : "regress_estat"
matrices:
                     e(b) :  1 x 5
                     e(V) :  5 x 5
functions:
                e(sample)
```

Most of the items displayed above are recognizable from the regression output. Two that are not displayed in the regression output are `e(ll)` and `e(ll_0)`, which are, respectively, the values of the log-likelihood function for the fitted model and for the null model.[23] These values could be used to implement a likelihood-ratio test of the model's adequacy, similar to the Wald test provided by the ANOVA F.

Another result displayed above is `e(sample)`, which is listed as a `function` rather than a `scalar`, `macro`, or `matrix`. The `e(sample)` function returns 1 if an observation was included in the estimation sample and 0 otherwise. The `regress` command honors any `if` *exp* and `in` *range* qualifiers and then does casewise deletion to remove any observations with missing values from the data $(\mathbf{y}, \mathbf{X})$. Thus the observations actually used in generating the regression estimates may be fewer than those specified in the `regress` command. A subsequent command, such as `summarize` *regressors* `if` *exp* (or `in` *range*), will not necessarily use the same observations as the previous regression. But we can easily restrict the set of observations to those used in estimation with the qualifier `if e(sample)`. For example,

```
. summarize regressors if e(sample)
```

will yield the summary statistics from the regression sample. The estimation sample may be retained for later use by placing it in a new variable:

```
. generate byte reg1sample = e(sample)
```

where we use the `byte` data type to save memory since `e(sample)` is an indicator {0,1} variable.

The `estat` command displays several items after any estimation command. Some of those items (`ic`, `summarize`, and `vce`) are common to all estimation commands, whereas others depend on the specific estimation command that precedes `estat`. After `regress`, the `estat summarize` command produces summary statistics, computed over the estimation sample, for the response variable and all regressors from the previous `regress` command.

23. These values are identical to those discussed above in the output of `estat ic`.

```
. estat summarize
Estimation sample regress                    Number of obs =      506

    Variable         Mean      Std. Dev.         Min          Max

      lprice     9.941057       .409255      8.51719      10.8198
        lnox     1.693091      .2014102      1.34807      2.16447
       ldist     1.188233       .539501      .122218      2.49568
       rooms     6.284051      .7025938         3.56         8.78
     stratio     18.45929       2.16582         12.6           22
```

In the following example, we use the `matrix list` command to display the coefficient matrix generated by our regression: `e(b)`, the k-element row vector of estimated coefficients. Like all Stata matrices, this array bears row and column labels, so an element may be addressed by either its row and column number[24] or by its row and column names.

```
. matrix list e(b)
e(b)[1,5]
          lnox        ldist        rooms      stratio        _cons
y1  -.95354002   -.13434015    .25452706   -.05245119    11.083865
```

We can use the `estat vce` command to display the estimated variance–covariance (VCE) matrix.[25] This command provides several options to control the display of the matrix:

```
. estat vce
Covariance matrix of coefficients of regress model
        e(V)        lnox        ldist        rooms      stratio        _cons

        lnox   .01362865
       ldist   .00426247    .00185789
       rooms   .00035279    .00003043    .00034337
     stratio   9.740e-07    .00002182    .00003374    .00003478
       _cons  -.03037429   -.01001835   -.00341397   -.00088151    .10119496
```

The diagonal elements of the VCE matrix are the squares of the estimated standard errors (`_se[]`) of the respective coefficients.

4.3.7 Detecting collinearity in regression

If the sample $(\mathbf{X}'\mathbf{X})$ matrix is numerically singular, not all the regression parameter estimates are numerically identified. $(\mathbf{X}'\mathbf{X})$ will be singular, or noninvertible, when one variable is *perfectly collinear* with some of the other variables. That variable can be represented as a linear combination of the other regressors. Stata automatically detects perfect collinearity, but near-collinearity is more difficult to diagnose. Both perfect and near-collinearity change how we may interpret regression estimates.

24. Stata matrices' rows and columns are numbered starting from 1.
25. In earlier versions of Stata, the `vce` command provided this functionality.

First, consider perfect collinearity. When Stata determines that $(\mathbf{X}'\mathbf{X})$ is numerically singular, it drops variables until the resulting regressor matrix is invertible, marking their coefficients with `(dropped)` in place of a value.[26] If two variables are perfectly collinear, only one of those variables can be included in the model and the estimated coefficient is the sum of the two coefficients on the original variables.

Near-collinearity arises when pairwise correlations of regressors are high, or in general, in the presence of near-linear dependencies in the regressor matrix. Failure of the full rank condition on $\mathbf{X}$ is a problem of the sample. The information in the estimation sample does not numerically identify all the regression parameters, but a different or expanded sample might.

With near-collinearity, small changes in the data matrix may cause large changes in the parameter estimates since they are nearly unidentified. Although the overall fit of the regression (as measured by R^2 or $\overline{R}^2$) may be very good, the coefficients may have very high standard errors and perhaps even incorrect signs or implausibly large magnitudes. If we consider a k-variable regression model containing a constant and $(k-1)$ regressors, we may write the kth diagonal element of the VCE as

$$\frac{s^2}{(1-R_k^2)\mathbf{S}_{kk}}$$

where R_k^2 is the partial R^2 from the regression of variable k on all other variables in the model and $\mathbf{S}_{kk}$ is the variation in the kth variable about its mean. Some observations about this expression:

- the greater the correlation of $\mathbf{x}_k$ with the other regressors (including the constant term), ceteris paribus, the higher the estimated variance will be;
- the greater the variation in $\mathbf{x}_k$ about its mean, ceteris paribus, the lower the estimated variance will be; and
- the better the overall fit of the regression, the lower the estimated variance will be.

This expression is the rationale for the VIF, or variance inflation factor, $(1-R_k^2)^{-1}$. VIF_k measures the degree to which the variance has been inflated because regressor k is not orthogonal to the other regression. After fitting a model with `regress`, the VIF measures may be calculated with the `estat vif` command. A rule of thumb states that there is evidence of collinearity if the largest VIF is greater than 10. We can be comfortable with the conditioning of our housing price regression model, as the maximum VIF is less than four:

26. This situation commonly arises with the *dummy variable trap*, in which a complete set of dummies and a constant term are included in the model. See chapter 7.

```
. regress lprice lnox ldist rooms stratio
      Source |       SS       df       MS              Number of obs =     506
-------------+------------------------------           F(  4,   501) =  175.86
       Model |  49.3987735     4  12.3496934           Prob > F      =  0.0000
    Residual |  35.1834974   501  .070226542           R-squared     =  0.5840
-------------+------------------------------           Adj R-squared =  0.5807
       Total |  84.5822709   505  .167489645           Root MSE      =    .265

------------------------------------------------------------------------------
      lprice |      Coef.   Std. Err.      t    P>|t|     [95% Conf. Interval]
-------------+----------------------------------------------------------------
        lnox |    -.95354   .1167418    -8.17   0.000    -1.182904   -.7241762
       ldist |  -.1343401   .0431032    -3.12   0.002    -.2190255   -.0496548
       rooms |   .2545271   .0185303    13.74   0.000     .2181203    .2909338
     stratio |  -.0524512   .0058971    -8.89   0.000    -.0640373   -.0408651
       _cons |   11.08387   .3181115    34.84   0.000     10.45887    11.70886
------------------------------------------------------------------------------

. estat vif
    Variable |       VIF       1/VIF
-------------+----------------------
        lnox |      3.98    0.251533
       ldist |      3.89    0.257162
       rooms |      1.22    0.820417
     stratio |      1.17    0.852488
-------------+----------------------
    Mean VIF |      2.56
```

How do we detect near-collinearity in an estimated regression? A summary measure for the regression equation is the *condition number* of $(\mathbf{X}'\mathbf{X})$, which measures the sensitivity of the estimates to changes in $\mathbf{X}$.[27] A large condition number indicates that small changes in $\mathbf{X}$ can cause large changes in the estimated coefficients. Belsley (1991), in an update of the seminal work on collinearity, recommends that the condition number be calculated from a transformed data matrix in which each regressor has unit length. A condition number for Belsley's normalized $(\mathbf{X}'\mathbf{X})$ in excess of 20 might be cause for concern. But just as there is no objective measure of how small the determinant of $\mathbf{X}'\mathbf{X}$ might be to trigger instability in the estimates, it is difficult to come up with a particular value that would indicate a problem: to some degree, it depends. Although we assume that the identification condition is satisfied and $\mathbf{X}$ is of full (numerical) rank to run a regression, there is no basis for a statistical test of the adequacy of that condition.

Official Stata does not have a command to generate the conditioning diagnostics of Belsley (1991), which beyond the computation of condition numbers include the variance decomposition matrix, which may be used to identify the regressors that are involved in near-linear dependencies. The `coldiag2` routine, contributed by John Hendrickx and available from the SSC archive, implements several of the diagnostic measures proposed in Belsley (1991), with options to select various features of the measures.

27. The condition number is the positive square root of the ratio of the largest to the smallest eigenvalue of $(\mathbf{X}'\mathbf{X})$. A large condition number indicates a nearly singular matrix, because a matrix that is near singular will have at least one very small eigenvalue. Large condition numbers of such a matrix will be large relative to the value of unity that would apply for $\mathbf{I}$.

How should we proceed if the conditioning diagnostics or VIFs show collinearity in an estimated equation? You can safely ignore near-collinearity that does not affect your key parameters. Because near-collinearity inflates standard errors, significant coefficients would become more significant if the sample contained fewer collinear regressors. Most microeconomic variables are intercorrelated—some of them highly—but that in itself may not be a concern.

If collinearity adversely affects your regression equation, you have two options: respecify the model to reduce the near-linear dependencies among the regressors or acquire a larger or better sample. Sometimes, near-collinearity reflects the homogeneity of a sample, so a broader sample from that population would be helpful.

For more details on near-collinearity, see Hill and Adkins (2003).

4.4 Presenting regression estimates

The `estimates` command makes it easy to store and present different sets of estimation results. The `estimates store` command stores results (up to 300 sets of estimates in memory for the current session) under a name and, optionally, a descriptive title.

You can organize several equations' estimates into a table by using `estimates table`. You specify that the table include several sets of results, and Stata automatically aligns the coefficients into the appropriate rows of a table. Options allow you to add estimated standard errors (`se`), t-values (`t`), p-values (`p`), or significance stars (`star`). You can assign each of these quantities its own display `format` if the default is not appropriate so that the coefficients, standard errors, and t- and p-values need not be rounded by hand. You can change the order of coefficients in the table by using the `keep()` option rather than relying on the order in which they appear in the list of estimates' contents. You can use `drop()` to remove certain parameter estimates from the coefficient table. You can add any result left in `e()` (see [P] **ereturn**) to the table with the `stat()` option, as well as several other criteria such as the AIC and BIC. Consider an example using several specifications from the housing-price model:

```
. use http://www.stata-press.com/data/imeus/hprice2a, clear
(Housing price data for Boston-area communities)
. generate rooms2 = rooms^2
. quietly regress lprice rooms                          // Fit model 1
. estimates store model1                                // Store estimates as model1
. quietly regress lprice rooms rooms2 ldist             // Fit model 2
. estimates store model2                                // Store estimates as model2
. quietly regress lprice ldist stratio lnox             // Fit model 3
. estimates store model3                                // Store estimates as model3
. quietly regress lprice lnox ldist rooms stratio // Fit model 4
. estimates store model4                                // Store estimates as model4
```

```
. estimates table model1 model2 model3 model4, stat(r2_a rmse)
> b(%7.3g) se(%6.3g) p(%4.3f)
```

Variable	model1	model2	model3	model4
rooms	.369	-.821		.255
	.0201	.183		.0185
	0.000	0.000		0.000
rooms2		.0889		
		.014		
		0.000		
ldist		.237	-.157	-.134
		.0255	.0505	.0431
		0.000	0.002	0.002
stratio			-.0775	-.0525
			.0066	.0059
			0.000	0.000
lnox			-1.22	-.954
			.135	.117
			0.000	0.000
_cons	7.62	11.3	13.6	11.1
	.127	.584	.304	.318
	0.000	0.000	0.000	0.000
r2_a	.399	.5	.424	.581
rmse	.317	.289	.311	.265

legend: b/se/p

After fitting and storing four different models of median housing price, we use **estimates table** to present the coefficients, estimated standard errors, and *p*-values in tabular form. The **stats()** option adds summary statistics from the **e()** results. Using the **star** option presents the results in the form

```
. estimates table model4 model1 model3 model2, stat(r2_a rmse ll)
> b(%7.3g) star title("Models of median housing price")
```

Models of median housing price

Variable	model4	model1	model3	model2
lnox	-.954***		-1.22***	
ldist	-.134**		-.157**	.237***
rooms	.255***	.369***		-.821***
stratio	-.0525***		-.0775***	
rooms2				.0889***
_cons	11.1***	7.62***	13.6***	11.3***
r2_a	.581	.399	.424	.5
rmse	.265	.317	.311	.289
ll	-43.5	-136	-124	-88.6

legend: * p<0.05; ** p<0.01; *** p<0.001

I chose to suppress the standard errors and display significance stars for the estimates. We add the log-likelihood value for each model with the **stats()** option.

We can use the **estimates** commands after any Stata estimation command, including multiple-equation commands. Ben Jann's **estout** is a full-featured solution for preparing publication-quality tables in various output formats (Jann 2005). This routine, which he describes as a wrapper for **estimates table**, reformats stored estimates in a variety of formats, combines summary statistics from model estimation, and produces output in several formats, such as tab-delimited (for word processors or spreadsheets), LaTeX, and HTML. A companion program, **estadd**, lets you add specific statistics to the e() arrays accessible by **estimates**. These useful programs are available from ssc.

As an example, we format the four models of median housing price for inclusion in a LaTeX document. This rather involved example using **estout** places the LaTeX headers and footers in the file and ensures that all items are in proper format for that typesetting language (e.g., using _cons would cause a formatting error unless it were modified).

```
. estout model1 model2 model3 model4 using ch3.19b_est.tex,
> style(tex) replace title("Models of median housing price")
> prehead(\\begin{table}[htbp]\\caption{{\sc @title}}\\centering\\medskip
> \begin{tabular}{l*{@M}{r}})
> posthead("\hline") prefoot("\hline")
> varlabels(rooms2 "rooms$^2$" _cons "constant") legend
> stats(N F r2_a rmse, fmt(%6.0f %6.0f %8.3f %6.3f)
> labels("N" "F" "$\bar{R}^2$" "RMS error"))
> cells(b(fmt(%8.3f)) se(par fmt(%6.3f)))
> postfoot(\hline\end{tabular}\end{table}) notype
```

You can insert the LaTeX fragment produced by this command directly in a research paper.

(*Continued on next page*)

Table 4.1: Models of median housing price

Parameter	model1 b/se	model2 b/se	model3 b/se	model4 b/se
rooms	0.369	−0.821		0.255
	(0.020)	(0.183)		(0.019)
rooms^2		0.089		
		(0.014)		
ldist		0.237	−0.157	−0.134
		(0.026)	(0.050)	(0.043)
stratio			−0.077	−0.052
			(0.007)	(0.006)
lnox			−1.215	−0.954
			(0.135)	(0.117)
constant	7.624	11.263	13.614	11.084
	(0.127)	(0.584)	(0.304)	(0.318)
N	506	506	506	506
F	337	169	125	176
$\overline{R}^2$	0.399	0.500	0.424	0.581
RMS error	0.317	0.289	0.311	0.265

You can change virtually every detail of the table by using `estout` directives. Since LaTeX, like HTML, is a markup language, you can program formatting changes. This flexibility is lacking from other `estout` output options, such as tab-delimited text for inclusion in a word-processing document or spreadsheet.

4.4.1 Presenting summary statistics and correlations

Most empirical papers containing regression results also provide one or more tables of summary statistics and, possibly, correlations. In more complex data structures, such as pooled cross-section time-series data or panel data, summary statistics for particular categories are often reported. You can use Stata commands and user-written commands to produce publication-quality tables with appropriate precision (number of decimal places) in tab-delimited, LaTeX, and HTML formats.

To produce summary statistics over categories, you could use Stata's `tabstat` command, but retrieving its results is hard. You could generate the same summary statistics across categories with Cox and Baum's `statsmat` routine, which places the statistics in one Stata matrix.[28] Here we use the egen function cut() to create crimelevel as a variable with five integer values defined by the smallest $(N/5)$ to largest $(N/5)$ values of crime. We then compute descriptive statistics for price for each of these five subsets of the data with the stat smat command, placing the results in a Stata matrix.

```
. label define crlev 0 "v.low" 1 "low" 2 "medium" 3 "high" 4 "v.high"
. egen crimelevel = cut(crime), group(5)
. label values crimelevel crlev
. statsmat price, stat(n mean p50) by(crimelevel)
> matrix(price_crime) format(%9.4g) title("Housing price by quintile of crime")
price_crime[5,3]:  Housing price by quintile of crime
               n    mean     p50
 v,low       101   27273   24499
   low       101   24806   22800
medium       101   23374   21600
  high       101   22222   19900
v,high       102   14957   13350
```

Ian Watson's tabout routine, available from ssc, provides another approach to this problem. tabout provides publication-quality output of cross tabulations in several output formats.

Another useful routine is Nicholas Cox's makematrix routine (available from ssc), which can execute any r-class (nonestimation) statistical command and produce a matrix of results. For instance, you could use this routine to display an oblong subset of a full correlation matrix. Here we generate the correlations of three of the variables in the dataset with median housing price. These are pairwise correlations (see pwcorr in [R] **correlate**), invoked with the listwise option.

```
. makematrix pc, from(r(rho) r(N)) label cols(price)
> title("Correlations with median housing price") listwise:
> corr crime nox dist
pc[3,2]:  Correlations with median housing price
                                      rho          N
  crimes committed per capita  -.38791912        506
nitrous oxide, parts per 100m  -.42603704        506
                         dist   .24933944        506
```

4.5 Hypothesis tests, linear restrictions, and constrained least squares

Researchers often apply regression methods in economics and finance to test hypotheses that are implied by a specific theoretical model. This section discusses hypothesis tests and interval estimates assuming that the model is properly specified and that the errors are i.i.d. Estimators are random variables, and their sampling distributions depend on those of the error process. In chapter 5, we extend many of these techniques to cases in which the errors are not i.i.d.

28. If you want LaTeX output, Baum and de Azevedo's outtable routine will generate a LaTeX table. Both statsmat and outtable are available from ssc by using the findit command.

Three tests are commonly used in econometrics: Wald tests, Lagrange multiplier (LM) tests, and likelihood-ratio (LR) tests. These tests share the same large-sample distribution, so choosing a test is usually a matter of convenience. Any hypothesis involving the coefficients of a regression equation can be expressed as one or more restrictions on the coefficient vector, reducing the dimensionality of the estimation problem. In the *unrestricted model*, the restrictions are not imposed on the estimation procedure. In the *restricted model*, they are imposed. The Wald test uses the point and VCE estimates from the unrestricted model to evaluate whether there is evidence that the restrictions are false. Loosely speaking, the LM test evaluates whether the restricted point estimates would be produced by the unrestricted estimator. I discuss several LM tests in chapter 6. The LR test compares the objective-function values from the unrestricted and restricted models. I present several LR tests in chapter 10.

Almost all the tests I present here are Wald tests. Let us consider the general form of the Wald test statistic. Given the population regression equation

$$y = \mathbf{x}\boldsymbol{\beta} + u$$

any set of linear restrictions on the coefficient vector may be expressed as

$$\mathbf{R}\boldsymbol{\beta} = \mathbf{r}$$

where $\mathbf{R}$ is a $q \times k$ matrix and $\mathbf{r}$ is a q-element column vector, with $q < k$. The q restrictions on the coefficient vector $\boldsymbol{\beta}$ imply that $(k-q)$ parameters are to be estimated in the restricted model. Each row of $\mathbf{R}$ imposes one restriction on the coefficient vector; one restriction can involve multiple coefficients. For instance, given the regression equation

$$y = \beta_1 x_1 + \beta_2 x_2 + \beta_3 x_3 + \beta_4 x_4 + u$$

we might want to test the hypothesis $H_0\colon \beta_2 = 0$. This restriction on the coefficient vector implies $\mathbf{R}\boldsymbol{\beta} = r$, where

$$\begin{aligned} \mathbf{R} &= (0\ 1\ 0\ 0) \\ \mathbf{r} &= (0) \end{aligned}$$

A test of $H_0\colon \beta_2 = \beta_3$ would imply the single restriction

$$\begin{aligned} \mathbf{R} &= (0\ 1\ -\ 1\ 0) \\ \mathbf{r} &= (0) \end{aligned}$$

whereas the ANOVA F test presented in section 4.3.2, that all the slope coefficients are zero, implies three restrictions: $H_0\colon \beta_2 = \beta_3 = \beta_4 = 0$, or

$$\mathbf{R} = \begin{pmatrix} 0 & 1 & 0 & 0 \\ 0 & 0 & 1 & 0 \\ 0 & 0 & 0 & 1 \end{pmatrix}$$

with

$$\mathbf{r} = \begin{pmatrix} 0 \\ 0 \\ 0 \end{pmatrix}$$

Given a hypothesis expressed as $H_0 : \mathbf{R}\boldsymbol{\beta} = \mathbf{r}$, we can construct the Wald statistic as

$$W = (\mathbf{R}\widehat{\boldsymbol{\beta}} - \mathbf{r})' \left\{ \mathbf{R}(\widehat{\mathbf{V}})^{-1}\mathbf{R}' \right\}^{-1} (\mathbf{R}\widehat{\boldsymbol{\beta}} - \mathbf{r})$$

This quadratic form uses the vector of estimated coefficients, $\widehat{\boldsymbol{\beta}}$, and the estimated VCE, $\widehat{\mathbf{V}}$, and evaluates the degree to which the restrictions fail to hold: the magnitude of the elements of the vector $(\mathbf{R}\widehat{\boldsymbol{\beta}} - \mathbf{r})$. The Wald statistic evaluates the sums of squares of that vector, each weighted by a measure of their precision.

The assumptions used to derive the large-sample distribution of the OLS estimator imply that w has a large-sample χ^2 distribution when H_0 is true. In small samples, the distribution of w/q may be better approximated by an F distribution with q and $(N-k)$ degrees of freedom. When $q = 1$, $\sqrt{w}$ has a large-sample normal distribution, which is sometimes better approximated by a Student t distribution with $(N - k)$ degrees of freedom.[29]

Now that we know the distribution of w when H_0 is true, we can set up standard hypothesis tests, which begin by specifying that

$$\Pr(\text{Reject } H_0 \mid H_0) = \alpha$$

where α is the significance level of the test.[30] Then we use the distribution of w to identify a critical value for the rejection region at a specific significance level.

Rather than reporting these critical values, Stata presents p-values, which measure the evidence against H_0. A p-value is the largest significance level at which a test can be conducted without rejecting H_0. The smaller the p-value, the more evidence there is against H_0.

Suppose that the estimates of a coefficient and its standard error are -96.17418 and 81.51707, respectively. These estimates imply that a t statistic of the null hypothesis—in which the population coefficient is zero—is -1.18. The Student t approximation to the distribution of this Wald test produces a two-sided p-value (`P>|t|`) of 0.242. We cannot reject H_0 at the conventional levels of 0.1, 0.05, or 0.01.[31] However, the p-value for the analogous test on another coefficient in the same model is 0.013, so we could reject H_0 at the 10% and the 5% levels, but not at the 1% level.

29. The small-sample approximations are exact when the errors are i.i.d. draws from a normal distribution. As above, the exact results are not highlighted because the normality assumption is too strong.
30. This section provides only a cursory introduction to hypothesis testing. See Wooldridge (2006, appendix C) for a more complete introduction.
31. These levels are commonly referred to as the 10%, 5%, and 1% levels, respectively.

Most Stata commands report p-values for *two-tailed* tests: a `P>|t|` value of 0.242 implies that 12.1% of the distribution lies in each tail of the distribution.[32] What if we have a one-tailed hypothesis, such as $H_0: \beta > 0$? In the example above, we would compute the one-tailed p-value by dividing the two-tailed p-value in half. If we place 5% of the mass of the distribution in one tail rather than 2.5% in each tail, the critical values become smaller in absolute value. Thus, under $H_0: \beta > 0$, the coefficient reported above would have a one-tailed p-value of 0.121, so we still fail to reject H_0 at the conventional levels.

As we have seen in `regress`'s output, Stata automatically generates several test statistics and their p-values: the ANOVA F and the t statistics for each coefficient, with the null hypothesis that the coefficients equal zero in the population. If we want to test more hypotheses after a regression equation, three Stata commands are particularly useful: `test`, `testparm`, and `lincom`. The first syntax for the `test` command is

`test` *coeflist*

where *coeflist* contains the names of one or more variables in the regression model. A second syntax is

`test` *exp*=*exp*

where *exp* is an algebraic expression in the names of the regressors.[33] The `testparm` command works similarly but allows wildcards in the coefficient list

`testparm` *varlist*

where the *varlist* may contain `*` or a hyphenated range expression such as `ind1-ind9`. The `lincom` command evaluates linear combinations of coefficients

`lincom` *exp*

where *exp* is any linear combination of coefficients that is valid in the second syntax of `test`. For `lincom`, the *exp* must not contain an equal sign.

We begin the discussion of hypothesis tests with the simplest case: a hypothesis that involves one regression coefficient.

4.5.1 Wald tests with test

If we want to test the hypothesis $H_0: \beta_j = 0$, the ratio of the estimated coefficient to its estimated standard error has an approximate t distribution under the null hypothesis

32. An exception is the `ttest` command, which presents both two-tailed and one-tailed tests of the difference of two means.

33. We can repeat the arguments of `test` in parentheses, as shown below in discussing joint tests. More syntaxes for `test` are available for multiple-equation models; see [R] **test** or type `help test`.

that the population coefficient equals zero. `regress` displays that ratio as the t column of the coefficient table. Returning to our median-housing-price equation, we could produce a test statistic for the significance of a coefficient by using the commands

```
. use http://www.stata-press.com/data/imeus/hprice2a, clear
(Housing price data for Boston-area communities)

. regress lprice lnox ldist rooms stratio

      Source |       SS       df       MS              Number of obs =     506
-------------+------------------------------           F(  4,   501) =  175.86
       Model |  49.3987735     4  12.3496934           Prob > F      =  0.0000
    Residual |  35.1834974   501  .070226542           R-squared     =  0.5840
-------------+------------------------------           Adj R-squared =  0.5807
       Total |  84.5822709   505  .167489645           Root MSE      =    .265

------------------------------------------------------------------------------
      lprice |      Coef.   Std. Err.      t    P>|t|     [95% Conf. Interval]
-------------+----------------------------------------------------------------
        lnox |   -.95354    .1167418    -8.17   0.000    -1.182904   -.7241762
       ldist |  -.1343401   .0431032    -3.12   0.002    -.2190255   -.0496548
       rooms |   .2545271   .0185303    13.74   0.000     .2181203    .2909338
     stratio |  -.0524512   .0058971    -8.89   0.000    -.0640373   -.0408651
       _cons |   11.08387   .3181115    34.84   0.000     10.45887    11.70886
------------------------------------------------------------------------------

. test rooms

 ( 1)  rooms = 0

       F(  1,   501) =  188.67
            Prob > F =    0.0000
```

which in Stata's shorthand is equivalent to the command `test _b[rooms] = 0` (and much easier to type). The `test` command displays the statistic as $F(1, N-k)$ rather than in the t_{N-k} form of the coefficient table. Because many hypotheses to which `test` can be applied involve more than one restriction on the coefficient vector—and thus more than one degree of freedom—Stata routinely displays an F statistic.[34] If we cannot reject the hypothesis $H_0\colon \beta_j = 0$ and wish to restrict the equation accordingly, we remove that variable from the list of regressors.

More generally, we may want to test the hypothesis $\beta_j = \beta_j^0 = \theta$, where θ is any constant value. If theory suggests that the coefficient on variable `rooms` should be 0.33, we can specify that hypothesis in `test`:

```
. quietly regress lprice lnox ldist rooms stratio

. test rooms = 0.33

 ( 1)  rooms = .33

       F(  1,   501) =   16.59
            Prob > F =    0.0001
```

Thus we can strongly reject the null hypothesis that the population coefficient is 0.33.

34. Recall that $(t_{N-k})^2 = F^1_{N-k}$: the square of a Student's t with $(N-k)$ degrees of freedom is an F statistic with $(1, N-k)$ degrees of freedom.

4.5.2 Wald tests involving linear combinations of parameters

We might want to compute a point and interval estimate for the sum of several coefficients. We can do that with the `lincom` (linear combination) command, which lets us specify any linear expression in the coefficients. With our median-housing-price equation, consider an arbitrary restriction: that the coefficients on `rooms`, `ldist`, and `stratio` sum to zero, so we can write

$$H_0 : \beta_{\texttt{rooms}} + \beta_{\texttt{ldist}} + \beta_{\texttt{stratio}} = 0$$

Although this hypothesis involves *three* estimated coefficients, it involves only *one* restriction on the coefficient vector. Here we have unitary coefficients on each term, but that need not be so.

```
. quietly regress lprice lnox ldist rooms stratio
. lincom rooms + ldist + stratio
 ( 1)  ldist + rooms + stratio = 0
```

lprice	Coef.	Std. Err.	t	P>|t|	[95% Conf.	Interval]
(1)	.0677357	.0490714	1.38	0.168	-.0286753	.1641468

The sum of the three estimated coefficients is 0.068, with an interval estimate including zero. The t statistic provided by `lincom` provides the same p-value that `test` would produce.

We can use `test` to consider the equality of two of the coefficients or to test that their ratio equals a particular value:

```
. quietly regress lprice lnox ldist rooms stratio
. test ldist = stratio
 ( 1)  ldist - stratio = 0
       F(  1,   501) =    3.63
            Prob > F =    0.0574
. test lnox  = 10 * stratio
 ( 1)  lnox - 10 stratio = 0
       F(  1,   501) =   10.77
            Prob > F =    0.0011
```

Whereas we cannot reject the hypothesis that the coefficients on `ldist` and `stratio` are equal at the 5% level, we can reject the hypothesis that the ratio of the `lnox` and `stratio` coefficients equals 10 at the 1% level. Stata rewrites both expressions into a normalized form. Although the ratio of two coefficients would appear to be a nonlinear expression, we can test this assumption by rewriting it as shown above. We cannot use the same strategy to evaluate a hypothesis involving a product of coefficients. We will discuss a solution to that problem in section 4.5.4.

In the estimates above, we cannot reject the hypothesis that the sum of the coefficients on `rooms`, `ldist`, and `stratio` is zero, but we estimate that sum to be slightly positive. To estimate the equation subject to that restriction, we have two options. First, we could substitute the restriction(s) into the model algebraically and fit the restricted model. Here that would be simple enough, but in more complicated models, it may be cumbersome. Second, we could use Stata's `constraint` command to define each constraint to be imposed on the equation and estimate with the `cnsreg` (constrained regression) command. The `constraint` command has the following syntax:

`constraint` [`define`] # [*exp*=*exp* | *coeflist*]

where # is the number of the constraint, which may be expressed either in an algebraic expression or as a *coeflist*. Using the latter syntax, the regressors in *coeflist* are removed from the equation. We can use the constraints in `cnsreg`:

`cnsreg` *depvar indepvars* [*if*] [*in*] [*weight*], `constraints(`*numlist*`)`

The command's syntax echoes that of `regress`, but it requires the `constraints()` option with the constraints to be imposed listed by number (# above).

To illustrate the latter strategy, we use `constraint`:

```
. constraint def 1 ldist + rooms + stratio = 0
. cnsreg lprice lnox ldist rooms stratio, constraint(1)
Constrained linear regression                       Number of obs =      506
                                                    F(  3,    502) =   233.42
                                                    Prob > F      =   0.0000
                                                    Root MSE      =   .26524
 ( 1)  ldist + rooms + stratio = 0
```

lprice	Coef.	Std. Err.	t	P>|t|	[95% Conf.	Interval]
lnox	-1.083392	.0691935	-15.66	0.000	-1.219337	-.9474478
ldist	-.1880712	.0185284	-10.15	0.000	-.2244739	-.1516684
rooms	.2430633	.01658	14.66	0.000	.2104886	.2756381
stratio	-.0549922	.0056075	-9.81	0.000	-.0660092	-.0439752
_cons	11.48651	.1270377	90.42	0.000	11.23691	11.7361

This format displays all three coefficients' estimates, so we need not generate new variables to impose the constraint. You should not perform a `test` of the restrictions that have been imposed on the equation. By construction, the restrictions will be satisfied (within your machine's precision) in these estimates, so this hypothesis cannot be tested. Also the `Root MSE` has marginally increased in the constrained equation. Estimation subject to linear restrictions cannot improve the fit of the equation relative to the unrestricted counterpart and will increase `Root MSE` to the degree that the restrictions are binding.

4.5.3 Joint hypothesis tests

All the tests illustrated above involve only one restriction on the coefficient vector. Often we wish to test a hypothesis involving multiple restrictions on the coefficient vector. Multiple restrictions on the coefficient vector imply a *joint test*, the result of which is not simply a box score of individual tests. Every user of regression is familiar with this concept. You may often encounter a regression in which each slope coefficient has a t statistic falling short of significance, but nevertheless the ANOVA F is significant. The ANOVA F is a joint test that all the regressors are jointly uninformative. In the presence of a high degree of collinearity, we often encounter exactly this result. The data cannot attribute the explanatory power to one regressor or another, but the combination of regressors can explain much of the variation in the response variable.

We can construct a joint test in Stata by listing each hypothesis to be tested in parentheses on the `test` command.[35] The joint F test statistic will have as many numerator degrees of freedom as there are restrictions on the coefficient vector. As presented above, the first syntax of the `test` command, `test` *coeflist*, performs the joint test that two or more coefficients are jointly zero, such as H_0: $\beta_2 = 0$ and $\beta_3 = 0$. This joint hypothesis is not the same as H_0': $\beta_2 + \beta_3 = 0$. The latter hypothesis will be satisfied by a locus of $\{\beta_2, \beta_3\}$ values: all pairs that sum to zero. The former hypothesis will be satisfied only at the point where *each coefficient* equals zero. We can test the joint hypothesis for our median-housing-price equation with

```
. quietly regress lprice lnox ldist rooms stratio
. test lnox ldist
 ( 1)  lnox = 0
 ( 2)  ldist = 0
       F(  2,   501) =   58.95
            Prob > F =    0.0000
```

The data overwhelmingly reject the joint hypothesis that the model excluding `lnox` and `ldist` is correctly specified relative to the full model.

We can formulate a joint hypothesis that combines two restrictions on the equation above as follows:

```
. quietly regress lprice lnox ldist rooms stratio
. test (lnox = 10 * stratio) (ldist = stratio)
 ( 1)  lnox - 10 stratio = 0
 ( 2)  ldist - stratio = 0
       F(  2,   501) =    5.94
            Prob > F =    0.0028
```

Here we have imposed two restrictions on the coefficient vector so that the final F statistic has two numerator degrees of freedom. The joint test rejects the two hypotheses at the 1% level.

35. We can combine separate hypotheses by using the `accumulate` option on the `test` command.

Just as we cannot algebraically restrict the model or apply constraints that are inconsistent linear equations (i.e., `ldist - stratio = 0` and `ldist - stratio = 1`) we cannot `test` inconsistent linear hypotheses.

4.5.4 Testing nonlinear restrictions and forming nonlinear combinations

All the hypotheses discussed above are linear in that they may be written in the form

$$H_0\colon \mathbf{R}\boldsymbol{\beta} = \mathbf{r}$$

where we can express a set of $q < k$ linear restrictions on $\boldsymbol{\beta}_{k\times 1}$ as the $q \times k$ matrix $\mathbf{R}$ and the q-vector $\mathbf{r}$. Indeed, the constrained least-squares method implemented in `cnsreg` may be expressed as solving the constrained optimization problem

$$\begin{aligned} \widehat{\boldsymbol{\beta}} = \arg\min_{\boldsymbol{\beta}}\ \widehat{\mathbf{u}}'\widehat{\mathbf{u}} &= \arg\min_{\boldsymbol{\beta}}\ (\mathbf{y}-\mathbf{X}\boldsymbol{\beta})'(\mathbf{y}-\mathbf{X}\boldsymbol{\beta}) \\ s.t.\ \mathbf{R}\boldsymbol{\beta} &= \mathbf{r} \end{aligned}$$

and all the tests above may be expressed as an appropriate choice of $\mathbf{R}$ and $\mathbf{r}$.

Suppose that the hypothesis tests to be conducted cannot be written in this linear form, for example, if theory predicts a certain value for the product of two coefficients in the model or for an expression such as $(\beta_2/\beta_3+\beta_4)$. Two Stata commands are analogues to those we have used above.

`testnl` lets us specify nonlinear hypotheses on the β values, but unlike with `test`, we must use the syntax `_b[`*varname*`]` to refer to each coefficient value. For a joint test, we must write the equations defining each nonlinear restriction in parentheses, as illustrated below.

`nlcom` lets us compute nonlinear combinations of the estimated coefficients in point and interval form, similar to `lincom`. Both commands use the *delta method*, an approximation to the distribution of a nonlinear combination of random variables appropriate for large samples that constructs Wald-type tests.[36] Unlike tests of linear hypotheses, nonlinear Wald-type tests based on the delta method are sensitive to the scale of the $\mathbf{y}$ and $\mathbf{X}$ data.

The median-housing-price regression illustrates these two commands.

```
. quietly regress lprice lnox ldist rooms stratio
. testnl _b[lnox] * _b[stratio] = 0.06
  (1)  _b[lnox] * _b[stratio] = 0.06
               F(1, 501) =        1.44
                Prob > F =        0.2306
```

This example considers a restriction on the product of the coefficients of `lnox` and `stratio`. The product of these coefficients cannot be distinguished from 0.06. Just as

36. See Wooldridge (2002) for a discussion of the delta method.

we can use `lincom` to evaluate a linear expression in the coefficients, we can use `nlcom` in the nonlinear context.

We can also test a joint nonlinear hypothesis:

```
. quietly regress lprice lnox ldist rooms stratio

. testnl (_b[lnox] * _b[stratio] = 0.06)
>        (_b[rooms] / _b[ldist] = 3 * _b[lnox])

  (1)  _b[lnox] * _b[stratio] = 0.06
  (2)  _b[rooms] / _b[ldist] = 3 * _b[lnox]

               F(2, 501) =        5.13
                Prob > F =        0.0062
```

We can reject the joint hypothesis at the 1% level.

4.5.5 Testing competing (nonnested) models

How do we compare two regression models that attempt to explain the same response variable but that differ in their regressor lists? If one of the models is strictly nested within the other, we can use the `test` command to apply a Wald test to the original or unconstrained model to evaluate whether the data reject the restrictions implied by the constrained model. This approach works well for classical hypothesis testing where the parameters of one model are a proper subset of another. But economic theories often are cast in the form of competing hypotheses, where neither model may be nested within the other. Furthermore, no proposed theory may correspond to the *supermodel* that encompasses all elements of both theories by artificially nesting both models' unique elements in one structure. Tests of competing hypotheses versus a supermodel pit one model against a hybrid model that contains elements of both that are not proposed by either theory. If we have competing hypotheses such as

$$H_0: y = \mathbf{x}\boldsymbol{\beta} + \epsilon_0 \tag{4.7}$$

$$H_1: y = \mathbf{z}\gamma + \epsilon_1 \tag{4.8}$$

where some of the elements of both $\mathbf{x}$ and $\mathbf{z}$ are unique (not included in the other regressor matrix), we must use a different strategy.[37] Examining goodness of fit by comparing `Root MSE` or noting that one of these models has a higher R^2 or $\overline{R}^2$ is not likely to yield conclusive results and lacks a statistical rationale.

Davidson and MacKinnon (1981) proposed their J test as a solution to this problem.[38] This test relies on a simple approach: if model 0 has better explanatory power than model 1, model 0 is superior, and vice versa. We perform the J test by generating the predicted values of each series and including them in an augmented regression of the other model. Let $\widehat{\mathbf{y}}_1$ and $\widehat{\mathbf{y}}_2$ be the predicted values of $\mathbf{y}$ using the estimates of

37. A Bayesian econometrician would have no difficulty in this context; she would merely ask "which of these hypotheses is more likely to have generated the data?"

38. Do not confuse this test with Hansen's J test in the generalized method-of-moments (GMM) literature described in chapter 8.

the parameters in (4.7) and (4.8), respectively. We include the $\widehat{y}$ from the alternative hypothesis above in the null hypothesis' model (4.7). If the coefficient on $\widehat{y}_1$ is significant, we reject the model of the null hypothesis. We now reverse the definitions of the two models and include the $\widehat{y}_0$ from the null hypothesis in the alternative hypothesis' model (4.8). If the coefficient on $\widehat{y}_0$ is significant, we reject the model of the alternative hypothesis. Unfortunately, all four possibilities can arise: H_0 may stand against H_1, H_1 may stand against H_0, both hypotheses may be rejected, or neither hypothesis may be rejected. Only in the first two cases does the J test deliver a definitive verdict. Official Stata does not implement this test, but you can install Gregorio Impavido's command `nnest` (from `ssc describe nnest`).

Similar tests are those of Cox (1961), Cox (1962), extended by Pesaran (1974) and Pesaran and Deaton (1978). These tests are based on likelihood-ratio tests that can be constructed from the fitted values and sums of squared residuals of the nonnested models. The Cox–Pesaran–Deaton tests are also performed by Impavido's `nnest` package.

We illustrate these tests with our median-housing-price regression by specifying two forms of the equation, one including `crime` and `proptax` but excluding pollution levels (`lnox`), the other vice versa. The command uses an unusual syntax in which the first regression specification is given (as it would be for `regress`) and the regressors of the second specification are listed in parentheses. The dependent variable should not appear in the parenthesized list.

```
. nnest lprice lnox ldist rooms stratio (crime proptax ldist rooms stratio)
M1 : Y = a + Xb with X = [lnox ldist rooms stratio]
M2 : Y = a + Zg with Z = [crime proptax ldist rooms stratio]
J test for non-nested models
H0 : M1  t(500)      10.10728
H1 : M2  p-val        0.00000
H0 : M2  t(499)       7.19138
H1 : M1  p-val        0.00000
Cox-Pesaran test for non-nested models
H0 : M1  N(0,1)     -20.07277
H1 : M2  p-val        0.00000
H0 : M2  N(0,1)     -17.63186
H1 : M1  p-val        0.00000
```

Here the Davidson–MacKinnon test and the Cox–Pesaran–Deaton test reject both H_0 and H_1, indicating that a model excluding any of these three explanatory variables is misspecified relative to the supermodel including all of them. For our microeconomics research project, these findings cast doubt on the specification excluding the `crime` and `proptax` explanatory factors and suggest that we revisit our specification of the model with respect to those factors.

4.6 Computing residuals and predicted values

After fitting a linear regression model with **regress**, we can compute the regression residuals or the predicted values. Computing the residuals for each observation allows us to assess how well the model explains the value of the response variable for that observation. Is the in-sample prediction $\widehat{y}_i$ much larger or smaller than the actual value y_i? Computing predicted values lets us generate in-sample predictions: the values of the response variable generated by the fitted model. We may also want to generate out-of-sample predictions, that is, apply the estimated regression function to observations that were not used to generate the estimates. We may need to use hypothetical values of the regressors or actual values. In the latter case, we may want to apply the estimated regression function to a separate sample (e.g., to Springfield-area communities rather than Boston-area communities) to evaluate its applicability beyond the regression sample. A well-specified regression model should generate reasonable predictions for any sample from the population. If out-of-sample predictions are poor, the model's specification may be too specific to the original sample.

`regress` does not calculate the residuals or predicted values, but we can compute either after `regress` with the `predict` command,

`predict` [*type*] *newvar* [*if*] [*in*] [, *choice*]

where *choice* specifies the quantity to be computed for each observation. For linear regression, `predict` computes predicted values by default so that

```
. predict double lpricehat predict double lpricehat, xb
```

will yield identical results. The choice `xb` refers to the matrix form of the regression model (4.10) in which $\widehat{\mathbf{y}} = \mathbf{X}\widehat{\boldsymbol{\beta}}$. These are known as the *point predictions*. If you need the residuals, use

```
. predict double lpriceeps, residual
```

The regression estimates are available only to `predict` until another estimation command (e.g., `regress`) is issued, so if you need these series, you should compute them as early as possible. Using `double` as the optional *type* in these commands ensures that the series will be generated with full numerical precision and is strongly recommended.

In both instances, `predict` works like `generate`: the named series must not already exist, and the series will be computed for the entire available dataset, not merely the estimation sample. You may use the qualifier `if e(sample)` to restrict the computation to the estimation sample and compute only in-sample predictions.[39]

To evaluate the quality of the regression fit graphically, we can use a plot of actual and predicted values of y_i versus x_i with one regressor. In multiple regression, the natural analogue is a plot of actual y_i versus the predicted $\widehat{y}_i$ values. The commands

39. When applied to a model fitted with time-series data, `predict` can generate only static or one-step-ahead forecasts. See Baum (2005) if you want dynamic or k-step-ahead forecasts.

below generate such a graph (figure 4.2) illustrating the fit of our regression model. The aspect ratio has been constrained to unity so that points on the 45° line represent perfect predictions. The model systematically overpredicts the (log) price of relatively low-priced houses, which may give cause for concern about the applicability of the model to lower-income communities. There are also several very high median prices that appear to be seriously underpredicted by the model.

```
. use http://www.stata-press.com/data/imeus/hprice2a, clear
(Housing price data for Boston-area communities)
. quietly regress lprice lnox ldist rooms stratio
. predict double lpricehat, xb
. label var lpricehat "Predicted log price"
. twoway (scatter lpricehat lprice, msize(small) mcolor(black) msize(tiny)) ||
> (line lprice lprice if lprice <., clwidth(thin)),
> ytitle("Predicted log median housing price")
> xtitle("Actual log median housing price") aspectratio(1) legend(off)
```

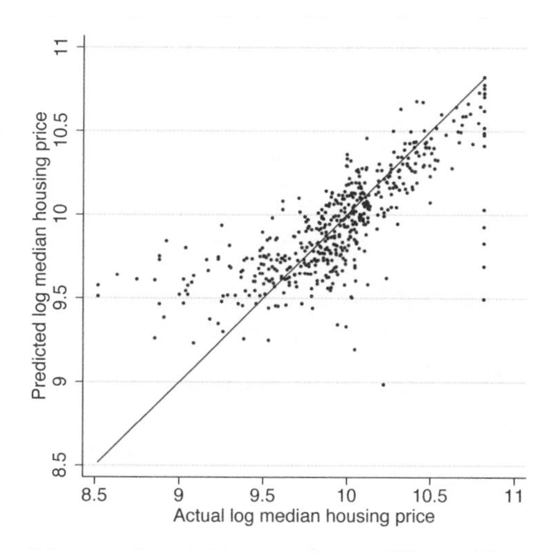

Figure 4.2: Actual versus predicted values from regression model

4.6.1 Computing interval predictions

Besides predicted values and least-squares residuals, we may want to use `predict` to compute other observation-specific quantities from the fitted model.[40] I will discuss some of the more specialized quantities available after `regress` in the next section. First, I discuss how `predict` may provide interval predictions to complement the point predictions.

40. Documentation for the capabilities of `predict` after `regress` is presented in [R] **regress postestimation**.

The interval prediction is simply a confidence interval for the prediction. There are two commonly used definitions of "prediction", the *predicted value* and the *forecast*. The predicted value estimates the average value of the dependent variable for given values of the regressors. The forecast estimates the value of the dependent variable for a given set of regressors. The mechanics of OLS implies that the point estimates are the same, but the variances of the predicted value and the forecast are different. As is intuitive, the variance of the forecast is higher than the variance of the predicted value.

Given regressor values $\mathbf{x}_0$, the predicted value is

$$E[y|\mathbf{x}_0] = \widehat{y}_0 = \mathbf{x}_0\boldsymbol{\beta}$$

A consistent estimator of the variance of the predicted value is

$$\widehat{V}_p = s^2 * \mathbf{x}_0(\mathbf{X}'\mathbf{X})^{-1}\mathbf{x}_0'$$

Given the regressor values x_0, the forecast error for a particular y_0 is

$$\widehat{e}_0 = y_0 - \widehat{y}_0 = \mathbf{x}_0\beta + u_0 - \widehat{y}_0$$

`predict` performs this calculation for each observation when the `stdp` option is specified.

The zero covariance between u_0 and $\widehat{\boldsymbol{\beta}}$ implies[41] that

$$\text{Var}[\widehat{e}_0] = \text{Var}[\widehat{y}_0] + \text{Var}[u_0]$$

for which

$$\widehat{V}_f = s^2 * \mathbf{x}_0(\mathbf{X}'\mathbf{X})^{-1}\mathbf{x}_0' + s^2$$

is a consistent estimator. `predict` performs this calculation for each observation when the `stdf` option is specified. As one would expect, the variance of the forecast is higher than the variance of the predicted value.

An interval prediction is an upper and lower bound that contain the true value with a given probability in repeated samples.[42] Here I present a method for finding the bounds for the forecast. Given that the standardized-prediction error has an approximate Student t distribution, the interval prediction begins by choosing bounds that enclose it with probability $1 - \alpha$:

$$\Pr\left\{-t_{1-\alpha/2} < \frac{y_0 - \widehat{y}_0}{\sqrt{\text{Var}[\widehat{e}_0]}} < t_{1-\alpha/2}\right\} = 1 - \alpha$$

where α is the significance level[43] and $t_{1-\alpha/2}$ is the inverse of the Student t at $1 - \alpha/2$. Standard manipulations of this condition yield

$$\Pr\left\{\widehat{y}_0 - t_{1-\alpha/2} * \sqrt{\text{Var}[\widehat{e}_0]} < y < \widehat{y}_0 + t_{1-\alpha/2} * \sqrt{\text{Var}[\widehat{e}_0]}\right\} = 1 - \alpha$$

41. See Wooldridge (2006) for a discussion of this point.
42. See Wooldridge (2006, section 6.4) for more about forming and interpreting interval predictions.
43. Loosely speaking, the significance level is the error rate that we are willing to tolerate in repeated samples. The often-chosen significance level of 5% yields a 95% confidence interval.

Plugging in our consistent estimators yields the bounds

$$\widehat{y}_0 \pm t_{1-\alpha/2} * \sqrt{\widehat{V}_f}$$

Substituting $E[y|\mathbf{x}_0]$ for y_0 and using the variance of predicted value presented in the text yields a prediction interval for the predicted value $\widehat{y}_0$.

The variance of the predicted value increases as we consider an x value farther from the mean of the estimation sample. The interval predictions for the predicted value lie on a pair of parabolas with the narrowest interval at $\overline{\mathbf{x}}$, widening as we diverge from the sample point of means. To compute this confidence interval, we use `predict`'s `stdp` option (see [R] **regress postestimation**). An appropriate confidence interval may be constructed from [$\pm t$ `stdp`], where t would be 1.96 for a 95% confidence interval for a sample with a large N. You may then construct two more variables to hold the lower-limit and upper-limit values and graph the point and interval predictions.

We consider a bivariate regression of log median housing price on `lnox`. For illustration, we fit only the model to 100 communities of the 506 in the dataset. The two `predict` commands generate the predicted values of `lprice` as `xb` and the standard error of prediction and `stdpred`, respectively:

```
. use http://www.stata-press.com/data/imeus/hprice2a, clear
(Housing price data for Boston-area communities)
. quietly regress lprice lnox if _n<=100
. predict double xb if e(sample)
(option xb assumed; fitted values)
(406 missing values generated)
. predict double stdpred if e(sample), stdp
(406 missing values generated)
```

To calculate the prediction interval, we use the `invttail()` function to generate the correct t-value for the sample size and a 95% prediction interval as a scalar. The variables `uplim` and `lowlim` can then be computed:

```
. scalar tval = invttail(e(df_r),0.025)
. generate double uplim = xb + tval * stdpred
(406 missing values generated)
. generate double lowlim = xb - tval * stdpred
(406 missing values generated)
```

Finally, we want to highlight the mean value of `lnox` (calculated by the `summarize` command, storing that value as `local` macro `lnoxbar`) and label the variables appropriately for the graph:

```
. summarize lnox if e(sample), meanonly
. local lnoxbar = r(mean)
. label var xb "Pred"
. label var uplim "95% prediction interval"
. label var lowlim "95% prediction interval"
```

We may now generate the figure by using three **graph twoway** types: **scatter** for the scatterplot, **connected** for the predicted values, and **rline** for the prediction interval limits:[44,45]

```
. twoway (scatter lprice lnox if e(sample),
> sort ms(Oh) xline(`lnoxbar'))
> (connected xb lnox if e(sample), sort msize(small))
> (rline uplim lowlim lnox if e(sample), sort),
> ytitle(Actual and predicted log price) legend(cols(3))
```

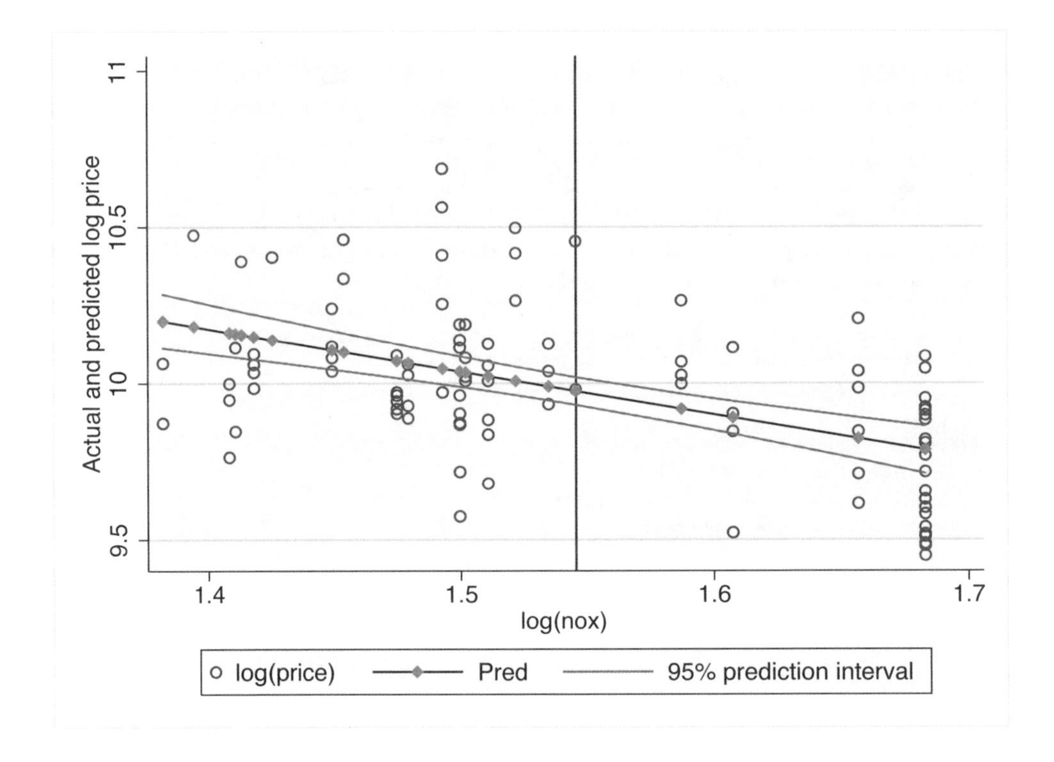

Figure 4.3: Point and interval predictions from bivariate regression

Figure 4.3 plots the actual values of the response variable against their point and interval predictions. The prediction interval is narrowest at the mean value of the regressor. The vertical line (calculated by `summarize lnox if e(sample)`, storing that value as `local` macro `lnoxbar`), marks the sample mean of `lnox` observations used in the regression.

44. For an introduction to Stata graphics, please see [G] **graph intro** and **help graph intro**. For an in-depth presentation of Stata's graphics capabilities, please see *A Visual Guide to Stata Graphics* (Mitchell 2004).

45. See [G] **graph twoway lfitci** or **help graph twoway lfitci** for another way to plot regression lines and prediction intervals.

We can compute residuals and predicted values of the dependent variable from the data and the regression point estimates. The residuals and in-sample predictions are used to assess how well the model explains the dependent variable. Whereas the goal of some studies is to obtain out-of-sample predictions for the dependent variable, using either actual or hypothetical values for the regressors, in other cases these out-of-sample predictions can be used to evaluate a model's usefulness. For example, we may apply the estimated coefficients to a separate sample (e.g., the Springfield-area communities rather than the Boston-area communities) to evaluate its out-of-sample applicability. If a regression model is well specified, it should generate reasonable predictions for any sample from the population. If out-of-sample predictions are poor, the model's specification may be too specific to the original sample.

A prediction interval for the forecast may be computed with `predict`'s `stdf` (standard error of forecast) option (see [R] **regress postestimation**). Unlike `stdp`, which calculates an interval around the expected value of y for a given set of X values (in or out of sample), `stdf` accounts for the additional uncertainty associated with the prediction of one y value (i.e., σ_u^2). We can use a confidence interval formed with `stdf` to evaluate an out-of-sample data point, y_0, and formally test whether it could have been generated by the process generating the fitted model. The null hypothesis for that test implies that data point should lie within the interval $\widehat{y}_0 \pm t$ `stdf`.

4.7 Computing marginal effects

One of Stata's most powerful statistical features is the `mfx` command, which computes marginal effects or elasticities after estimation in point and interval form:

`mfx` [*if*] [*in*] [, *options*]

`mfx` calculates the marginal effect that a change in a regressor has on those quantities computed with `predict` after estimation. It automatically uses the default prediction option, for instance, for `regress`, the `xb` option that computes $\widehat{\mathbf{y}}$.

If you use `mfx` (with the default `dydx` option) after a regression equation, the results merely reproduce the `regress` coefficient table with one change: the mean of each regressor is displayed. For regression, the coefficient estimates calculate the marginal effects, and they do not vary across the sample space. Of greater interest to economists are the elasticity and semielasticity measures, which we obtain with `mfx` options `eyex`, `dyex`, and `eydx`. The first is the elasticity of y with respect to x_j, equivalent to $\partial \log(y)/\partial \log(x_j)$. By default, these are evaluated at the multivariate point of means of the data, but they can be evaluated at any point using the `at()` option. The second, `dyex`, would be appropriate if the response variable was already in logarithmic form, but the regressor was not; this is the semielasticity $\partial y/\partial \log(x_j)$ of a log-linear model. The third form. `eydx`, would be appropriate if the regressor was in logarithmic form, but the respons variable was not; this is the semielasticity $\partial \log(y)/\partial x_j$.

The following example shows some of these options, using a form of the median price regression in levels rather than logarithms to illustrate. We compute elasticities (by default, at the point of means) with the `eyex` option for each explanatory variable.

```
. use http://www.stata-press.com/data/imeus/hprice2a, clear
(Housing price data for Boston-area communities)

. regress price nox dist rooms stratio proptax

      Source |       SS       df       MS              Number of obs =     506
-------------+------------------------------           F(  5,   500) =  165.85
       Model |  2.6717e+10     5  5.3434e+09           Prob > F      =  0.0000
    Residual |  1.6109e+10   500  32217368.7           R-squared     =  0.6239
-------------+------------------------------           Adj R-squared =  0.6201
       Total |  4.2826e+10   505    84803032           Root MSE      =    5676

------------------------------------------------------------------------------
       price |      Coef.   Std. Err.      t    P>|t|     [95% Conf. Interval]
-------------+----------------------------------------------------------------
         nox |  -2570.162    407.371    -6.31   0.000    -3370.532   -1769.793
        dist |  -955.7175   190.7124    -5.01   0.000    -1330.414    -581.021
       rooms |   6828.264   399.7034    17.08   0.000     6042.959    7613.569
     stratio |  -1127.534   140.7653    -8.01   0.000    -1404.099   -850.9699
     proptax |  -52.24272   22.53714    -2.32   0.021    -96.52188   -7.963555
       _cons |   20440.08   5290.616     3.86   0.000      10045.5    30834.66
------------------------------------------------------------------------------

. mfx, eyex

Elasticities after regress
      y  = Fitted values (predict)
         =    22511.51
------------------------------------------------------------------------------
variable |      ey/ex    Std. Err.     z    P>|z|  [    95% C.I.   ]      X
---------+--------------------------------------------------------------------
     nox |  -.6336244      .10068   -6.29   0.000  -.830954 -.436295   5.54978
    dist |  -.1611472      .03221   -5.00   0.000  -.224273 -.098022   3.79575
   rooms |   1.906099       .1136   16.78   0.000   1.68344  2.12876   6.28405
 stratio |  -.9245706      .11589   -7.98   0.000  -1.15171 -.697429   18.4593
 proptax |  -.0947401      .04088   -2.32   0.020  -.174871 -.014609   40.8237
------------------------------------------------------------------------------
```

The significance levels of the elasticities are identical to those of the original coefficients.[46] The regressor `rooms` is elastic, with an increase in `rooms` having almost twice as large an effect on `price` in proportional terms. The other three regressors are inelastic, with estimated elasticities within the unit interval, but the 95% confidence interval for `stratio` includes values less than -1.0.

The `at()` option of `mfx` can compute point and interval estimates of the marginal effects or elasticities at any point in the sample space. We can specify that one variable take on a specific value while all others are held at their (estimation sample) means or medians to trace out the effects of that regressor. For instance, we may calculate a house price elasticity over the range of values of `lnox` in the sample. The command also handles the discrete changes appropriate for indicator variables.

46. `mfx` uses a large-sample normal, whereas `regress` uses a Student t, thus causing the small difference in the output.

The example below evaluates the variation in the elasticity of median housing price with respect to the community's student–teacher ratio in both point and interval form. We first run the regression and compute selected percentiles of `stratio` by using the `detail` option of `summarize`, saving them in variable `x_val`.

```
. // run regression
. quietly regress price nox dist rooms stratio
. // compute appropriate t-statistic  for 95% confidence interval
. scalar tmfx = invttail(e(df_r),0.975)
. generate y_val = .                // generate variables needed
(506 missing values generated)
. generate x_val = .
(506 missing values generated)
. generate eyex_val = .
(506 missing values generated)
. generate seyex1_val = .
(506 missing values generated)
. generate seyex2_val = .
(506 missing values generated)
. // summarize, detail computes percentiles of stratio
. quietly summarize stratio if e(sample), detail
. local pct  1 10 25 50 75 90 99
. local i = 0
. foreach p of local pct {
  2.         local pc`p'=r(p`p')
  3.         local ++i
  4. // set those percentiles into x_val
.         quietly replace x_val = `pc`p'' in `i'
  5. }
```

To produce the graph, we must compute elasticities at the selected percentiles and store the `mfx` results in variable `y_val`. The `mfx` command, like all estimation commands, leaves behind results described in `ereturn list`. The saved quantities include scalars such as `e(Xmfx_y)`, the predicted value of y generated from the regressors, and matrices containing the marginal effects or elasticities. The example above uses `eyex` to compute the elasticities, which are returned in the matrix `e(xMfx_eyex)` with standard errors returned in the matrix `e(xMfx_se_eyex)`. The do-file extracts the appropriate values from those matrices and uses them to create variables containing the percentiles of `stratio`, the corresponding predicted values of `price`, the elasticity estimates, and their confidence interval bounds.

(Continued on next page)

```
. local i = 0

. foreach p of local pct {
  2. // compute elasticities at those points
.    quietly mfx compute, eyex at(mean stratio=`pc`p'')
  3.    local ++i
  4. // save predictions at these points in y_val
.    quietly replace y_val = e(Xmfx_y) in `i'
  5. // retrieve elasticities
.    matrix Meyex = e(Xmfx_eyex)
  6.    matrix eta = Meyex[1,"stratio"]              // for the stratio column
  7.    quietly replace eyex_val = eta[1,1] in `i'    // and save in eyex_val
  8. // retrieve standard errors of the elasticities
.    matrix Seyex = e(Xmfx_se_eyex)
  9.    matrix se = Seyex[1,"stratio"]               // for the stratio column
 10. // compute upper and lower bounds of confidence interval
.    quietly replace seyex1_val = eyex_val + tmfx*se[1,1] in `i'
 11.    quietly replace seyex2_val = eyex_val - tmfx*se[1,1] in `i'
 12. }
```

I graph these series in figure 4.4, combining three `twoway` graph types: `scatter` for the elasticities, `rline` for their standard errors, and `connected` for the predicted values, with a second axis labeled with their magnitudes.[47]

```
. label variable x_val "Student/teacher ratio (percentiles `pct')"

. label variable y_val "Predicted median house price"

. label variable eyex_val "Elasticity"

. label variable seyex1_val "95% c.i."

. label variable seyex2_val "95% c.i."

. // graph the scatter of elasticities vs. percentiles of stratio
. // as well as the predictions with rline
. // and the 95% confidence bands with connected
. twoway (scatter eyex_val x_val, ms(Oh) yscale(range(-0.5 -2.0)))
> (rline seyex1_val seyex2_val x_val)
> (connected y_val x_val, yaxis(2) yscale(axis(2) range(20000 35000))),
> ytitle(Elasticity of price vs. student/teacher ratio)

. drop y_val x_val eyex_val seyex1_val seyex2_val  // discard graph's variables
```

47. For more about Stata's graphics capabilities, including overlaying several plot types, see *A Visual Guide to Stata Graphics* (Mitchell 2004).

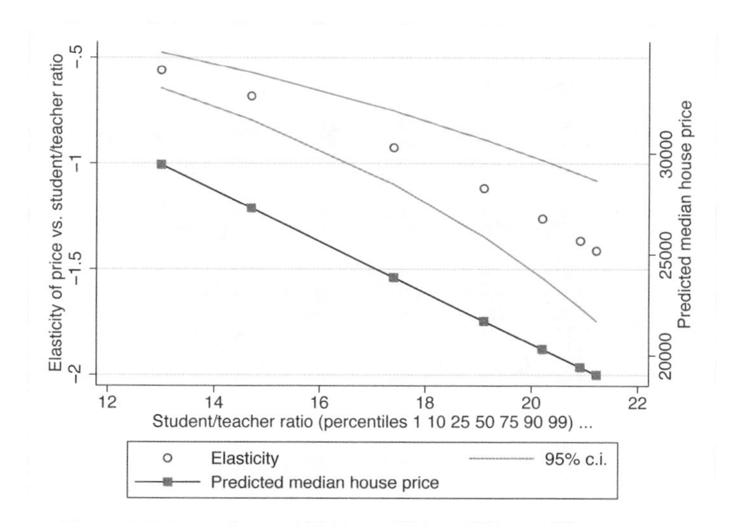

Figure 4.4: Point and interval elasticities computed with `mfx`

The model's predictions for various levels of the student–teacher ratio demonstrate that more crowded schools are associated with lower housing prices, ceteris paribus. The elasticities vary considerably over the range of `stratio` values.

These do-files demonstrate how much you can automate generating a table of point and interval elasticity estimates, in this case to present them graphically, by using values stored in the `r()` and `e()` structures. You could adapt the do-files to generate similar estimates for a different regressor or from a different regression equation. We choose the x-axis points from the percentiles of the regressor and specify the list of percentiles as a local macro. Although many users will use `mfx` just for its results, you can also use those results to produce a table or graph showing the variation in marginal effects or elasticities over a range of regressor values.

Exercises

1. Regress $y = (2,1,0)$ on $X = (0,1,2)$ without a constant term, and calculate the residuals. Refit the model with a constant term, and calculate the residuals. Compare the residual sum of squares from this model with those from the model with a constant term included. What do you conclude about the model fitted without a constant term?

2. Fit the regression of section 4.5.2, and use `test` to evaluate the hypothesis $H_0: 2\ \beta_{\texttt{ldist}} = -\beta_{\texttt{rooms}}$. Compute the linear combination $2\ b_{\texttt{ldist}} - b_{\texttt{rooms}}$ by using `lincom`. Why do these two commands yield the same p-values? What is the relationship between the F statistic reported by `test` and the t statistic reported by `lincom`?

3. Fit the regression of section 4.5.2. Refit the model subject to the linear restriction that $2\ \beta_{\texttt{ldist}} = -\beta_{\texttt{rooms}}$. Do the results change appreciably? Why or why not?

4. Using the regression equation estimated in the example of section 4.7, compute the elasticities of `price` with respect to `dist` at each decile of the `price` distribution (hint: see `xtile`) and produce a table containing the 10 deciles of `price` and the corresponding elasticities.

4.A Appendix: Regression as a least-squares estimator

We can express the linear regression problem in matrix notation with $\mathbf{y}$ as an N vector, $\mathbf{X}$ a $N \times k$ matrix, and $\mathbf{u}$ an N vector as

$$\mathbf{y} = \mathbf{X}\boldsymbol{\beta} + \mathbf{u} \tag{4.9}$$

Using the *least-squares* approach to estimation, we want to solve the sample analogue to this problem as

$$\mathbf{y} = \mathbf{X}\widehat{\boldsymbol{\beta}} + \widehat{\mathbf{u}} \tag{4.10}$$

where $\widehat{\boldsymbol{\beta}}$ is the k-element vector of estimates of $\boldsymbol{\beta}$ and $\widehat{\mathbf{u}}$ is the N-vector of least-squares residuals. We want to choose the elements of $\widehat{\boldsymbol{\beta}}$ to achieve the minimum error sum of squares, $\widehat{\mathbf{u}}'\widehat{\mathbf{u}}$. We can write the least-squares problem as

$$\boldsymbol{\beta} = \arg\min_{\boldsymbol{\beta}}\ \widehat{\mathbf{u}}'\widehat{\mathbf{u}} = \arg\min_{\boldsymbol{\beta}}\ (\mathbf{y} - \mathbf{X}\boldsymbol{\beta})'(\mathbf{y} - \mathbf{X}\boldsymbol{\beta})$$

Assuming $N > k$ and linear independence of the columns of $\mathbf{X}$ (i.e., $\mathbf{X}$ must have full column rank), this problem has the unique solution

$$\widehat{\boldsymbol{\beta}} = (\mathbf{X}'\mathbf{X})^{-1}\mathbf{X}'\mathbf{y} \tag{4.11}$$

The values calculated by least squares in (4.11) are identical to those computed by the method of moments in (4.4) since the first-order conditions used to derive the least-squares solution above define the moment conditions used by the method of moments.

4.B Appendix: The large-sample VCE for linear regression

The sampling distribution of an estimator describes the estimates produced by applying that estimator to repeated samples from the underlying population. If the size of each sample N is large enough, the sampling distribution of the estimator may be approximately normal, whether or not the underlying stochastic disturbances are normally distributed. An estimator satisfying this property is said to be *asymptotically normal*. If we are consistently estimating one parameter, its sampling variance will shrink to zero as $N \rightarrow \infty$. An estimated parameter may be biased in small samples, but that bias will disappear with large N if the estimator is consistent. In the multivariate context, the variability of the estimates is described by the variance–covariance matrix of the large-sample normal distribution. We call this matrix the variance–covariance matrix of our estimator, or VCE. To evaluate the variability of our estimates, we need a consistent estimator of the VCE.

If the regressors are "well behaved" with finite second moments, we can write the probability limit, or *plim*, of their moments matrix, scaled by sample size N, as

$$\text{plim}\ \frac{\mathbf{X}'\mathbf{X}}{N} = \mathbf{Q} \tag{4.12}$$

where $\mathbf{Q}$ is a positive-definite matrix.[48] We can then derive the distribution of the random estimates $\widehat{\boldsymbol{\beta}}$ as

$$\sqrt{N}\ (\widehat{\boldsymbol{\beta}} - \beta) \xrightarrow{d} N\left(0, \sigma_u^2 \mathbf{Q}^{-1}\right) \tag{4.13}$$

where $\xrightarrow{d}$ denotes convergence in distribution as the sample size $N \rightarrow \infty$. For $\widehat{\boldsymbol{\beta}}$ itself, we can write

$$\widehat{\boldsymbol{\beta}} \overset{a}{\sim} N\left(\beta, \frac{\sigma_u^2}{N}\mathbf{Q}^{-1}\right) \tag{4.14}$$

where $\overset{a}{\sim}$ denotes the large-sample distribution. To estimate the large-sample VCE of $\widehat{\boldsymbol{\beta}}$, we must estimate the two quantities in (4.14): σ_u^2 and $(1/N)\mathbf{Q}^{-1}$. We can consistently estimate the first quantity, σ_u^2, as shown in (4.5) by $\mathbf{e}'\mathbf{e}/(N-k)$, where $\mathbf{e}$ is the regression residual vector. We can estimate the second quantity consistently from the sample by $(\mathbf{X}'\mathbf{X})^{-1}$. Thus we can estimate the large-sample VCE of $\widehat{\boldsymbol{\beta}}$ from the sample as

$$\text{VCE}(\widehat{\boldsymbol{\beta}}) = s^2(\mathbf{X}'\mathbf{X})^{-1} = \frac{\widehat{\mathbf{u}}'\widehat{\mathbf{u}}}{N-k}(\mathbf{X}'\mathbf{X})^{-1} \tag{4.15}$$

48. A sequence of random variables $\widehat{\theta}_N$ *converges* in probability to the constant a if for $\epsilon > 0$, $\Pr(|\widehat{\theta}_N - a| > \epsilon) \rightarrow 0$ as $N \rightarrow \infty$. a is the *plim* of $\widehat{\theta}_N$. If $\widehat{\theta}_N$ is an estimator of the population parameter θ and $a = \theta$, $\widehat{\theta}_N$ is a *consistent* estimator of θ.

5 Specifying the functional form

5.1 Introduction

A key assumption maintained in the previous chapter is that the functional form was correctly specified. Here we discuss some methods for checking the validity of this assumption. If the *zero-conditional-mean* assumption

$$E[u \mid x_1, x_2, \ldots, x_k] = 0 \tag{5.1}$$

is violated, the coefficient estimates are inconsistent.

The three main problems that cause the zero-conditional-mean assumption to fail in a regression model are

- improper specification of the model;
- endogeneity of one or more regressors; or
- measurement error of one or more regressors.

The *specification* of a regression model may be flawed in its list of included regressors or in the functional form specified for the estimated relationship. *Endogeneity* means that one or more regressors may be correlated with the error term, a condition that often arises when those regressors are simultaneously determined with the response variable. *Measurement error* of a regressor implies that the underlying behavioral relationship includes one or more variables that the econometrician cannot accurately measure. This chapter discusses specification issues, whereas chapter 8 addresses endogeneity and measurement errors.

5.2 Specification error

The consistency of the linear regression estimator requires that the sample regression function correspond to the underlying population regression function or true model for the response variable y:

$$y_i = \mathbf{x}_i \boldsymbol{\beta} + u_i$$

Specifying a regression model often involves making a sequence of decisions about the model's contents. Economic theory often provides some guidance in model specification but may not explicitly indicate how a specific variable should enter the model, identify the functional form, or spell out precisely how the stochastic elements (u_i) enter the model. Comparative static results that provide expected signs for derivatives do not indicate which functional specification to use for the model. Should it be estimated in levels; as a log-linear structure; as a polynomial in one or more of the regressors; or in logarithms, implying a constant-elasticity relationship? Theory is often silent on such specifics, yet we must choose a specific functional form to proceed with empirical research.[1]

Let us assume that the empirical specification may differ from the population regression function in one of two ways (which both might be encountered in the same fitted model). Given the dependent variable y, we may omit relevant variables from the model, or we may include irrelevant variables in the model, making the fitted model "short" or "long", respectively, relative to the true model.

5.2.1 Omitting relevant variables from the model

Suppose that the true model is

$$y = \mathbf{x}_1\boldsymbol{\beta}_1 + \mathbf{x}_2\boldsymbol{\beta}_2 + u$$

with k_1 and k_2 regressors in the two subsets, but that we regress y on just the $\mathbf{x}_1$ variables:

$$y = \mathbf{x}_1\boldsymbol{\beta}_1 + u$$

This step yields the least-squares solution

$$\begin{aligned} \widehat{\boldsymbol{\beta}}_1 &= (\mathbf{X}_1'\mathbf{X}_1)^{-1}\mathbf{X}_1'\mathbf{y} && (5.2) \\ &= \boldsymbol{\beta}_1 + (\mathbf{X}_1'\mathbf{X}_1)^{-1}\mathbf{X}_1'\mathbf{X}_2\boldsymbol{\beta}_2 + (\mathbf{X}_1'\mathbf{X}_1)^{-1}\mathbf{X}_1'\mathbf{u} && (5.3) \end{aligned}$$

Unless $\mathbf{X}_1'\mathbf{X}_2 = \mathbf{0}$ or $\boldsymbol{\beta}_2 = 0$, the estimate $\widehat{\boldsymbol{\beta}}_1$ is biased, since

$$E[\widehat{\boldsymbol{\beta}}_1|\mathbf{X}] = \boldsymbol{\beta}_1 + \mathbf{P}_{1\cdot 2}\,\boldsymbol{\beta}_2$$

where $\mathbf{P}_{1\cdot 2} = (\mathbf{X}_1'\mathbf{X}_1)^{-1}\mathbf{X}_1'\mathbf{X}_2$ is the $k_1 \times k_2$ matrix reflecting the regression of each column of $\mathbf{X}_2$ on the columns of $\mathbf{X}_1$. If $k_1 = k_2 = 1$ and the single variable in $\mathbf{X}_2$ is correlated with the single variable in $\mathbf{X}_1$, we can derive the direction of bias. Generally, with multiple variables in each set, we can make no statements about the nature of the bias of the $\widehat{\boldsymbol{\beta}}_1$ coefficients.

We may conclude that the cost of omitting relevant variables is high. If $E[\mathbf{x}_1'\mathbf{x}_2] \neq \mathbf{0}$, (5.3) would have showed that the estimator was inconsistent. If the population correlations between elements of $\mathbf{x}_1$ and $\mathbf{x}_2$ are zero, regression estimates would be consistent but probably biased in finite samples. In economic research, a variable mistakenly excluded from a model is unlikely to be uncorrelated in the population or in the sample with the regressors.

1. This requirement holds unless one chooses to use nonparametric methods that are beyond the scope of this book. See Hardle (1990) for an introduction to nonparametric methods.

Specifying dynamics in time-series regression models

A related concern arises in models for time-series data, in which theory rarely fully specifies the *time form* of the dynamic relationship. For instance, consumer theory may specify the ultimate response of an individual's saving to a change in her after-tax income. However, theory may fail to indicate how rapidly the individual will adjust her saving to a permanent change in her salary. Will that adjustment take place within one, two, three, or more biweekly pay periods? From our analysis of the asymmetry of specification error, we know that the advice to the modeler should be "do not underfit the dynamics." If we do not know the time form of a dynamic relationship with certainty, we should include several lagged values of the regressor. We can then use the "test down" strategy discussed below to determine whether the longer lags are necessary. Moreover, omitting higher-order dynamic terms may cause apparent nonindependence of the regression errors, as signaled by residual independence tests.

5.2.2 Graphically analyzing regression data

With Stata's graphics, you can easily perform exploratory data analysis on the regression sample, even with large datasets. In specification analysis, we may want to examine the simple bivariate relationships between y and the regressors in $\mathbf{x}$. Although multiple linear regression coefficients are complicated functions of the various bivariate regression coefficients among these variables, we still often find it useful to examine a set of bivariate plots. We use `graph matrix` to generate a set of plots illustrating the bivariate relationships underlying our regression model of median housing prices:

(*Continued on next page*)

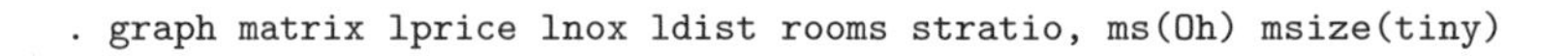

```
. graph matrix lprice lnox ldist rooms stratio, ms(Oh) msize(tiny)
```

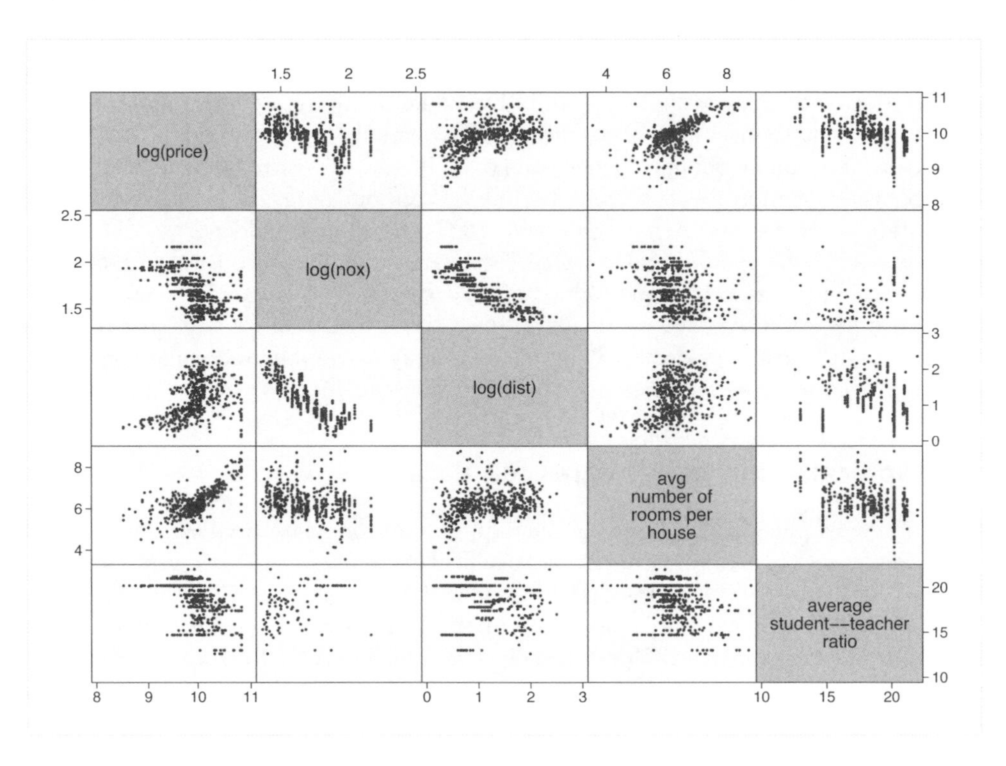

Figure 5.1: `graph matrix` of regression variables

The first row (or column) of the plot matrix in figure 5.1 illustrates the relationships between the variable to be explained (the log of median housing price) and the four causal factors. These plots are the $y - x$ planes, in which a simple regression of log housing price on each of these factors in turn would determine the line of best fit.

The other bivariate graphs below the main diagonal are also illustrative. If any of these relationships could be fitted well by a straight line, the intercorrelation among those regressors would be high, and we would expect collinearity problems. For instance, the scatter of points between `lnox` and `ldist` appears to be compact and linear. The `correlate` command shows that those two variables have a simple correlation of -0.86.

5.2.3 Added-variable plots

The added-variable plot identifies the important variables in a relationship by decomposing the multivariate relationship into a set of two-dimensional plots.[2] Taking each regressor in turn, the added-variable plot is based on two residual series. The first series, e_1, contains the residuals from the regression of $\mathbf{x}_g$ on all other $\mathbf{x}$, whereas the second series, e_2, contains the residuals from the regression of y on all x variables except $\mathbf{x}_g$. That is, e_1 represents the part of $\mathbf{x}_g$ that cannot be linearly related to those other regressors, whereas e_2 represents the information in y that is not explained by all other regressors (excluding $\mathbf{x}_g$). The added-variable plot for $\mathbf{x}_g$ is then the scatter of e_2 (on the y-axis) versus e_1 (on the x-axis). Two polar cases (as discussed by Cook and Weisberg [1994, 194]) are of interest. If most points are clustered around a horizontal line at ordinate zero in the added-variable plot, $\mathbf{x}_g$ is irrelevant. On the other hand, if most points are clustered around a vertical line with abscissa zero, the plot would indicate near-perfect collinearity. Here as well, adding $\mathbf{x}_g$ to the model would not be helpful.

The strength of a linear relationship between e_1 and e_2 (that is, the slope of a least-squares line through this scatter of points) represents the marginal value of $\mathbf{x}_g$ in the full model. If the slope is significantly different from zero, $\mathbf{x}_g$ makes an important contribution to the model beyond that of the other regressors. The more closely the points are grouped around a straight line in the plot, the more important is the contribution of $\mathbf{x}_g$ at the margin. As an added check, if the specification of the full model (including $\mathbf{x}_g$) is correct, the plot of e_1 versus e_2 must exhibit linearity. Significant departures from linearity in the plot cast doubt on the appropriate specification of $\mathbf{x}_g$ in the model.

After `regress`, the command to generate an added-variable plot is given as

```
avplot varname
```

where *varname* is the variable on which the plot is based, which can be a regressor or a variable not included in the regression model. Alternatively,

```
avplots
```

produces one graph with all added-variable plots from the last regression, as we now illustrate.

2. For details about the added-variable plot, see Cook and Weisberg (1994, 191–194). See [R] **regress postestimation** for more details about its implementation in Stata.

```
. use http://www.stata-press.com/data/imeus/hprice2a, clear
(Housing price data for Boston-area communities)

. generate rooms2 = rooms^2

. regress lprice lnox ldist rooms rooms2 stratio lproptax

      Source |       SS       df       MS              Number of obs =     506
-------------+------------------------------           F(  6,   499) =  138.41
       Model |  52.8357813     6  8.80596356           Prob > F      =  0.0000
    Residual |  31.7464896   499   .06362022           R-squared     =  0.6247
-------------+------------------------------           Adj R-squared =  0.6202
       Total |  84.5822709   505  .167489645           Root MSE      =  .25223

------------------------------------------------------------------------------
      lprice |      Coef.   Std. Err.      t    P>|t|     [95% Conf. Interval]
-------------+----------------------------------------------------------------
        lnox |  -.6615694   .1201606    -5.51   0.000    -.8976524   -.4254864
       ldist |   -.095087   .0421435    -2.26   0.024    -.1778875   -.0122864
       rooms |  -.5625662   .1610315    -3.49   0.001    -.8789496   -.2461829
      rooms2 |   .0634347   .0124621     5.09   0.000     .0389501    .0879193
     stratio |  -.0362928   .0060699    -5.98   0.000    -.0482185   -.0243671
    lproptax |  -.2211125   .0410202    -5.39   0.000     -.301706   -.1405189
       _cons |   14.15454   .5693846    24.86   0.000     13.03585    15.27323
------------------------------------------------------------------------------

. avplots, ms(Oh) msize(small) col(2)
```

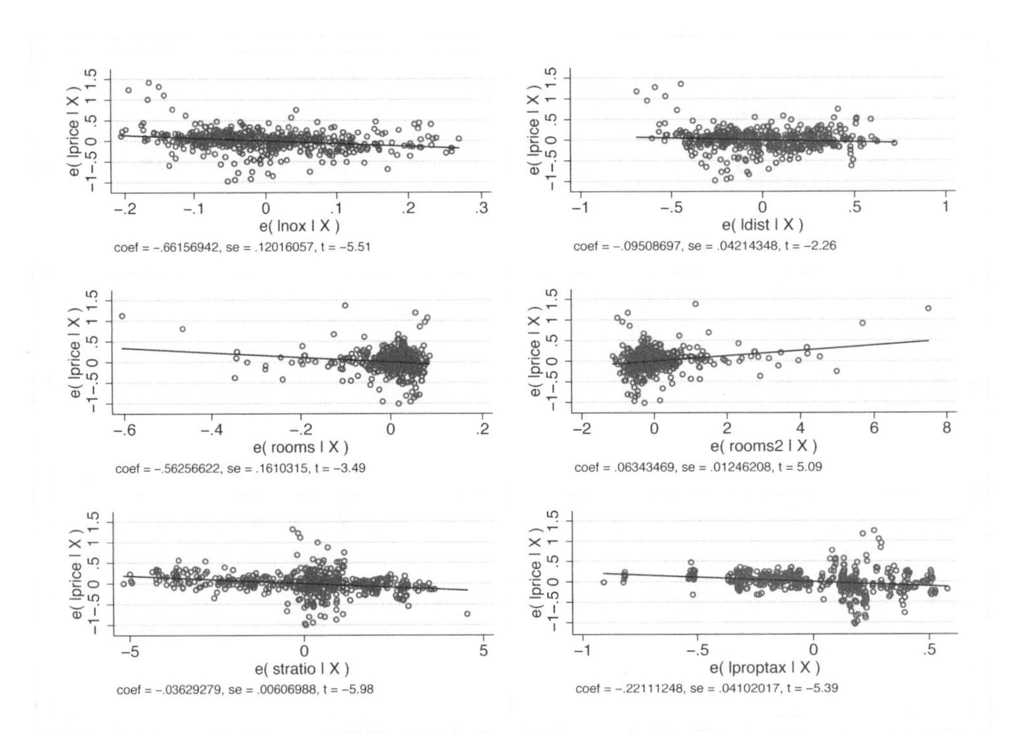

Figure 5.2: Added-variable plots

In each pane of figure 5.2, we see several observations that are far from the straight line linking the response variable and that regressor. The outlying values are particularly evident in the graphs for `lnox` and `ldist`, where low values of $E[\texttt{lnox}|X]$ and $E[\texttt{ldist}|X]$ are associated with prices much higher than those predicted by the model. The t statistics shown in each panel test the hypothesis that the least-squares line has a slope significantly different from zero. These test statistics are identical to those of the original regression, shown above.

5.2.4 Including irrelevant variables in the model

Including irrelevant regressors does not violate the zero-conditional-mean assumption. Since their population coefficients are zero, including them in the regressor list does not cause the conditional mean of the u process to differ from zero. Suppose that the true model is

$$y = \mathbf{x}_1\boldsymbol{\beta}_1 + u \tag{5.4}$$

but we mistakenly include several $\mathbf{x}_2$ variables in our regression model. In that case, we fail to impose the restrictions that $\boldsymbol{\beta}_2 = 0$. Since $\boldsymbol{\beta}_2 = 0$ in the population, including $\mathbf{x}_2$ leaves our estimates of $\boldsymbol{\beta}_1$ unbiased and consistent, as is the estimate of σ_u^2. Overfitting the model and including the additional variables causes a loss of *efficiency* (see section 4.2.3). By ignoring that the $\mathbf{x}_2$ variables do not belong in the model, our estimates of $\boldsymbol{\beta}_1$ are less precise than they would be with the correctly specified model, and the estimated standard errors of $\boldsymbol{\beta}_1$ will be larger than those fitted from the correct model of (5.4). This property is especially apparent if we have $k_1 = k_2 = 1$ and the correlation between x_1 and x_2 is high. Mistakenly including $\mathbf{x}_2$ will lead to imprecise estimates of $\boldsymbol{\beta}_1$.

Clearly, overfitting the model costs much less than underfitting, as discussed earlier. The long model delivers unbiased and consistent estimates of all its parameters, including those of the irrelevant regressors, which tend to zero.

5.2.5 The asymmetry of specification error

We may conclude that the costs of these two types of specification error are asymmetric. We would much rather err on the side of caution (including additional variables) to avoid the inconsistent estimates that would result from underfitting the model. Given this conclusion, a model selection strategy that starts with a simple specification and seeks to refine it by adding variables is flawed. The opposite approach, starting with a general specification and seeking to refine it by imposing appropriate restrictions, has much more to recommend it.[3] Although a general specification may be plagued by collinearity, a recursive simplification strategy is much more likely to yield a usable model at the end of the specification search. Ideally, we would not need to search for a specification. We would merely write down the fitted model that theory propounds, run one regression,

3. The *general-to-specific* approach proposed by econometrician David Hendry in several of his works implements such a refinement strategy. See http://ideas.repec.org/e/phe33.html for more information.

and evaluate its results. Unfortunately, most applied work is not that straightforward. Most empirical investigations contain some amount of specification searching.

In considering such a research strategy, we also must be aware of the limits of statistical inference. We might run 20 regressions in which the regressors do not appear in the true model, but at the 5% level, we would expect one of those 20 estimates to erroneously show a relationship between the response variable and regressors. Many articles in the economic literature decry "data mining" or "fishing for results". The rationale for fitting a variety of models in search of the true model is to avoid using statistical inference to erroneously reject a theory because we have misspecified the relationship. If we write down one model that bears little resemblance to the true model, fit that model, and conclude that the data reject the theory, we are resting our judgment on the *maintained hypothesis* that we have correctly specified the population model. But if we used a transformation of that model, or added omitted variables to the model, our inference might reach a different conclusion.

5.2.6 Misspecification of the functional form

A model that includes the appropriate regressors may be misspecified because the model may not reflect the algebraic form of the relationship between the response variable and those regressors. For instance, suppose that the true model specifies a nonlinear relationship between y and x_j—such as a polynomial relationship—and we omit the squared term.[4] Doing so would be misspecifying the functional form. Likewise, if the true model expresses a constant-elasticity relationship, the model fitted to logarithms of y and $\mathbf{x}$ could render conclusions different from those of a model fitted to levels of the variables. In one sense, this problem may be easier to deal with than the omission of relevant variables. In a misspecification of the functional form, we have all the appropriate variables at hand and only have to choose the appropriate form in which they enter the regression function. Ramsey's omitted-variable regression specification error test (RESET) implemented by Stata's `estat ovtest` may be useful in this context.

5.2.7 Ramsey's RESET

Linear regression of y on the levels of various x's restricts the effects of each x_j to be strictly linear. If the functional relationship linking y to x_j is nonlinear, a linear function may work reasonably well for some values of x_j but will eventually break down. Ramsey's RESET is based on this simple notion. RESET runs an augmented regression that includes the original regressors, powers of the predicted values from the original regression, and powers of the original regressors. Under the null hypothesis of no misspecification, the coefficients on these additional regressors are zero. RESET is simply a Wald test of this null hypothesis. The test works well because polynomials in $\widehat{y}$ and x_j can approximate a variety of functional relationships between y and the regressors $\mathbf{x}$.

4. Distinguish between a model linear in the *parameters* and a nonlinear relationship between y and x. $y_i = \beta_1 + \beta_2 x_i + \beta_3 x_i^2 + u_i$ is linear in the $\boldsymbol{\beta}$ parameters but defines a nonlinear function, $E[y|x] = f(x)$.

As discussed in [R] **regress postestimation**, to compute the RESET after `regress`, we use the following command syntax:[5]

`estat ovtest` [, `rhs`]

The parsimonious flavor of the test, computed by default, augments the regression with the second, third, and fourth powers of the $\widehat{y}$ series. With the `rhs` option, powers of the individual regressors themselves are used. This option may considerably reduce the power of the test in small samples because it will include many regressors. For example, if we perform RESET after our regression model of log housing prices,

```
. quietly regress lprice lnox ldist rooms stratio
. estat ovtest
Ramsey RESET test using powers of the fitted values of lprice
       Ho:  model has no omitted variables
                 F(3, 498) =      9.69
                  Prob > F =      0.0000
. estat ovtest, rhs
Ramsey RESET test using powers of the independent variables
       Ho:  model has no omitted variables
                F(12, 489) =     11.79
                  Prob > F =      0.0000
```

we can reject RESET's null hypothesis of no omitted variables for the model using either form of the test. We respecify the equation to include the square of `rooms` and include another factor, `lproptax`, the log of property taxes in the community:

```
. regress lprice lnox ldist rooms rooms2 stratio lproptax
      Source |       SS       df       MS              Number of obs =     506
-------------+------------------------------           F(  6,   499) =  138.41
       Model |  52.8357813     6  8.80596356           Prob > F      =  0.0000
    Residual |  31.7464896   499   .06362022           R-squared     =  0.6247
-------------+------------------------------           Adj R-squared =  0.6202
       Total |  84.5822709   505  .167489645           Root MSE      =  .25223

------------------------------------------------------------------------------
      lprice |      Coef.   Std. Err.      t    P>|t|     [95% Conf. Interval]
-------------+----------------------------------------------------------------
        lnox |  -.6615694   .1201606    -5.51   0.000    -.8976524   -.4254864
       ldist |   -.095087   .0421435    -2.26   0.024    -.1778875   -.0122864
       rooms |  -.5625662   .1610315    -3.49   0.001    -.8789496   -.2461829
      rooms2 |   .0634347   .0124621     5.09   0.000     .0389501    .0879193
     stratio |  -.0362928   .0060699    -5.98   0.000    -.0482185   -.0243671
    lproptax |  -.2211125   .0410202    -5.39   0.000     -.301706   -.1405189
       _cons |   14.15454   .5693846    24.86   0.000     13.03585    15.27323
------------------------------------------------------------------------------
. estat ovtest
Ramsey RESET test using powers of the fitted values of lprice
       Ho:  model has no omitted variables
                 F(3, 496) =      1.64
                  Prob > F =      0.1798
```

5. A more general command that implements several flavors of the RESET, and may be applied after instrumental-variables estimation, is Mark Schaffer's `ivreset`, available from `ssc`.

This model's predicted values no longer reject the RESET. The relationship between `rooms` and housing values appears to be nonlinear (although the pattern of signs on the `rooms` and `rooms2` coefficients is not that suggested by theory). But as theory suggests, communities with higher property tax burdens have lower housing values, ceteris paribus.

5.2.8 Specification plots

Many plots based on the residuals have been developed to help you evaluate the specification of the model because certain patterns in the residuals indicate misspecification. We can graph the residuals versus the predicted values with `rvfplot` (residual-versus-fitted plot) or plot them against a specific regressor with `rvpplot` (residual-versus-predictor plot) by using the regression model above:

```
. quietly regress lprice lnox ldist rooms rooms2 stratio lproptax
. rvpplot ldist, ms(Oh) yline(0)
```

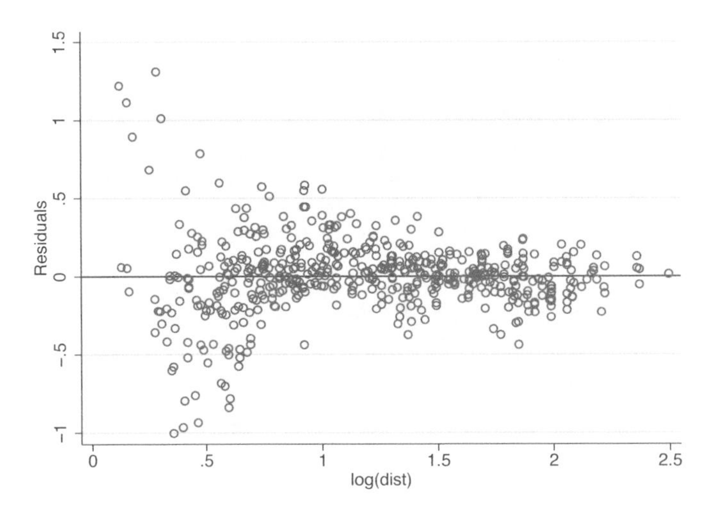

Figure 5.3: Residual-versus-predictor plot

The latter plot is displayed in figure 5.3. Any pattern in this graph indicates a problem with the model. For instance, the residuals appear much more variable for low levels versus high levels of log of distance (`ldist`), so the assumption of homoskedasticity (a constant variance of the disturbance process) is untenable.

A variety of other graphical techniques for identifying specification problems have been proposed, and several are implemented in Stata; see [R] **regress postestimation**.

5.2.9 Specification and interaction terms

We might also encounter specification error with respect to interactions among the regressors. If the true model implies that $\partial y/\partial x_j$ is a function of x_ℓ, we should fit the model

$$y = \beta_1 + \beta_2 X_2 + \cdots + \beta_j x_j + \beta_\ell X_\ell + \beta_p(x_j \cdot x_\ell) + \cdots + u \quad (5.5)$$

in which the regressor $(x_j \cdot x_\ell)$ is an *interaction term*. With this term added to the model, we find that $\partial y/\partial x_j = \beta_j + \beta_p x_\ell$. The effect of x_j then depends on x_ℓ. For example, in a regression of housing prices on attributes of the dwelling, the effect of adding a bedroom to the house may depend on the house's square footage.[6] If the coefficient β_p is constrained to equal zero [that is, if we estimate (5.5) without interaction effects], the partial derivatives of both x_j and x_ℓ are constrained to be constant rather than varying, as they would be for the equation including the interaction term. If the interaction term or terms are irrelevant, their t statistics will indicate that you can safely omit them. But here the correct specification of the model requires that you enter the regressors in the proper form in the fitted model.

As an example of misspecification due to interaction terms, we include `taxschl`—an interaction term between `lproptax`, the logarithm of average property taxes in the community, and `stratio`, the student–teacher ratio—in our housing-price model.[7] Both are negative factors, in the sense that buyers would prefer to pay lower taxes and enjoy schools with larger staff and would not be willing to pay as much for a house in a community with high values for either attribute.

```
. generate taxschl = lproptax * stratio
. regress lprice lnox ldist lproptax stratio taxschl
      Source |       SS       df       MS              Number of obs =     506
-------------+------------------------------           F(  5,   500) =   84.47
       Model |  38.7301562     5  7.74603123           Prob > F      =  0.0000
    Residual |  45.8521148   500   .09170423           R-squared     =  0.4579
-------------+------------------------------           Adj R-squared =  0.4525
       Total |  84.5822709   505  .167489645           Root MSE      =  .30283

------------------------------------------------------------------------------
      lprice |      Coef.   Std. Err.      t    P>|t|     [95% Conf. Interval]
-------------+----------------------------------------------------------------
        lnox |  -.9041103   .1441253    -6.27   0.000    -1.187276   -.6209444
       ldist |  -.1430541   .0501831    -2.85   0.005    -.2416499   -.0444583
    lproptax |   -1.48103   .5163117    -2.87   0.004    -2.495438   -.4666219
     stratio |  -.4388722   .1538321    -2.85   0.005    -.7411093   -.1366351
     taxschl |   .0641648    .026406     2.43   0.015     .0122843    .1160452
       _cons |   21.47905   2.952307     7.28   0.000      15.6786    27.27951
------------------------------------------------------------------------------
```

The interaction term is evidently significant, so a model excluding that term can be considered misspecified for the reasons discussed in section 5.2.1, although the omitted variable is algebraically related to the included regressors. The interaction term has

6. And vice versa: in (5.5), $\partial y/\partial x_\ell$ is a function of the level of x_j.
7. We exclude the `rooms` and `rooms2` regressors from this example for illustration.

a positive sign, so the negative partial derivative of `lprice` with respect to `lproptax` (`stratio`) becomes less negative (closer to zero) for higher levels of `stratio` (`lproptax`).

5.2.10 Outlier statistics and measures of leverage

To evaluate the adequacy of the specification of an fitted model, we must also consider evidence relating to the model's robustness to *influential data*. The OLS estimator is designed to fit the regression sample as well as possible. However, our objective in fitting the model often includes inference about the population from which the sample was drawn or computing out-of-sample forecasts. Evidence that the model's coefficients have been strongly influenced by a few data points or of structural instability over subsamples casts doubt on the fitted model's worth in any broader context. For this reason, we consider tests and plots designed to identify influential data.

A variety of statistics have been designed to evaluate influential data and the relationship between those data and the fitted model. A pioneering work in this field is Belsley, Kuh, and Welsch (1980) and the later version, Belsley (1991). An *outlier* in a regression relationship is a data point with an unusual value, such as a value of housing price twice as high as any other or a community with a student–teacher ratio 3 standard deviations below the mean. An outlier may be an observation associated with a large residual (in absolute terms), a data point that the model fits poorly.

On the other hand, an unusual data point that is far from the center of mass of the x_j distribution may also be an outlier, although the residual associated with that data point will often be small because the least-squares process attaches a squared penalty to the residual in forming the least-squares criterion. Just as the arithmetic mean (a least-squares estimator) is sensitive to extreme values (relative to the sample median), the least-squares regression fit will attempt to prevent such an unusual data point from generating a sizable residual. We say that this unusual point has a high degree of *leverage* on the estimates because including it in the sample alters the estimated coefficients by a sizable amount. Data points with large residuals may also have high leverage. Those with low leverage may still have a large effect on the regression estimates. Measures of influence and the identification of influential data points take their leverage into account.

You can calculate a measure of each data point's leverage after `regress` with

```
. predict double lev if e(sample), leverage
```

These leverage values are computed from the diagonal elements of the "hat matrix", $h_j = \mathbf{x}_j(\mathbf{X}'\mathbf{X})^{-1}\mathbf{x}_j'$, where $\mathbf{x}_j$ is the jth row of the regressor matrix.[8] You can use `lvr2plot` to graphically display leverage values versus the (normalized) squared residuals. Points with very high leverage or very large squared residuals may deserve scrutiny. We can also examine those statistics directly. Consider our housing-price regression model, for which we compute leverage and squared residuals. The `town` variable identifies the community.

8. The formulas for `predict` options are presented in [R] **regress postestimation**.

```
. quietly regress lprice lnox ldist rooms rooms2 stratio lproptax
. generate town = _n
. predict double lev if e(sample), leverage
. predict double eps if e(sample), res
. generate double eps2 = eps^2
. summarize price lprice
```

Variable	Obs	Mean	Std. Dev.	Min	Max
price	506	22511.51	9208.856	5000	50001
lprice	506	9.941057	.409255	8.517193	10.8198

We then list the five largest values of the leverage measure, using **gsort** to produce the descending-sort order:

```
. gsort -lev
. list town price lprice lev eps2 in 1/5
```

	town	price	lprice	lev	eps2
1.	366	27499	10.2219	.17039262	.61813718
2.	368	23100	10.04759	.11272637	.30022048
3.	365	21900	9.994242	.10947853	.33088957
4.	258	50001	10.8198	.08036068	.06047061
5.	226	50001	10.8198	.0799096	.03382768

We can also examine the towns with the largest squared residuals:

```
. gsort -eps2
. list town price lprice lev eps2 in 1/5
```

	town	price	lprice	lev	eps2
1.	369	50001	10.8198	.02250047	1.7181195
2.	373	50001	10.8198	.01609848	1.4894088
3.	372	50001	10.8198	.02056901	1.2421055
4.	370	50001	10.8198	.0172083	1.0224558
5.	406	5000	8.517193	.00854955	1.0063662

As these results show, a large value of leverage does not imply a large squared residual, and vice versa. Several of the largest values of leverage or the squared residuals correspond to the extreme values of median housing prices recorded in the dataset, which range from \$5,000 to \$50,001. These data may have been coded with observations outside that range equal to that minimum or maximum value, respectively.

The DFITS statistic

A summary of the leverage values and magnitude of residuals is provided by the DFITS statistic of Welsch and Kuh (1977),

$$\text{DFITS}_j = r_j \sqrt{\frac{h_j}{1-h_j}}$$

where r_j is a studentized (standardized) residual,

$$r_j = \frac{e_j}{s_{(j)}\sqrt{1-h_j}}$$

with $s_{(j)}$ referring to the root mean squared error (s) of the regression equation with the jth observation removed. Working through the algebra shows that either a large value of leverage (h_j) or a large absolute residual (e_j) will generate a large $|\text{DFITS}_j|$. The DFITS measure is a scaled difference between the in-sample and out-of-sample predicted values for the jth observation. DFITS evaluates the result of fitting the regression model including and excluding that observation. Belsley, Kuh, and Welsch (1980) suggest that a cutoff value of $|\text{DFITS}_j| > 2\sqrt{k/N}$ indicates highly influential observations. We now compute DFITS in our housing-price regression model:[9]

```
. predict double dfits if e(sample), dfits
```

We then sort the calculated DFITS statistic in descending order and calculate the recommended cutoff value as an indicator variable, `cutoff`, equal to 1 if the absolute value of DFITS is large and zero otherwise. Consider the values of DFITS for which `cutoff` = 1:

9. See [R] **regress postestimation** for more details.

```
. gsort -dfits
. quietly generate cutoff = abs(dfits) > 2*sqrt((e(df_m)+1)/e(N)) & e(sample)
. list town price lprice dfits if cutoff
```

	town	price	lprice	dfits
1.	366	27499	10.2219	1.5679033
2.	368	23100	10.04759	.82559867
3.	369	50001	10.8198	.8196735
4.	372	50001	10.8198	.65967704
5.	373	50001	10.8198	.63873964
6.	371	50001	10.8198	.55639311
7.	370	50001	10.8198	.54354054
8.	361	24999	10.12659	.32184327
9.	359	22700	10.03012	.31516743
10.	408	27901	10.23642	.31281326
11.	367	21900	9.994242	.31060611
12.	360	22600	10.02571	.28892457
13.	363	20800	9.942708	.27393758
14.	358	21700	9.985067	.24312885
490.	386	7200	8.881836	-.23838749
491.	388	7400	8.909235	-.25909393
492.	491	8100	8.999619	-.26584795
493.	400	6300	8.748305	-.28782824
494.	416	7200	8.881836	-.29288953
495.	402	7200	8.881836	-.29595696
496.	381	10400	9.249561	-.29668364
497.	258	50001	10.8198	-.30053391
498.	385	8800	9.082507	-.302916
499.	420	8400	9.035987	-.30843965
500.	490	7000	8.853665	-.3142718
501.	401	5600	8.630522	-.33273658
502.	417	7500	8.922658	-.34950136
503.	399	5000	8.517193	-.36618139
504.	406	5000	8.517193	-.37661853
505.	415	7012	8.855378	-.43879798
506.	365	21900	9.994242	-.85150064

About 6% of the observations are flagged by the DFITS cutoff criterion. Many of those observations associated with large positive DFITS have the top-coded value of $50,001 for median housing price, and the magnitude of positive DFITS is considerably greater than that of negative DFITS. The identification of top-coded values that represent an arbitrary maximum recorded price suggests that we consider a different estimation technique for this model. The `tobit` regression model, presented in section 10.3.2, can properly account for the *censored* nature of the median housing price.

The DFBETA statistic

We may also want to focus on one regressor and consider its effect on the estimates by computing the DFBETA series with the **dfbeta** command after a regression.[10] The jth observation's DFBETA measure for regressor ℓ may be written as

$$\text{DFBETA}_j = \frac{r_j v_j}{\sqrt{v^2(1-h_j)}}$$

where the v_j are the residuals obtained from the partial regression of x_ℓ on the remaining columns of $\mathbf{X}$, and v^2 is their sum of squares. The DFBETAs for regressor ℓ measure the distance that this regression coefficient would shift when the jth observation is included or excluded from the regression, scaled by the estimated standard error of the coefficient. One rule of thumb suggests that a DFBETA value greater than unity in absolute value might be reason for concern since this observation might shift the estimated coefficient by more than one standard error. Belsley, Kuh, and Welsch (1980) suggest a cutoff of $|\text{DFBETA}_j| > 2/\sqrt{N}$.

We compute DFBETAs for one of the regressors, `lnox`, in our housing-price regression model:

```
. quietly regress lprice lnox ldist rooms rooms2 stratio lproptax
. dfbeta lnox
                          DFlnox:  DFbeta(lnox)
. quietly generate dfcut = abs(DFlnox) > 2/sqrt(e(N)) & e(sample)
. sort DFlnox
. summarize lnox
    Variable |       Obs        Mean    Std. Dev.       Min        Max
-------------+--------------------------------------------------------
        lnox |       506    1.693091    .2014102   1.348073   2.164472
```

10. As discussed in [R] **regress postestimation**, we can calculate one `dfbeta` series with `predict`, whereas you can use one `dfbeta` command to compute one or all of these series and automatically name them.

```
. list town price lprice lnox DFlnox if dfcut
```

	town	price	lprice	lnox	DFlnox
1.	369	50001	10.8198	1.842136	-.4316933
2.	372	50001	10.8198	1.842136	-.4257791
3.	373	50001	10.8198	1.899118	-.3631822
4.	371	50001	10.8198	1.842136	-.2938702
5.	370	50001	10.8198	1.842136	-.2841335
6.	365	21900	9.994242	1.971299	-.2107066
7.	408	27901	10.23642	1.885553	-.1728729
8.	368	23100	10.04759	1.842136	-.1309522
9.	11	15000	9.615806	1.656321	-.1172723
10.	410	27499	10.2219	1.786747	-.1117743
11.	413	17900	9.792556	1.786747	-.0959273
12.	437	9600	9.169518	2.00148	-.0955826
13.	146	13800	9.532424	2.164472	-.0914387
490.	154	19400	9.873029	2.164472	.0910494
491.	463	19500	9.87817	1.964311	.0941472
492.	464	20200	9.913438	1.964311	.0974507
493.	427	10200	9.230143	1.764731	.1007114
494.	406	5000	8.517193	1.93586	.1024767
495.	151	21500	9.975808	2.164472	.1047597
496.	152	19600	9.883285	2.164472	.1120427
497.	460	20000	9.903487	1.964311	.1142668
498.	160	23300	10.05621	2.164472	.1165014
499.	491	8100	8.999619	1.806648	.1222368
500.	362	19900	9.898475	2.04122	.1376445
501.	363	20800	9.942708	2.04122	.1707894
502.	490	7000	8.853665	1.806648	.1791869
503.	358	21700	9.985067	2.04122	.1827834
504.	360	22600	10.02571	2.04122	.2209745
505.	361	24999	10.12659	2.04122	.2422512
506.	359	22700	10.03012	2.04122	.2483543

Compared to the DFITS measure, we see a similar pattern for the DFBETA for `lnox`, with roughly 6% of the sample exhibiting large values of this measure. As with DFITS, the large positive values exceed their negative counterparts in magnitude. Many of the positive values are associated with the top-coded median house value of $50,001. These (presumably wealthy) communities have values of `lnox` well in excess of its minimum or mean. In contrast, many of the communities with large negative DFBETA values have extremely high values (or the maximum recorded value) of that pollutant.

How should we react to this evidence of many data points with a high degree of leverage? For this research project, we might consider that the price data have been improperly coded, particularly on the high end. Any community with a median housing value exceeding $50,000 has been coded as $50,001. These observations in particular have been identified by the DFITS and DFBETA measures.

Removing the bottom-coded and top-coded observations from the sample would remove communities from the sample nonrandomly, affecting the wealthiest and poorest communities. To resolve this problem of *censoring* (or coding of extreme values) we could use the `tobit` model (see section 10.3.2). A version of the tobit model, two-limit tobit, can handle censoring at both lower and upper limits.

5.3 Endogeneity and measurement error

In econometrics, a regressor is endogenous if it violates the zero-conditional-mean assumption $E[u \mid X] = 0$: that is, if the variable is correlated with the error term, it is *endogenous*. I deal with this problem in chapter 8.

We often must deal with measurement error, meaning that the variable that theory tells us belongs in the relationship cannot be precisely measured in the available data. For instance, the exact marginal tax rate that an individual faces will depend on many factors, only some of which we might be able to observe. Even if we knew the individual's income, number of dependents, and homeowner status, we could only approximate the effect of a change in tax law on her tax liability. We are faced with using an approximate measure, including some error of measurement, whenever we try to formulate and implement such a model.

This issue is similar to a proxy variable problem, but here it is not an issue of a latent variable such as aptitude or ability. An observable magnitude does exist, but the econometrician cannot observe it. Why is this measurement error of concern? Because the economic behavior we want to model—that of individuals, firms, or industries—presumably is driven by the *actual* measures, not our mismeasured approximations of those factors. If we fail to capture the actual measure, we may misinterpret the behavioral response.

Mathematically, measurement error (commonly termed errors-in-variables) has the same effect on an OLS regression model as endogeneity of one or more regressors (see chapter 8).

Exercises

1. Using the `lifeexpW` dataset, regress `lifeexp` on `popgrowth` and `lgnppc`. Generate an added-value plot by using `avplot safewater`. What do you conclude about the regression estimates?
2. Refit the model, including `safewater`. Use Ramsey's RESET to evaluate the specification. What do you conclude?
3. Generate the `dfits` series from the regression, and list the five countries with the largest absolute value of the DFITS measure. Which of these countries stand out?
4. Refit the model, omitting Haiti, and apply the RESET. What do you conclude about the model's specification?

6 Regression with non-i.i.d. errors

As discussed in section 4.2.2, if the regression errors are i.i.d., OLS produces consistent estimates; the large-sample distribution in large samples is normal with a mean at the true coefficient values, and the VCE is consistently estimated by (4.15). If the zero-conditional-mean assumption holds but the errors are not i.i.d., OLS produces consistent estimates whose sampling distribution in large samples is still normal with a mean at the true coefficient values but whose VCE cannot be consistently estimated by (4.15).

We have two options when the errors are not i.i.d. First, we can use the consistent OLS point estimates with a different estimator of the VCE that accounts for non-i.i.d. errors. Or, if we can specify how the errors deviate from i.i.d. in our regression model, we can use a different estimator that produces consistent and more efficient point estimates.

The tradeoff between these two methods is *robustness* versus *efficiency*. A robust approach places fewer restrictions on the estimator: the idea is that the consistent point estimates are good enough, although we must correct our estimator of their VCE to account for non-i.i.d. errors. The efficient approach incorporates an explicit specification of the non-i.i.d. distribution into the model. If this specification is appropriate, the additional restrictions it implies will produce a more efficient estimator than that of the robust approach.

The i.i.d. assumption fails when the errors are either not *identically* distributed or not *independently* distributed (or both). When the variance of the errors, conditional on the regressors, changes over the observations, the identically distributed assumption fails. This problem is known as *heteroskedasticity* (unequal variance), with its opposite being *homoskedasticity* (common variance). The i.i.d. case assumes that the errors are *conditionally homoskedastic*: there is no information in the regressors about the variance of the disturbances.

When the errors are correlated with each other, they are not *independently* distributed. In this chapter, we allow the errors to be correlated with *each other* (violating the i.i.d. assumption) but not with the regressors. We still maintain the zero-conditional-mean assumption, which implies that there is no correlation between the regressors and the errors. The case of nonindependent errors is different from the case in which the regressors are correlated with the errors.

After introducing some common causes for failure of the assumption of i.i.d. errors, we present the robust approach. We then discuss the general form of the efficient approach, the estimation and testing in the most common special cases, and the testing for i.i.d. errors in these subsections because efficient tests require that we specify the form of the deviation.

6.1 The generalized linear regression model

The popularity of the least-squares regression technique is linked to its generality. If we have a model linking a response variable to several regressors that satisfy the zero-conditional-mean assumption of (5.1), OLS will yield consistent point estimates of the β parameters. We need not make any further assumptions on the distribution of the u process and specifically need not assume that it is distributed multivariate normal.[1]

Here I present the *generalized linear regression model* (GLRM) that lets us consider the consequences of non-i.i.d. errors on the estimated covariance matrix of the estimated parameters $\widehat{\boldsymbol{\beta}}$. The GLRM is

$$\begin{aligned} \mathbf{y} &= \mathbf{X}\beta + \mathbf{u} \\ E[\mathbf{u}|\mathbf{X}] &= \mathbf{0} \\ E[\mathbf{uu}'|\mathbf{X}] &= \boldsymbol{\Sigma}_u \end{aligned}$$

where $\boldsymbol{\Sigma}_u$ is a positive-definite matrix of order $N \times N$.[2] This is a generalization of the i.i.d. error model in which $\boldsymbol{\Sigma}_u = \sigma^2 I_N$.

When $\boldsymbol{\Sigma}_u \neq \sigma^2 I_N$ the OLS estimator of β is still unbiased, consistent, and normally distributed in large samples, but it is no longer efficient, as demonstrated by

$$\begin{aligned} \widehat{\boldsymbol{\beta}} &= (\mathbf{X}'\mathbf{X})^{-1}\mathbf{X}'\mathbf{y} \\ &= (\mathbf{X}'\mathbf{X})^{-1}\mathbf{X}'(\mathbf{X}\beta + \mathbf{u}) \\ &= \beta + (\mathbf{X}'\mathbf{X})^{-1}\mathbf{X}'\mathbf{u} \\ E[\widehat{\boldsymbol{\beta}} - \beta] &= 0 \end{aligned}$$

given the assumption of zero-conditional mean of the errors. That assumption implies that the sampling variance of the linear regression estimator (conditioned on $\mathbf{X}$) will be

$$\begin{aligned} \text{Var}[\widehat{\boldsymbol{\beta}}|\mathbf{X}] &= E[(\mathbf{X}'\mathbf{X})^{-1}\mathbf{X}'\mathbf{uu}'\mathbf{X}(\mathbf{X}'\mathbf{X})^{-1}] \\ &= (\mathbf{X}'\mathbf{X})^{-1}(\mathbf{X}'\boldsymbol{\Sigma}_u\mathbf{X})(\mathbf{X}'\mathbf{X})^{-1} \end{aligned} \tag{6.1}$$

The VCE computed by `regress` is merely $\sigma_u^2(\mathbf{X}'\mathbf{X})^{-1}$, with σ_u^2 replaced by its estimate, s^2.

When $\boldsymbol{\Sigma}_u \neq \sigma^2 I_N$, this simple estimator of the VCE is not consistent and the usual inference procedures are inappropriate. Hypothesis tests and confidence intervals using the simple estimator of the VCE after `regress` will not be reliable.

6.1.1 Types of deviations from i.i.d. errors

The GLRM lets us consider models in which $\boldsymbol{\Sigma}_u \neq \sigma^2 I_N$. Three special cases are of interest. First, consider the case of pure *heteroskedasticity* in which $\boldsymbol{\Sigma}_u$ is a diagonal

---

1. We do need to place some restrictions on the higher moments of u. But we can safely ignore those technical regularity conditions.

2. $\mathbf{y}$ is an $N \times 1$ vector of observations on y, $\mathbf{X}$ is an $N \times K$ matrix of observations on $\mathbf{x}$, and $\mathbf{u}$ is an $N \times 1$ disturbance vector.

matrix. This case violates the *identically distributed* assumption. When the diagonal elements of $\mathbf{\Sigma}_u$ differ, as in

$$\mathbf{\Sigma}_u = \begin{pmatrix} \sigma_1^2 & 0 & \dots & 0 \\ 0 & \sigma_2^2 & \dots & 0 \\ & & \vdots & \\ 0 & 0 & \dots & \sigma_N^2 \end{pmatrix}$$

the model allows the variance of u, conditional on $\mathbf{X}$, to vary across the observations. For instance, using a household survey, we could model consumer expenditures as a function of household income. We might expect the error variance of high-income individuals to be much greater than that of low-income individuals because high-income individuals have much more discretionary income.

Second, we can separate the observations into several groups or *clusters* within which the errors are correlated. For example, when we are modeling households' expenditures on housing as a function of their income, the errors may be correlated over the households within a neighborhood.

Clustering—correlation of errors within a cluster of observations—causes the $\mathbf{\Sigma}_u$ matrix to be *block-diagonal* because the errors in different groups are independent of one another. This case drops the *independently distributed* assumption in a particular way. Since each cluster of observations may have its own error variance, the *identically distributed* assumption is relaxed as well.

$$\mathbf{\Sigma}_u = \begin{pmatrix} \mathbf{\Sigma}_1 & \mathbf{0} & & & \mathbf{0} \\ \mathbf{0} & \ddots & & & \\ & & \mathbf{\Sigma}_m & & \\ & & & \ddots & \mathbf{0} \\ \mathbf{0} & & & \mathbf{0} & \mathbf{\Sigma}_M \end{pmatrix} \tag{6.2}$$

In this notation, $\mathbf{\Sigma}_m$ represents an intracluster covariance matrix. For cluster (group) m with τ_m observations, $\mathbf{\Sigma}_m$ will be $\tau_m \times \tau_m$. Zero covariance between observations in the M different clusters gives the covariance matrix $\mathbf{\Sigma}_u$ a block-diagonal form.

Third, the errors in time-series regression models may show *serial correlation*, in which the errors are correlated with their predecessor and successor. In the presence of serial correlation, the error covariance matrix becomes

$$\mathbf{\Sigma}_u = \sigma_u^2 \begin{pmatrix} 1 & \rho_1 & \dots & \rho_{N-1} \\ \rho_1 & 1 & \dots & \rho_{2N-3} \\ \vdots & & \ddots & \\ \rho_{N-1} & \rho_{2N-3} & \dots & 1 \end{pmatrix} \tag{6.3}$$

where the unknown parameters $\rho_1, \rho_2, \dots, \rho_{\{N(N-1)\}/2}$ represent the correlations between successive elements of the error process. This case also drops the *independently distributed* assumption but parameterizes the correlations differently.

6.1.2 The robust estimator of the VCE

If the errors are conditionally heteroskedastic and we want to apply the robust approach, we use the Huber–White–sandwich estimator of the variance of the linear regression estimator. Huber and White independently derived this estimator, and the *sandwich* aspect helps you understand the robust approach. We need to estimate $\text{Var}[\widehat{\boldsymbol{\beta}}|X]$, which according to (6.1) is of the form

$$\begin{aligned}\text{Var}[\widehat{\boldsymbol{\beta}}|\mathbf{X}] &= (\mathbf{X}'\mathbf{X})^{-1}(\mathbf{X}'\boldsymbol{\Sigma}_u\mathbf{X})(\mathbf{X}'\mathbf{X})^{-1} \\ &= (\mathbf{X}'\mathbf{X})^{-1}(\mathbf{X}'E[\mathbf{uu}'|\mathbf{X}]\mathbf{X})(\mathbf{X}'\mathbf{X})^{-1} \qquad (6.4)\end{aligned}$$

The term that we must estimate, $\{\mathbf{X}'E[\mathbf{uu}'|\mathbf{X}]\mathbf{X}\}$, is sandwiched between the $(\mathbf{X}'\mathbf{X})^{-1}$ terms. Huber (1967) and White (1980) showed that

$$\widehat{S}_0 = \frac{1}{N}\sum_{i=1}^{N}\widehat{u}_i^2\mathbf{x}_i'\mathbf{x}_i \qquad (6.5)$$

consistently estimates $(\mathbf{X}'E[\mathbf{uu}'|\mathbf{X}]\mathbf{X})$ when the u_i are conditionally heteroskedastic. In this expression, $\widehat{u}_i$ is the ith regression residual, and $\mathbf{x}_i$ is the ith row of the regressor matrix: a $1 \times k$ vector of sample values. Substituting the consistent estimator from (6.5) for its population equivalent in (6.4) yields the robust estimator of the VCE[3]

$$\text{Var}[\widehat{\boldsymbol{\beta}}|\mathbf{X}] = \frac{N}{N-k}(\mathbf{X}'\mathbf{X})^{-1}\left(\sum_{i=1}^{N}\widehat{u}_i^2\mathbf{x}_i'\mathbf{x}_i\right)(\mathbf{X}'\mathbf{X})^{-1} \qquad (6.6)$$

The `robust` option available with most Stata estimation commands, including the `regress` command, implements the sandwich estimator described above. Calculating robust standard errors affects only the coefficients' standard errors and interval estimates and does not affect the point estimates $\widehat{\boldsymbol{\beta}}$. The ANOVA F table will be suppressed, as will the adjusted R^2 measure because neither is valid when robust standard errors are being computed. The title `Robust` will be displayed above the standard errors of the coefficients to remind you that a robust estimator of the VCE is in use. After `regress`, Wald tests produced by `test` and `lincom`, which use the robust estimator of the VCE, will be robust to conditional heteroskedasticity of unknown form.[4] See [U] **20.14 Obtaining robust variance estimates** for more detail.

If the assumption of homoskedasticity is valid, the simple estimator of the VCE is more efficient than the robust version. If we are working with a sample of modest size and the assumption of homoskedasticity is tenable, we should rely on the simple estimator of the VCE. But because the robust estimator of the VCE is easily calculated in

3. Equation (6.6) is correct. As in the appendix to chapter 4, we define $\text{Var}[\widehat{\boldsymbol{\beta}}|\mathbf{X}]$ to be a large-sample approximation to the variance of our estimator. The large-sample calculations cause the $1/N$ factor in (6.5) to be dropped from (6.6). The factor $N/(N-k)$ improves the approximation in small samples.

4. Davidson and MacKinnon (1993) recommend using a different divisor that improves the performance of the robust estimator of the VCE estimator in small samples. Specifying the `hc3` option on `regress` will produce this robust estimator of the VCE.

Stata, it is simple to estimate both VCEs for a particular equation and consider whether inference based on the nonrobust standard errors is fragile. In large datasets, it has become increasingly common to report results using the robust estimator of the VCE.

To illustrate the use of the robust estimator of the VCE, we use a dataset (fertil2) that contains data on 4,361 women from a developing country. We want to model the number of children ever born (ceb) to each woman based on their age, their age at first birth (agefbrth), and an indicator of whether they regularly use a method of contraception (usemeth).[5] I present the descriptive statistics for the dataset with summarize based on those observations with complete data for a regression:

```
. use http://www.stata-press.com/data/imeus/fertil2, clear
. quietly regress ceb age agefbrth usemeth
. estimates store nonRobust
. summarize ceb age agefbrth usemeth children if e(sample)
```

Variable	Obs	Mean	Std. Dev.	Min	Max
ceb	3213	3.230003	2.236836	1	13
age	3213	29.93931	7.920432	15	49
agefbrth	3213	19.00498	3.098121	10	38
usemeth	3213	.6791161	.4668889	0	1
children	3213	2.999378	2.055579	0	13

The average woman in the sample is 30 years old, first bore a child at 19, and has had 3.2 children, with just under three children in the household. We expect that the number of children ever born is increasing in the mother's current age and decreasing in her age at the birth of her first child. The use of contraceptives is expected to decrease the number of children ever born.

For later use, we use estimates store to preserve the results of this (undisplayed) regression. We then fit the same model with a robust estimator of the VCE, saving those results with estimates store. We then use the estimates table command to display the two sets of coefficient estimates, standard errors, and t statistics.

```
. regress ceb age agefbrth usemeth, robust
Linear regression                                      Number of obs =    3213
                                                       F(  3,  3209) =  874.06
                                                       Prob > F      =  0.0000
                                                       R-squared     =  0.5726
                                                       Root MSE      =   1.463
```

ceb	Coef.	Robust Std. Err.	t	P>|t|	[95% Conf.	Interval]
age	.2237368	.0046619	47.99	0.000	.2145962	.2328775
agefbrth	-.2606634	.0095616	-27.26	0.000	-.2794109	-.2419159
usemeth	.1873702	.0606446	3.09	0.002	.0684642	.3062762
_cons	1.358134	.1675624	8.11	0.000	1.029593	1.686674

```
. estimates store Robust
```

5. Since the dependent variable is an integer, this model would be properly fitted with Poisson regression. For pedagogical reasons, we use linear regression.

```
. estimates table nonRobust Robust, b(%9.4f) se(%5.3f) t(%5.2f)
> title(Estimates of CEB with OLS and Robust standard errors)
Estimates of CEB with OLS and Robust standard errors
```

Variable	nonRobust	Robust
age	0.2237	0.2237
	0.003	0.005
	64.89	47.99
agefbrth	-0.2607	-0.2607
	0.009	0.010
	-29.64	-27.26
usemeth	0.1874	0.1874
	0.055	0.061
	3.38	3.09
_cons	1.3581	1.3581
	0.174	0.168
	7.82	8.11

```
legend: b/se/t
```

Our prior results are borne out by the estimates, although the effect of contraceptive use appears to be marginally significant. The robust estimates of the standard errors are similar to the nonrobust estimates, suggesting that there is no conditional heteroskedasticity.

6.1.3 The cluster estimator of the VCE

Stata has implemented an estimator of the VCE that is robust to the correlation of disturbances within groups and to not identically distributed disturbances. It is commonly referred to as the cluster–robust–VCE estimator, because these groups are known as clusters. Within-cluster correlation allows the $\mathbf{\Sigma}_u$ in (6.2) to be *block-diagonal*, with nonzero elements within each block on the diagonal. This block-diagonal structure allows the disturbances within each cluster to be correlated with each other but requires that the disturbances from difference clusters be uncorrelated.

If the within-cluster correlations are meaningful, ignoring them leads to inconsistent estimates of the VCE. Since the `robust` estimate of the VCE assumes independently distributed errors, its estimate of $(\mathbf{X}'E[\mathbf{uu}'|\mathbf{X}]\mathbf{X})$ is not consistent. Stata's `cluster()` option, available on most estimation commands including `regress`, lets you account for such an error structure. Like the `robust` option (which it encompasses), application of the `cluster()` option does not affect the point estimates but only modifies the estimated VCE of the estimated parameters. The `cluster()` option requires you to specify a group- or cluster-membership variable that indicates how the observations are grouped.

The cluster–robust–VCE estimator is

$$\text{Var}[\widehat{\boldsymbol{\beta}}|\mathbf{X}] = \frac{N-1}{N-k}\frac{M}{M-1}(\mathbf{X}'\mathbf{X})^{-1}\left(\sum_{j=1}^{M}\widetilde{\mathbf{u}}_j'\widetilde{\mathbf{u}}_j\right)(\mathbf{X}'\mathbf{X})^{-1} \tag{6.7}$$

where M is the number of clusters, $\widetilde{\mathbf{u}}_j = \sum_{i=1}^{N_k} \widehat{u}_i \mathbf{x}_i$, N_j is the number of observations in the jth cluster, $\widehat{u}_i$ is the ith residual from the jth cluster, and $\mathbf{x}_i$ is the $1 \times k$ vector of regressors from the ith observation in the jth cluster.

Equation (6.7) has the same form as (6.6). Aside from the small-sample adjustments, the (6.7) differs from (6.6) only in that the "meat" of the sandwich is now the cluster–robust estimator of $(\mathbf{X}'E[\mathbf{uu}'|\mathbf{X}]\mathbf{X})$.

The goal of the robust and the cluster–robust–VCE estimators is to consistently estimate the $\text{Var}[\widehat{\boldsymbol{\beta}}|\mathbf{X}]$ in the presence of non-i.i.d. disturbances. Different violations of the i.i.d. disturbance assumption simply require distinct estimators of $(\mathbf{X}'E[\mathbf{uu}'|\mathbf{X}]\mathbf{X})$.

To illustrate the use of the cluster estimator of the covariance matrix, we revisit the model of fertility in a developing country that we estimated above via nonrobust and robust methods. The clustering variable is `children`: the number of living children in the household. We expect the errors from households of similar size to be correlated, while independent of those generated by households of different size.

```
. regress ceb age agefbrth usemeth, cluster(children)
Linear regression                                      Number of obs =    3213
                                                       F(  3,     13) =   20.91
                                                       Prob > F      =  0.0000
                                                       R-squared     =  0.5726
Number of clusters (children) = 14                     Root MSE      =   1.463

------------------------------------------------------------------------------
             |               Robust
         ceb |      Coef.   Std. Err.      t    P>|t|     [95% Conf. Interval]
-------------+----------------------------------------------------------------
         age |   .2237368   .0315086     7.10   0.000     .1556665    .2918071
    agefbrth |  -.2606634   .0354296    -7.36   0.000    -.3372045   -.1841224
     usemeth |   .1873702   .0943553     1.99   0.069     -.016472    .3912125
       _cons |   1.358134   .4248589     3.20   0.007     .4402818    2.275985
------------------------------------------------------------------------------
```

The cluster estimator, allowing for within-cluster correlation of errors, results in much more conservative standard errors (and smaller t statistics) than those displayed in the previous example.

6.1.4 The Newey–West estimator of the VCE

In the presence of heteroskedasticity and autocorrelation, we can use the *Newey–West* estimator of the VCE. This heteroskedastic and autocorrelation consistent (HAC) estimator of the VCE has the same form as the robust and cluster–robust estimators, but it uses a distinct estimator for $(\mathbf{X}'E[\mathbf{uu}'|\mathbf{X}]\mathbf{X})$. Rather than specifying a cluster variable, the Newey–West estimator requires that we specify the maximum order of any significant autocorrelation in the disturbance process—known as the maximum lag, denoted by L.

In addition to the term that adjusts for heteroskedasticity, the estimator proposed by Newey and West (1987) uses weighted cross products of the residuals to account for autocorrelation:

$$\widehat{\mathbf{Q}} = \widehat{\mathbf{S}}_0 + \frac{1}{T}\sum_{l=1}^{L}\sum_{t=l+1}^{T} w_l \, \widehat{u}_t\widehat{u}_{t-l}(\mathbf{x}_t'\mathbf{x}_{t-l} + \mathbf{x}_{t-l}'\mathbf{x}_t)$$

Here $\widehat{\mathbf{S}}_0$ is the robust estimator of the VCE from (6.5), $\widehat{u}_t$ is the tth residual, and $\mathbf{x}_t$ is the tth row of the regressor matrix. The Newey–West formula takes a specified number (L) of the sample autocorrelations into account, using the Bartlett kernel estimator,

$$w_l = 1 - \frac{l}{L+1}$$

to generate the weights.

The estimator is said to be HAC, allowing for any deviation of $\mathbf{\Sigma}_u$ from $\sigma_u^2 I$ up to Lth-order autocorrelation. The user must specify her choice of L, which should be large enough to encompass any likely autocorrelation in the error process. One rule of thumb is to choose $L = \sqrt[4]{N}$. This estimator is available in the Stata command `newey` (see [TS] **newey**), which you can use as an alternative to `regress` to estimate a regression with HAC standard errors. This command has the following syntax,

`newey` *depvar* [*indepvars*] [*if*] [*in*], `lag(#)`

where the number given for the `lag()` option is L above. Like the `robust` option, the HAC estimator does not modify the point estimates; it affects only the estimator of the VCE. Test statistics based on the HAC VCE are robust to arbitrary heteroskedasticity and autocorrelation.

We illustrate this estimator of the VCE by using a time-series dataset of monthly short-term and long-term interest rates on U.K. government securities (Treasury bills and gilts), 1952m3–1995m12. The descriptive statistics for those series are given by `summarize`:

```
. use http://www.stata-press.com/data/imeus/ukrates, clear
. summarize rs r20
    Variable |       Obs        Mean    Std. Dev.       Min        Max
-------------+--------------------------------------------------------
          rs |       526    7.651513    3.553109   1.561667      16.18
         r20 |       526    8.863726    3.224372       3.35      17.18
```

The model expresses the monthly change in the short rate `rs`, the Bank of England's monetary policy instrument, as a function of the prior month's change in the long-term rate **r20**. The regressor and regressand are created on the fly by Stata's time-series operators **D.** and **L.** The model represents a monetary policy reaction function.

We fit the model with and without HAC standard errors by using **regress** and **newey**, respectively, using **estimates store** to save the results and **estimates table** to juxtapose them. Since there are 524 observations, the rule of thumb for lag selection recommends five lags, which we specify in **newey**'s **lag()** option.

```
. quietly regress D.rs LD.r20
. estimates store nonHAC
. newey D.rs LD.r20, lag(5)
Regression with Newey-West standard errors          Number of obs  =       524
maximum lag: 5                                      F(  1,    522)  =     36.00
                                                    Prob > F        =    0.0000

------------------------------------------------------------------------------
             |             Newey-West
        D.rs |      Coef.   Std. Err.      t    P>|t|     [95% Conf. Interval]
-------------+----------------------------------------------------------------
         r20 |
         LD. |   .4882883   .0813867     6.00   0.000     .3284026     .648174
       _cons |   .0040183   .0254102     0.16   0.874    -.0459004    .0539371
------------------------------------------------------------------------------
. estimates store NeweyWest
. estimates table nonHAC NeweyWest, b(%9.4f) se(%5.3f) t(%5.2f)
> title(Estimates of D.rs with OLS and Newey--West standard errors)
Estimates of D.rs with OLS and Newey--West standard errors

----------------------------------------
    Variable |   nonHAC     NeweyWest
-------------+--------------------------
      LD.r20 |   0.4883       0.4883
             |    0.067        0.081
             |     7.27         6.00
       _cons |   0.0040       0.0040
             |    0.022        0.025
             |     0.18         0.16
----------------------------------------
                              legend: b/se/t
```

The HAC standard error estimate of the slope coefficient from **newey** is larger than that produced by **regress**, although the coefficient retains its significance.

Two issues remain with this HAC VCE estimator. First, although the Newey–West estimator is widely used, there is no justification for using the Bartlett kernel. We might use several alternative kernel estimators, and some may have better properties in specific instances. The only requirement is that the kernel delivers a positive-definite estimate of the VCE. Second, if there is no reason to question the assumption of homoskedasticity of u, we may want to deal with serial correlation under that assumption. We may want the AC without the H. The standard Newey–West procedure as implemented in **newey** does not allow for this. The **ivreg2** routine (Baum, Schaffer, and Stillman 2003) can estimate robust, AC, and HAC standard errors for regression models, and it provides a choice of alternative kernels. See chapter 8 for full details on this routine.

6.1.5 The generalized least-squares estimator

This section presents a class of estimators for estimating the coefficients of a GLRM when the zero-conditional-mean assumption holds, but the errors are not i.i.d. Known as *feasible generalized least squares* (FGLS), this technique relies on the insight that if we knew $\mathbf{\Sigma}_u$, we could algebraically transform the data so that the resulting errors were i.i.d. and then proceed with linear regression on the transformed data. We do not know $\mathbf{\Sigma}_u$, though, so this estimator is infeasible. The *feasible* alternative requires that we assume a structure that describes how the errors deviate from i.i.d. errors. Given that assumption, we can consistently estimate $\mathbf{\Sigma}_u$. Any consistent estimator of $\mathbf{\Sigma}_u$ may be used to transform the data to generate observations with i.i.d. errors.

Although both the robust estimator of the VCE approach and FGLS estimators account for non-i.i.d. disturbances, FGLS estimators place more structure on the estimation method to obtain more efficient point estimates and consistent estimators of the VCE. In contrast, the robust estimator of the VCE approach uses just the OLS point estimates and makes the estimator of the VCE robust to the non-i.i.d. disturbances.

First, consider the infeasible GLS estimator of

$$\begin{aligned} \mathbf{y} &= \mathbf{X}\boldsymbol{\beta} + \mathbf{u} \\ E[\mathbf{u}\mathbf{u}'|\mathbf{X}] &= \mathbf{\Sigma}_u \end{aligned}$$

The known $N \times N$ matrix $\mathbf{\Sigma}_u$ is symmetric and positive definite, which implies that it has an inverse $\mathbf{\Sigma}_u^{-1} = \mathbf{P}\mathbf{P}'$, where $\mathbf{P}$ is a triangular matrix. Premultiplying the model by $\mathbf{P}'$ yields

$$\begin{aligned} \mathbf{P}'\mathbf{y} &= \mathbf{P}'\mathbf{X}\boldsymbol{\beta} + \mathbf{P}'\mathbf{u} \\ \mathbf{y}_* &= \mathbf{X}_*\boldsymbol{\beta} + \mathbf{u}_* \end{aligned} \tag{6.8}$$

where[6]

$$\text{Var}[\mathbf{u}_*] = E[\mathbf{u}_*\mathbf{u}_*'] = \mathbf{P}'\mathbf{\Sigma}_u\mathbf{P} = \mathbf{I}_N$$

With a known $\mathbf{\Sigma}_u$ matrix, regression of $\mathbf{y}_*$ on $\mathbf{X}_*$ is asymptotically efficient by the Gauss–Markov theorem presented in section 4.2.3. That estimator merely represents standard linear regression on the transformed data:

$$\widehat{\boldsymbol{\beta}}_{\text{GLS}} = (\mathbf{X}_*'\mathbf{X}_*)^{-1}(\mathbf{X}_*'\mathbf{y}_*)$$

The VCE of the GLS estimator $\widehat{\boldsymbol{\beta}}_{\text{GLS}}$ is

$$\text{Var}[\widehat{\boldsymbol{\beta}}_{\text{GLS}}|\mathbf{X}] = (\mathbf{X}'\mathbf{\Sigma}_u^{-1}\mathbf{X})^{-1}$$

6. $E[\mathbf{P}'\mathbf{u}\mathbf{u}'\mathbf{P}] = \mathbf{P}'E[\mathbf{u}\mathbf{u}']\mathbf{P} = \mathbf{P}'\mathbf{\Sigma}_u\mathbf{P}$. But that expression equals $\mathbf{P}'(\mathbf{P}\mathbf{P}')^{-1}\mathbf{P} = \mathbf{P}'(\mathbf{P}')^{-1}\mathbf{P}^{-1}\mathbf{P} = \mathbf{I}_N$. See Davidson and MacKinnon (2004, 258).

The FGLS estimator

When $\mathbf{\Sigma}_u$ is unknown, we cannot apply the GLS estimator of (6.8). But if we have a consistent estimator of $\mathbf{\Sigma}_u$, denoted $\widehat{\mathbf{\Sigma}}_u$, we may apply the FGLS estimator, replacing $\mathbf{P}'$ with $\widehat{\mathbf{P}}'$ in (6.8). The FGLS estimator has the same large-sample properties as its infeasible counterpart.[7] That result does not depend on using an efficient estimator of $\mathbf{\Sigma}_u$, but merely any consistent estimator of $\mathbf{\Sigma}_u$.

The challenge in devising a consistent estimator of $\mathbf{\Sigma}_u$ lies in its dimension. $\mathbf{\Sigma}_u$ is a square symmetric matrix of order N with $\{N(N+1)\}/2$ distinct elements. Fortunately, the most common departures from i.i.d. errors lead to parameterizations of $\widehat{\mathbf{\Sigma}}_u$ with many fewer parameters. As I discuss in the next sections, heteroskedasticity and autocorrelation can often be modeled with a handful of parameters. All we need for consistency of these estimates is a fixed number of parameters in $\widehat{\mathbf{\Sigma}}_u$ as $N \rightarrow \infty$.

The gain from using FGLS depends on the degree to which $\mathbf{\Sigma}_u$ diverges from $\sigma^2 I_N$, the covariance matrix for i.i.d. errors. If that divergence is small, the FGLS estimates will be similar to those of standard linear regression, and vice versa.

The following two sections discuss the most common violations of the i.i.d. errors assumption—heteroskedasticity and serial correlation—and present the FGLS estimator appropriate for each case.

6.2 Heteroskedasticity in the error distribution

In cross-sectional datasets representing individuals, households, or firms, the disturbance variances are often related to some measure of scale. For instance, in modeling consumer expenditures, the disturbance for variance of high-income households is usually larger than that of poorer households. For the FGLS estimator described above, the diagonal elements of the $\mathbf{\Sigma}_u$ matrix for these errors will be related to that scale measure.

We may instead have a dataset in which we may reasonably assume that the disturbances are homoskedastic *within* groups of observations but potentially heteroskedastic *between* groups. For instance, in a labor market survey, self-employed individuals or workers paid by salary and commission (or salary and tips) may have a greater variance around their conditional-mean earnings than salaried workers. For the FGLS estimator, there will be several distinct values of σ_u^2, each common to those individuals in a group but differing between groups.

As a third potential cause of heteroskedasticity, consider the use of *grouped data*, in which each observation is the average of microdata (e.g., state-level data for the United States, where the states have widely differing populations). Since means computed from larger samples are more accurate, the disturbance variance for each observation is known up to a factor of proportionality. Here we are certain (by the nature of grouped

7. See Davidson and MacKinnon (2004).

data) that heteroskedasticity exists, and we can construct the appropriate $\widehat{\mathbf{\Sigma}}_u$. In the two former cases, we are not so fortunate.

We may also find heteroskedasticity in time-series data, especially *volatility clustering*, which appears in high-frequency financial-market data. I will not discuss this type of conditional heteroskedasticity at length, but the use of the autoregressive conditional heteroskedasticity (ARCH) and generalized ARCH (GARCH) models for high-frequency time-series data is based on the notion that the errors in these contexts are *conditionally* heteroskedastic and that the evolution of the conditional variance of the disturbance process may be modeled.[8]

6.2.1 Heteroskedasticity related to scale

We often use an economic rationale to argue that the variance of the disturbance process is related to some measure of *scale* of the individual observations. For instance, if the response variable measures expenditures on food by individual households, the disturbances will be denominated in dollars (or thousands of dollars). No matter how well the estimated equation fits, the dollar dispersion of wealthy households' errors around their predicted values will likely be much greater than those of low-income households.[9] Thus a hypothesis of

$$\sigma_i^2 \propto z_i^\alpha \tag{6.9}$$

is often made, where z_i is some scale-related measure for the ith unit. The notion of proportionality comes from the definition of FGLS: we need only estimate $\widehat{\mathbf{\Sigma}}_u$ up to a factor of proportionality. It does not matter whether z is one of the regressors or merely more information we have about each unit in the sample.

We write z_i^α in (6.9) since we must indicate the nature of this proportional relationship. For instance, if $\alpha = 2$, we are asserting that the *standard deviation* of the disturbance process is proportional to the level of z_i (e.g., to household income or a firm's total assets). If $\alpha = 1$, we imply that the *variance* of the disturbance process is proportional to the level of z_i, so that the standard deviation is proportional to $\sqrt{z_i}$.

Given a plausible choice of z_i, why is the specification of α so important? If we are to use FGLS to deal with heteroskedasticity, our choices of z_i and α in (6.9) will define the FGLS estimator to be used. Before I discuss correcting for heteroskedasticity related to scale, you must understand how to detect the presence of heteroskedasticity.

---

8. The development of ARCH models was a major factor in the award of the Bank of Sweden Prize in Economic Sciences in Memory of Alfred Nobel to Robert F. Engle in 2003. He shared the prize with fellow time-series econometrician Clive Granger. A bibliography of Engle's published and unpublished works may be found at http://ideas.repec.org/e/pen9.html.

9. With firm data, the same logic applies. If we are explaining a firm's capital investment expenditures, the degree to which spending differs from the model's predictions could be billions of dollars for a huge multinational but much smaller for a firm of modest size.

Testing for heteroskedasticity related to scale

After fitting a regression model, we can base a test for heteroskedasticity on the regression residuals. Why is this approach reasonable, if the presence of heteroskedasticity renders the standard errors unusable? The consistent point estimates $\widehat{\boldsymbol{\beta}}$ produce estimated residuals that may be used to make inferences about the distribution of u. If the assumption of homoskedasticity conditional on the regressors holds, it can be expressed as follows:

$$H_0\colon \text{Var}\left[u|\mathbf{X}\right] = \sigma_u^2 \tag{6.10}$$

Under this null hypothesis the conditional variance of the error process does not depend on the explanatory variables. Given that $E[u] = 0$, this null hypothesis is equivalent to requiring that $E[u^2|\mathbf{X}] = \sigma_u^2$. The conditional mean of the squared disturbances should not be a function of the regressors, so a regression of the squared residuals on any candidate $\mathbf{z}_i$ should have no meaningful explanatory power.[10,11]

One of the most common tests for heteroskedasticity is derived from this line of reasoning: the *Breusch–Pagan* (BP) test (Breusch and Pagan 1979).[12] The BP test, an LM test, involves regressing the squared residuals on a set of variables in an auxiliary regression:[13]

$$\widehat{u}_i^2 = d_1 + d_2 z_{i2} + d_3 z_{i3} + \ldots\ d_\ell z_{i\ell} + v_i \tag{6.11}$$

We could use the original regressors from the fitted model as the $\mathbf{z}$ variables,[14] use a subset of them, or add measures of scale as discussed above. If the magnitude of the squared residual is not systematically related to any of the $\mathbf{z}$ variables, then this auxiliary regression will have no explanatory power. Its R^2 will be small, and its ANOVA F statistic will indicate that it fails to explain any meaningful fraction of the variation of $\widehat{u}_i^2$ around its own mean.[15]

The BP test can be conducted by using either the F or LM statistic from the auxiliary regression (6.11). Under the null hypothesis of (6.10), LM $\sim \chi_\ell^2$, where there are ℓ regressors in the auxiliary regression. We can obtain the BP test with `estat hettest` after `regress`. If no regressor list (of z's) is provided, `hettest` uses the fitted values from the previous regression (the $\widehat{y}_i$ values). As mentioned above, the variables specified in the set of z's could be chosen as measures that did not appear in the original regressor list.

---

10. $\mathbf{z}_i$ must be a function of the regressor.
11. $\mathbf{z}_i$ has been generalized to be a vector.
12. The Stata manuals document this test as that of Cook and Weisberg. Breusch and Pagan (1979), Godfrey (1978), and Cook and Weisberg (1983) separately derived (and published) the same test statistic. It should not be confused with a different test devised by Breusch and Pagan implemented in `sureg`.
13. An LM test statistic evaluates the results of a restricted regression model. In the BP test, the restrictions are those implied by homoskedasticity, which implies that the squared regression disturbances should be uncorrelated with any measured characteristics in the regression. For more details, see Wooldridge (2006, 185–186).
14. Although the residuals are uncorrelated by construction with each of the regressors of the original model, that condition need not hold for their squares.
15. Although the regression residuals from a model with a constant term have mean zero, the mean of their squares must be positive unless $R^2 = 1$.

The BP test with $\mathbf{z} = \mathbf{x}$ is a special case of *White's general test* (White 1980) for heteroskedasticity, which takes the list of regressors $(x_2, x_3, \ldots, x_k)$ and augments it with squares and cross products of each of these variables. The White test then runs an auxiliary regression of $\widehat{u}_i^2$ on the regressors, their squares, and their cross products, removing duplicate elements. For instance, if `crime` and `crime`-squared were in the original regression, only one instance of the squares term will enter the list of Zs. Under the null hypothesis, none of these variables should have any explanatory power for the squared residual series. The White test is another LM test of the $N \times R^2$ form but involves many more regressors in the auxiliary regression (especially for a regression in which k is sizable). The resulting test may have relatively low power because of the many degrees of freedom devoured by a lengthy regressor list. An alternate form of White's test uses only the fitted values of the original regression and their squares. We can compute both versions of White's test with `whitetst` as described in Baum, Cox, and Wiggins (2000), which you can install by using `ssc`. The original version of White's test may also be computed by the `estat imtest` command, using the `white` option.

All these tests rest on the specification of the disturbance variance expressed in (6.9). A failure to reject the tests' respective null hypotheses of homoskedasticity does not indicate an absence of heteroskedasticity but implies that the heteroskedasticity is not likely to be of the specified form. In particular, if the heteroskedasticity arises from group membership (as discussed in section 6.2.2), we would not expect tests based on measures of scale to pick it up unless there was a strong correlation between scale and group membership.[16]

We consider the potential scale-related heteroskedasticity in our model of median housing prices where the scale can be thought of as the average size of houses in each community, roughly measured by number of rooms. After fitting the model, we calculate three test statistics: that computed by `estat hettest, iid` without arguments, which is the BP test based on fitted values; `estat hettest, iid` with a variable list, which uses those variables in the auxiliary regression; and White's general test statistic from `whitetst`.[17]

```
. use http://www.stata-press.com/data/imeus/hprice2a, clear
(Housing price data for Boston-area communities)
. regress lprice rooms crime ldist
      Source |       SS       df       MS              Number of obs =     506
-------------+------------------------------           F(  3,   502) =  219.03
       Model |  47.9496883     3  15.9832294           Prob > F      =  0.0000
    Residual |  36.6325827   502  .072973272           R-squared     =  0.5669
-------------+------------------------------           Adj R-squared =  0.5643
       Total |  84.5822709   505  .167489645           Root MSE      =  .27014
```

16. Many older textbooks discuss the Goldfeld–Quandt test, which is based on forming two groups of residuals defined by high and low values of one z variable. Because there is little to recommend this test relative to the BP or White test approaches, which allow for multiple z's, I do not discuss it further.

17. By default, `estat hettest` produces the original BP test, which assumes that the u_i are normally distributed. Typing `estat hettest, iid` yields the Koenker (1981) LM test, which assumes the u_i to be i.i.d. under the null hypothesis.

```
------------------------------------------------------------------------------
      lprice |      Coef.   Std. Err.      t    P>|t|     [95% Conf. Interval]
-------------+----------------------------------------------------------------
       rooms |   .3072343   .0178231    17.24   0.000     .2722172    .3422514
       crime |  -.0174486    .001591   -10.97   0.000    -.0205744   -.0143228
       ldist |    .074858   .0255746     2.93   0.004     .0246115    .1251045
       _cons |   7.984449   .1128067    70.78   0.000     7.762817     8.20608
------------------------------------------------------------------------------
. estat hettest, iid
Breusch-Pagan / Cook-Weisberg test for heteroskedasticity
         Ho: Constant variance
         Variables: fitted values of lprice
         chi2(1)      =    44.67
         Prob > chi2  =   0.0000
. estat hettest rooms crime ldist, iid
Breusch-Pagan / Cook-Weisberg test for heteroskedasticity
         Ho: Constant variance
         Variables: rooms crime ldist
         chi2(3)      =    80.11
         Prob > chi2  =   0.0000
. whitetst
White's general test statistic :  144.0052  Chi-sq( 9)  P-value =  1.5e-26
```

Each of these tests indicates that there is a significant degree of heteroskedasticity in this model.

FGLS estimation

To use FGLS on a regression equation in which the disturbance process exhibits heteroskedasticity related to scale, we must estimate the $\mathbf{\Sigma}_u$ matrix up to a factor of proportionality. We implement FGLS by transforming the data and running a regression on the transformed equation. For FGLS to successfully deal with the deviation from i.i.d. errors, the transformations must purge the heteroskedasticity from the disturbance process and render the disturbance process in the transformed equation i.i.d.

Say that we test for this form of heteroskedasticity and conclude, per (6.9), that the disturbance variance of the ith firm is proportional to z_i^2, with z defined as a measure of scale related to the covariates in the model. We assume that z_i is strictly positive or that it has been transformed to be strictly positive. The appropriate transformation to induce homoskedastic errors would be to divide each variable in $(y, \mathbf{X})$ (including $\boldsymbol{\iota}$, the first column of $\mathbf{X}$) by z_i. That equation will have a disturbance term u_i/z_i, and since z_i is a constant, $\text{Var}[u_i/z_i] = (1/z_i^2)\text{Var}[u_i]$. If the original disturbance variance is proportional to z_i, dividing it by z_i^2 will generate a constant value: homoskedasticity of the transformed equation's error process.

We could implement FGLS on the equation

$$y_i = \beta_1 + \beta_2 x_{i,2} + \cdots + \beta_k x_{i,k} + u_i \tag{6.12}$$

by specifying the transformed equation

$$\frac{y_i}{z_i} = \frac{\beta_1}{z_i} + \frac{\beta_2 x_{i,2}}{z_i} + \cdots + \frac{\beta_k x_{i,k}}{z_i} + \frac{u_i}{z_i} \tag{6.13}$$

or

$$y_i^* = \beta_1 \iota^* + \beta_2 x_{i,2}^* + \cdots + \beta_k x_{i,k}^* + u_i^* \tag{6.14}$$

where $i^* = 1/z_i$. The economic meaning of the coefficients in the transformed equation has not changed; β_2 and its estimate $\widehat{\beta}_2$ still represent $\partial y/\partial x_2$. Since we have changed the dependent variable, measures such as R^2 and `Root MSE` are not comparable to those of the original equation. In particular, the transformed equation does not have a constant term.

Although we could do these transformations by hand with `generate` statements followed by `regress` on the transformed (6.14), that approach is cumbersome. For instance, we will normally want to evaluate measures of goodness of fit based on the original data, not the transformed data. Furthermore, the transformed variables can be confusing. For example, if z_i were also regressor x_2 in (6.12),[18] the $\mathbf{x}^*$ variables would include $1/z_i$ and $\boldsymbol{\iota}$, a units vector. The coefficient on the former is really an estimate of the constant term of the equation, whereas the coefficient labeled as `_cons` by Stata is actually the coefficient on z, which could become confusing.

Fortunately, we need not perform FGLS by hand. FGLS in a heteroskedastic context can be accomplished by *weighted least squares*. The transformations we have defined above amount to *weighting* each observation (here by $1/z_i$). Observations with smaller disturbance variances receive a larger weight in the computation of the sums and therefore have greater weight in computing the weighted least-squares estimates. We can instruct Stata to perform this weighting when it estimates the original regression by defining $1/z_i^2$ as the so-called *analytical weight*. Stata implements several kinds of weights (see [U] **11 Language syntax** and [U] **20.16 Weighted estimation**), and this sort of FGLS involves the analytical weight (`aw`) variety. We merely estimate the regression specifying the weights,

```
. generate rooms2 = rooms^2
. regress lprice rooms crime ldist [aw=1/rooms2]
(sum of wgt is   1.3317e+01)
      Source |       SS       df       MS              Number of obs =     506
-------------+------------------------------           F(  3,   502) =  159.98
       Model |  39.6051883     3  13.2017294           Prob > F      =  0.0000
    Residual |   41.426616   502  .082523139           R-squared     =  0.4888
-------------+------------------------------           Adj R-squared =  0.4857
       Total |  81.0318042   505  .160459018           Root MSE      =  .28727

------------------------------------------------------------------------------
      lprice |      Coef.   Std. Err.      t    P>|t|     [95% Conf. Interval]
-------------+----------------------------------------------------------------
       rooms |   .2345368   .0194432    12.06   0.000     .1963367     .272737
       crime |  -.0175759   .0016248   -10.82   0.000    -.0207682   -.0143837
       ldist |   .0650916    .027514     2.37   0.018     .0110349    .1191483
       _cons |   8.450081   .1172977    72.04   0.000     8.219626    8.680536
------------------------------------------------------------------------------
```

18. We assume that z_i is strictly positive.

which indicates that the regression is to be performed using `1/rooms2` as the analytical weight,[19] where `rooms2` = rooms^2. These estimates are qualitatively similar to those obtained with `robust`, with slightly weaker measures of goodness of fit.

The coefficient estimates and standard errors from this weighted regression will be identical to those computed by hand if the $y^*, \mathbf{x}^*$ variables are generated. But unlike the regression output from (6.14), the regression with analytical weights produces the desired measures of goodness of fit (e.g., R^2 and `Root MSE`) and `predict` will generate predicted values or residuals in the units of the untransformed dependent variable. The FGLS *point estimates* differ from those generated by `regress` from the untransformed regression; see (6.12). However, both the standard regression and FGLS point estimates are consistent estimates of $\boldsymbol{\beta}$.

The series specified as the analytical weight (`aw`) must be the *inverse of the observation variance*, not its standard deviation, and the original data are multiplied by the analytical weight, not divided by it. Some other statistical packages that provide facilities for FGLS differ in how they specify the weighting variable, for instance, requiring you to provide the value that appears as the divisor in (6.13).

We often see empirical studies in which a regression equation has been specified in some ratio form. For instance, per capita dependent and independent variables for data on states or countries are often used, as are financial ratios for firm- or industry-level data. Although the study may not mention heteroskedasticity, these ratio forms probably have been chosen to limit the potential damage of heteroskedasticity in the fitted model. Heteroskedasticity in a per capita form regression on country-level data is much less likely to be a problem in that context than it would be if the levels of GDP were used in that model. Likewise, scaling firms' values by total assets, total revenues, or the number of employees can mitigate the difficulties caused by extremes in scale between large corporations and corner stores. Such models should still be examined for their errors' behavior, but the popularity of the ratio form in these instances is an implicit consideration of potential heteroskedasticity related to scale.

6.2.2 Heteroskedasticity between groups of observations

Between-group heteroskedasticity is often associated with *pooling* data across what may be nonidentically distributed sets of observations. For instance, a consumer survey conducted in Massachusetts (MA) and New Hampshire (NH) may give rise to a regression equation predicting the level of spending as a function of several likely factors. If we merely pool the sets of observations from MA and NH into one dataset (using `append`), we may want to test that any fitted model is *structurally stable* over the two states' observations: that is, are the same $\boldsymbol{\beta}$ parameters appropriate?[20] Even if the two states' observations share the same population parameter vector $\boldsymbol{\beta}$, they may have different σ_u^2 values. For instance, spending in MA may be more sensitive to the presence of sales tax on many nonfood items, whereas NH shoppers do not pay a sales tax. This difference

19. This is one of the rare instances in Stata syntax when the square brackets (`[]`) are used.
20. A discussion of testing for structural stability appears in section 7.4.

may affect not only the slope parameters of the model but also the error variance. If so, then the assumption of homoskedasticity is violated in a particular manner. We may argue that the intrastate (or more generally, intragroup) disturbance variance is constant but that it may differ *between* states (or groups).

This same situation may arise, as noted above, with other individual-level series. Earnings may be more variable for self-employed workers, or those who depend on commissions or tips than salaried workers. With firm data, we might expect that profits (or revenues or capital investment) might be much more variable in some industries than others. Capital-goods makers face a much more cyclical demand for their product than do, for example, electric utilities.

Testing for heteroskedasticity between groups of observations

How might we test for groupwise heteroskedasticity? With the assumption that each group's regression equation satisfies the classical assumptions (including that of homoskedasticity), the s^2 computed by `regress` is a consistent estimate of the group-specific variance of the disturbance process. For two groups, we can construct an F test, with the larger variance in the numerator; the degrees of freedom are the residual degrees of freedom of each group's regression. We can easily construct such a test if both groups' residuals are stored in one variable, with a group variable indicating group membership (here 1 or 2). We can then use the third form of `sdtest` (see [R] **sdtest**), with the `by(`*groupvar*`)` option, to conduct the F test.

What if there are more than two groups across which we wish to test for equality of disturbance variance: for instance, a set of 10 industries? We may then use the `robvar` command (see [R] **sdtest**), which like `sdtest` expects to find one variable containing each group's residuals, with a group membership variable identifying them. The `by(`*groupvar*`)` option is used here as well. The test conducted is that of Levene (1960), labeled as w_0, which is robust to nonnormality of the error distribution. Two variants of the test proposed by Brown and Forsythe (1992), which uses more robust estimators of central tendency (e.g., median rather than mean), w_{50} and w_{10}, are also computed.

I illustrate groupwise heteroskedasticity with state-level data from the `NEdata.dta`. These data comprise one observation per year for each of the six U.S. states in the New England region for 1981–2000. Descriptive statistics are generated by `summarize` for `dpipc`, state disposable personal income per capita.

```
. use http://www.stata-press.com/data/imeus/NEdata, clear
. summarize dpipc
    Variable |       Obs        Mean    Std. Dev.       Min        Max
-------------+--------------------------------------------------------
       dpipc |       120    18.15802    5.662848   8.153382   33.38758
```

We fit a linear trend model to `dpipc` by regressing that variable on `year`. The residuals are tested for equality of variances across states with `robvar`.

```
. regress dpipc year
      Source |       SS       df       MS              Number of obs =     120
-------------+------------------------------           F(  1,   118) =  440.17
       Model |  3009.33617     1  3009.33617           Prob > F      =  0.0000
    Residual |  806.737449   118  6.83675804           R-squared     =  0.7886
-------------+------------------------------           Adj R-squared =  0.7868
       Total |  3816.07362   119  32.0678456           Root MSE      =  2.6147

------------------------------------------------------------------------------
       dpipc |      Coef.   Std. Err.      t    P>|t|     [95% Conf. Interval]
-------------+----------------------------------------------------------------
        year |   .8684582   .0413941    20.98   0.000     .7864865    .9504298
       _cons |  -1710.508   82.39534   -20.76   0.000    -1873.673   -1547.343
------------------------------------------------------------------------------
. predict double eps, residual
. robvar eps, by(state)
            |      Summary of Residuals
      state |        Mean   Std. Dev.       Freq.
------------+------------------------------------
         CT |    4.167853   1.3596266          20
         MA |    1.618796   .86550138          20
         ME |  -2.9841056   .93797625          20
         NH |   .51033312   .61139299          20
         RI |   -.8927223   .63408722          20
         VT |  -2.4201543   .71470977          20
------------+------------------------------------
      Total |  -6.063e-14   2.6037101         120
W0  = 4.3882072   df(5, 114)     Pr > F = .00108562
W50 = 3.2989849   df(5, 114)     Pr > F = .00806752
W10 = 4.2536245   df(5, 114)     Pr > F = .00139064
```

The hypothesis of equality of variances is soundly rejected by all three `robvar` test statistics, with the residuals for Connecticut possessing a standard deviation considerably larger than those of the other three states.

FGLS estimation

If different groups of observations have different error variances, we can apply the GLS estimator using analytical weights, as described above in section 6.2.1. In the groupwise context, we define the analytical weight (`aw`) series as a constant value for each observation in a group. That value is calculated as the estimated variance of that group's OLS residuals. Using the residual series calculated above, we construct an estimate of its variance for each New England state with `egen` and generate the analytical weight series:

```
. by state, sort: egen sd_eps = sd(eps)
. generate double gw_wt = 1/sd_eps^2
. tabstat sd_eps gw_wt, by(state)
Summary statistics: mean
  by categories of: state
```

state	sd_eps	gw_wt
CT	1.359627	.5409545
MA	.8655014	1.334948
ME	.9379762	1.136623
NH	.611393	2.675218
RI	.6340872	2.48715
VT	.7147098	1.957675
Total	.8538824	1.688761

The `tabstat` command reveals that the standard deviations of New Hampshire and Rhode Island's residuals are much smaller than those of the other four states. We now reestimate the regression with FGLS, using the analytical weight series:

```
. regress dpipc year [aw=gw_wt]
(sum of wgt is   2.0265e+02)
```

Source	SS	df	MS
Model	2845.55409	1	2845.55409
Residual	480.921278	118	4.07560405
Total	3326.47537	119	27.9535745

```
Number of obs =      120
F(  1,    118) =   698.19
Prob > F      =   0.0000
R-squared     =   0.8554
Adj R-squared =   0.8542
Root MSE      =   2.0188
```

dpipc	Coef.	Std. Err.	t	P>|t|	[95% Conf.	Interval]
year	.8444948	.0319602	26.42	0.000	.7812049	.9077847
_cons	-1663.26	63.61705	-26.14	0.000	-1789.239	-1537.281

Compared with the unweighted estimates' `Root MSE` of 2.6147, FGLS yields a considerably smaller value of 2.0188.

6.2.3 Heteroskedasticity in grouped data

In section 6.2, I addressed a third case in which heteroskedasticity arises in cross-sectional data, where our observations are grouped or aggregated data, representing different numbers of microdata records. This situation arises when the variables in our dataset are averages or standard deviations of groups' observations, for instance, a set of 50 U.S. state observations. Because we know the population of each state, we know precisely how much more accurate California's observation (based on more than 30 million individuals) is than Vermont's (based on fewer than a million). This situation would also arise in the context of observations representing average attainment scores for individual schools or school districts, where we know that each school's (or school district's) student population is different. In these cases we know that heteroskedasticity will occur in the grouped or aggregated data, and we know Ω because it depends only on the N_g underlying each observation.

You could consider this a problem of nonrandom sampling. In the first example above, when 30 million California records are replaced by one state record, an individual has little weight in the average. In a smaller state, each individual would have a greater weight in her state's average values. If we want to conduct inference for a national random sample, we must equalize those weights, leading to a heavier weight being placed on California's observation and a lighter weight being placed on Vermont's. The weights are determined by the relative magnitudes of the states' populations. Each observation in our data stands for an integer number of records in the population (stored, for instance, in pop).

FGLS estimation

We can deal with the innate heteroskedasticity in an OLS regression on grouped data by considering that the precision of each group mean (i.e., its standard error) depends on the size of the group from which it is calculated. The analytical weight, proportional to the inverse of the observation's variance, must take the group size into account. If we have state-level data on per capita saving and per capita income, we could estimate

```
. regress saving income [aw=pop]
```

in which we specify that the analytical weight is pop. The larger states will have higher weights, reflecting the greater precision of their group means.

I illustrate this correction with a dataset containing 420 public school districts' characteristics. The districts' average reading score (read_scr) is modeled as a function of their expenditures per student (expn_stu), computers per student (comp_stu), and the percentage of students eligible for free school lunches (meal_pct, an indicator of poverty in the district). We also know the enrollment per school district (enrl_tot). The descriptive statistics for these variables are given by summarize:

```
. use http://www.stata-press.com/data/imeus/pubschl, clear
. summarize read_scr expn_stu comp_stu meal_pct enrl_tot
```

Variable	Obs	Mean	Std. Dev.	Min	Max
read_scr	420	654.9705	20.10798	604.5	704
expn_stu	420	5312.408	633.9371	3926.07	7711.507
comp_stu	420	.1359266	.0649558	0	.4208333
meal_pct	420	44.70524	27.12338	0	100
enrl_tot	420	2628.793	3913.105	81	27176

First, we estimate the parameters by using regress, ignoring the total enrollment per school district, which varies considerably over the districts. We expect that districts' average reading scores will be positively related to expenditures per student and computers per student and negatively related to poverty.

```
. regress read_scr expn_stu comp_stu meal_pct
      Source |       SS       df       MS              Number of obs =     420
-------------+------------------------------           F(  3,   416) =  565.36
       Model |  136046.267     3  45348.7558           Prob > F      =  0.0000
    Residual |  33368.3632   416  80.2124115           R-squared     =  0.8030
-------------+------------------------------           Adj R-squared =  0.8016
       Total |  169414.631   419  404.330861           Root MSE      =  8.9561

------------------------------------------------------------------------------
    read_scr |      Coef.   Std. Err.      t    P>|t|     [95% Conf. Interval]
-------------+----------------------------------------------------------------
    expn_stu |   .0046699   .0007204     6.48   0.000     .0032538     .006086
    comp_stu |   19.88584   7.168347     2.77   0.006     5.795143    33.97654
    meal_pct |   -.635131   .0164777   -38.54   0.000     -.667521    -.602741
       _cons |   655.8528   3.812206   172.04   0.000     648.3592    663.3464
------------------------------------------------------------------------------
```

Our prior results on the relationship between reading scores and these factors are borne out. We reestimate the parameters, using enrollment as an analytical weight.

```
. regress read_scr expn_stu comp_stu meal_pct [aw=enrl_tot]
(sum of wgt is   1.1041e+06)
      Source |       SS       df       MS              Number of obs =     420
-------------+------------------------------           F(  3,   416) =  906.75
       Model |  123692.671     3  41230.8903           Prob > F      =  0.0000
    Residual |  18915.9815   416  45.4711093           R-squared     =  0.8674
-------------+------------------------------           Adj R-squared =  0.8664
       Total |  142608.652   419  340.354779           Root MSE      =  6.7432

------------------------------------------------------------------------------
    read_scr |      Coef.   Std. Err.      t    P>|t|     [95% Conf. Interval]
-------------+----------------------------------------------------------------
    expn_stu |   .0055534   .0008322     6.67   0.000     .0039176    .0071892
    comp_stu |   27.26378   8.197228     3.33   0.001     11.15063    43.37693
    meal_pct |  -.6352229    .013149   -48.31   0.000    -.6610696   -.6093762
       _cons |    648.988   4.163875   155.86   0.000     640.8031    657.1728
------------------------------------------------------------------------------
```

Including the weights modifies the coefficient estimates and reduces the `Root MSE` of the estimated equation. Equally weighting very small and very large school districts places too much weight on the former and too little on the latter. For instance, the effect of increases in the number of computers per student is almost 50% larger in the weighted estimates, and the effect of expenditures per student is smaller in the OLS estimates. The weighting also yields more precise coefficient estimates.

6.3 Serial correlation in the error distribution

Our discussion of heteroskedasticity in the error process focused on the first i in i.i.d.: the notion that disturbances are *identically* distributed over the observations. As in the discussion of the cluster estimator, we also may doubt the second i, that the disturbances are *independently* distributed. With cross-sectional data, departures from independence may reflect neighborhood effects, as accounted for by the cluster–VCE estimator. Observations that are similar in some way share a correlation in their disturbances.

When we turn to time-series data, we see a similar rationale for departures from independence. Observations that are close in time may be correlated, with the strength of that correlation increasing with proximity. Although there is no natural measure of proximity in cross-sectional data, time-series data by its nature defines *temporal* proximity. The previous and subsequent observations are those closest to y_t chronologically. When correlations arise in a time series, we speak of the disturbance process exhibiting *serial correlation* or *autocorrelation*; it is literally correlated with itself.

We must be wary of specification issues, as apparent serial correlation in the errors may be nothing more than a reflection of one or more systematic factors mistakenly excluded from the regression model. As discussed in section 5.2, inadequate specification of dynamic terms may cause such a problem. But sometimes errors will be, by construction, serially correlated rather than independent across observations. Theoretical schemes such as partial-adjustment mechanisms and agents' adaptive expectations can give rise to errors that cannot be serially independent. Thus we also must consider this sort of deviation of $\mathbf{\Sigma}_u$ from $\sigma^2 I_N$, one that is generally more challenging to deal with than is pure heteroskedasticity.

6.3.1 Testing for serial correlation

How might we test for the presence of serially correlated errors? Just as for pure heteroskedasticity, we base tests of serial correlation on the regression residuals. In the simplest case, autocorrelated errors follow the AR(1) model: an *autoregressive process* of order one, also known as a first-order Markov process:

$$u_t = \rho u_{t-1} + v_t, \ |\rho| < 1 \tag{6.15}$$

where the v_t are uncorrelated random variables with mean zero and constant variance. We impose the restriction that $|\rho| < 1$ to ensure that the disturbance process u is stationary with a finite variance. If $\rho = 1$, we have a *random walk*, which implies that the variance of u is infinite, and u is termed a *nonstationary* series, or an integrated process of order one [often written as $I(1)$]. We assume that the u process is *stationary*, with a finite variance, which will imply that the effects of a shock, v_t, will dissipate over time.[21]

The larger (in absolute value) ρ is, the greater will be the *persistence* of that shock to u_t and the more highly *autocorrelated* will be the sequence of disturbances u_t. In fact, in the AR(1) model, the *autocorrelation function* of u will be the geometric sequence $\rho, \rho^2, \rho^3, \ldots$, and the correlation of disturbances separated by τ periods will be ρ^τ. In Stata, the autocorrelation function for a time series may be computed with the `ac` or `corrgram` commands ([TS] **corrgram** refers to the correlogram of the series).

If we suspect that there is autocorrelation in the disturbance process of our regression model, we could use the estimated residuals to diagnose it. The empirical counterpart

21. If there is reason to doubt the stationarity of a time series, a *unit root test* should be performed: see, for example, [TS] **dfgls**.

to u_t in (6.15) will be the $\widehat{u}_t$ series produced by `predict`. We estimate the auxiliary regression of $\widehat{u}_t$ on $\widehat{u}_{t-1}$ without a constant term because the residuals have mean zero. The resulting slope estimate is a consistent estimator of the first-order autocorrelation coefficient ρ of the u process from (6.15). Under the null hypothesis $\rho = 0$, so a rejection of this null hypothesis by this LM test indicates that the disturbance process exhibits AR(1) behavior.

A generalization of this procedure that supports testing for higher-order autoregressive disturbances is the LM test of Breusch and Godfrey (Godfrey 1988). In this test, the regression is augmented with p lagged residual series. The null hypothesis is that the errors are serially independent up to order p. The test evaluates the *partial correlations* of the regressors $\mathbf{x}$ partialled off.[22] The residuals at time t are orthogonal to the columns of $\mathbf{x}$ at time t, but that need not be so for the lagged residuals. This is perhaps the most useful test for nonindependence of time-series disturbances, since it allows the researcher to examine more than first-order serial independence of the errors in one test. The test is available in Stata as `estat bgodfrey` (see [R] **regress postestimation time series**).

A variation on the Breusch–Godfrey test is the Q test of Box and Pierce (1970), as refined by Ljung and Box (1979), which examines the first p sample autocorrelations of the residual series:

$$Q = T(T+2)\sum_{j=1}^{p} \frac{r_j^2}{T-j}$$

where r_j^2 is the jth autocorrelation of the residual series. Unlike the Breusch–Godfrey test, the Q test does not condition the autocorrelations on a particular x. Q is based on the simple correlations of the residuals rather than their partial correlations. Therefore, it is less powerful than the Breusch–Godfrey test when the null hypothesis (of no serial correlation in u up to order p) is false. However, the Q test may be applied to any time series whether or not it contains residuals from an estimated regression model. Under the null hypothesis, $Q \sim \chi^2(p)$. The Q test is available in Stata as `wntestq`, named such to indicate that it may be used as a general test for so-called *white noise*, a property of random variables that do not contain autocorrelation.

The oldest test (but still widely used and reported, despite its shortcomings) is the Durbin and Watson (1950) d statistic:

$$d = \frac{\sum_{t=2}^{T}(\widehat{u}_t - \widehat{u}_{t-1})^2}{\sum_{t=1}^{T}\widehat{u}_t^2} \simeq 2(1-\rho)$$

The Durbin–Watson (D–W) test proceeds from the principle that the numerator of the statistic, when expanded, contains twice the variance of the residuals minus twice the (first) autocovariance of the residual series. If $\rho = 0$, that autocovariance will be near zero, and d will equal 2.0. As $\rho \rightarrow 1$, $d \rightarrow 0$, whereas as $\rho \rightarrow -1$, $d \rightarrow 4$. However,

22. The partial autocorrelation function of a time series may be calculated with the `pac` command; see [TS] **corrgram**.

the exact distribution of the statistic depends on the regressor matrix (which must contain a constant term and must not contain a lagged dependent variable). Rather than having a set of critical values, the D–W test has two, labeled d_L and d_U. If the d statistic falls below d_L, we reject the null; above d_U, we do not reject; and in between, the statistic is inconclusive. (For negative autocorrelation, you test $4 - d$ against the same tabulated critical values.) The test is available in Stata as `estat dwstat` (see [R] **regress postestimation time series**) and is automatically displayed in the output of the `prais` estimation command.

In the presence of a lagged dependent variable or generally, predetermined regressors, the d statistic is biased toward 2.0, and Durbin's *alternative* (or h) test (Durbin 1970) must be used.[23] That test is an LM test, which is computed by regressing residuals on their lagged values and the original $\mathbf{X}$ matrix. The test is asymptotically equivalent to the Breusch–Godfrey test for $p = 1$ and is available in Stata as command `estat durbinalt` (see [R] **regress postestimation time series**).

I illustrate the diagnosis of autocorrelation with a time-series dataset of monthly short-term and long-term interest rates on U.K. government securities (Treasury bills and gilts), 1952m3–1995m12. `summarize` gives the descriptive statistics for these series:

```
. use http://www.stata-press.com/data/imeus/ukrates, clear
. summarize rs r20
    Variable |       Obs        Mean    Std. Dev.       Min        Max
-------------+--------------------------------------------------------
          rs |       526    7.651513    3.553109   1.561667      16.18
         r20 |       526    8.863726    3.224372       3.35      17.18
```

The model expresses the monthly change in the short rate `rs`, the Bank of England's monetary policy instrument, as a function of the prior month's change in the long-term rate `r20`. The regressor and regressand are created on the fly by Stata's time-series operators `D.` and `L.` The model represents a monetary policy reaction function. We save the model's residuals with `predict` so that we can use `wntestq`.

```
. regress D.rs LD.r20
      Source |       SS       df       MS              Number of obs =     524
-------------+------------------------------           F(  1,   522) =   52.88
       Model |  13.8769739     1  13.8769739           Prob > F      =  0.0000
    Residual |  136.988471   522  .262430021           R-squared     =  0.0920
-------------+------------------------------           Adj R-squared =  0.0902
       Total |  150.865445   523  .288461654           Root MSE      =  .51228

------------------------------------------------------------------------------
        D.rs |      Coef.   Std. Err.      t    P>|t|     [95% Conf. Interval]
-------------+----------------------------------------------------------------
         r20 |
         LD. |   .4882883   .0671484     7.27   0.000      .356374    .6202027
       _cons |   .0040183    .022384     0.18   0.858    -.0399555    .0479921
------------------------------------------------------------------------------
```

23. A variable x is predetermined if $E[x_t u_{t+s}] = 0$ for all t and s. See Davidson and MacKinnon (1993).

```
. predict double eps, residual
(2 missing values generated)
. estat bgodfrey, lags(6)
Breusch-Godfrey LM test for autocorrelation
```

lags(p)	chi2	df	Prob > chi2
6	17.237	6	0.0084

```
                        H0: no serial correlation
. wntestq eps
Portmanteau test for white noise

 Portmanteau (Q) statistic =    82.3882
 Prob > chi2(40)           =     0.0001
. ac eps
```

The Breusch–Godfrey test performed here considers the null of serial independence up to sixth order in the disturbance process, and that null is soundly rejected. That test is conditioned on the fitted model. The Q test invoked by **wntestq**, which allows for more general alternatives to serial independence of the residual series, confirms the diagnosis. To further analyze the nature of the residual series' lack of independence, we compute the autocorrelogram (displayed in figure 6.1). This graph indicates the strong presence of first-order autocorrelation—AR(1)—but also signals several other empirical autocorrelations outside the Bartlett confidence bands.

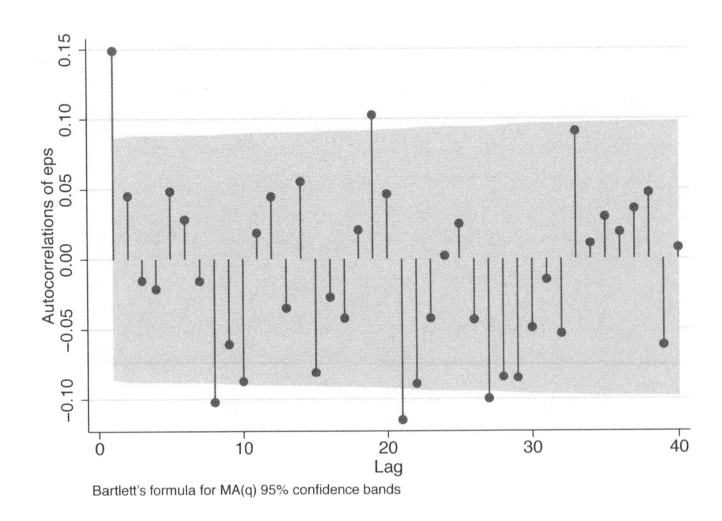

Figure 6.1: Autocorrelogram of regression residuals

6.3.2 FGLS estimation with serial correlation

For AR(1) disturbances of (6.15), if ρ were known, we could estimate the coefficients by GLS. The form of $\mathbf{\Sigma}_u$ displayed in (6.3) is simplified when we consider first-order serial correlation with one parameter ρ. An analytical inverse of $\mathbf{\Sigma}_u$ may be derived as

$$\mathbf{\Sigma}_u^{-1} = \sigma_u^{-2} \begin{pmatrix} \sqrt{1-\rho^2} & 0 & \dots & 0 \\ -\rho & 1 & \dots & 0 \\ & & \vdots & \\ 0 & -\rho & 1 & 0 \\ 0 & \dots & -\rho & 1 \end{pmatrix} \tag{6.16}$$

As with heteroskedasticity, we do not explicitly construct and apply this matrix. Rather, we can implement GLS by transforming the original data and running a regression on the transformed data. For observations 2, ..., T, we *quasidifference* the data: $y_t - \rho y_{t-1}$, $x_{j,t} - \rho x_{j,t-1}$, and so on. The first observation is multiplied by $\sqrt{1-\rho^2}$.

The GLS estimator is not feasible because ρ is an unknown population parameter just like $\boldsymbol{\beta}$ and σ_u^2. Replacing the unknown ρ values above with a consistent estimate and computing $\widehat{\mathbf{\Sigma}}_u$ yields the FGLS estimator. As with heteroskedasticity, the OLS residuals from the original model may be used to generate the necessary estimate. The Prais and Winsten (1954) estimator uses an estimate of ρ based on the OLS residuals to estimate $\widehat{\mathbf{\Sigma}}_u^{-1}$ by (6.16). The closely related Cochrane and Orcutt (1949) variation on that estimator differs only in its treatment of the first observation of the transformed data, given the estimate of ρ from the regression residuals. Either of these estimators may be iterated to convergence: essentially they operate by ping-ponging back and forth between estimates of $\boldsymbol{\beta}$ and ρ. Optional iteration refines the estimate of ρ, which is strongly recommended in small samples. Both estimators are available in Stata with the `prais` command.

Other approaches include maximum likelihood, which simultaneously estimates one parameter vector $(\boldsymbol{\beta}', \sigma^2, \rho)$, and the grid search approach of Hildreth and Lu (1960). Although you could argue for the superiority of a maximum likelihood approach, Monte Carlo studies suggest that the Prais–Winsten estimator is nearly as efficient in practice as maximum likelihood.

I illustrate the Prais–Winsten estimator by using the monetary policy reaction function displayed above. FGLS on this model finds a value of ρ of 0.19 and a considerably smaller coefficient on the lagged change in the long-term interest rate than that of our OLS estimate.

```
. prais D.rs LD.r20, nolog

Prais-Winsten AR(1) regression -- iterated estimates

      Source |       SS       df       MS              Number of obs =     524
-------------+------------------------------           F(  1,   522) =   25.73
       Model |  6.56420242     1  6.56420242           Prob > F      =  0.0000
    Residual |  133.146932   522   .25507075           R-squared     =  0.0470
-------------+------------------------------           Adj R-squared =  0.0452
       Total |  139.711134   523    .2671341           Root MSE      =  .50505

------------------------------------------------------------------------------
        D.rs |      Coef.   Std. Err.      t    P>|t|     [95% Conf. Interval]
-------------+----------------------------------------------------------------
         r20 |
         LD. |   .3495857    .068912     5.07   0.000     .2142067    .4849647
       _cons |   .0049985   .0272145     0.18   0.854    -.0484649    .0584619
-------------+----------------------------------------------------------------
         rho |   .1895324
------------------------------------------------------------------------------
Durbin-Watson statistic (original)    1.702273
Durbin-Watson statistic (transformed) 2.007414
```

In summary, although we may use FGLS to deal with autocorrelation, we should always be aware that this diagnosis may reflect misspecification of the model's dynamics or omission of one or more key factors from the model. We may mechanically correct for first-order serial correlation in a model, but we then attribute this persistence to some sort of clockwork in the error process rather than explaining its existence. Applying FGLS as described here is suitable for AR(1) errors but not for higher-order AR(p) errors or *moving-average* (MA) error processes, both of which may be encountered in practice. Regression equations with higher-order AR errors or MA errors can be modeled by using Stata's `arima` command.

Exercises

1. Use the `cigconsump` dataset, retaining only years 1985 and 1995. Regress `lpackpc` on `lavgprs` and `lincpc`. Use the Breusch–Pagan test (`hettest`) for variable `year`. Save the residuals, and use `robvar` to compute their variances by `year`. What do these tests tell you?
2. Use FGLS to refit the model, using analytical weights based on the residuals from each year. How do these estimates differ from the OLS estimates?
3. Use the `sp500` dataset installed with Stata (`sysuse sp500`), applying `tsset date`. Regress the first difference of `close` on two lagged differences and lagged `volume`. How do you interpret the coefficient estimates? Use the Breusch–Godfrey test to evaluate the errors' independence. What do you conclude?
4. Refit the model with FGLS (using `prais`). How do the FGLS estimates compare to those from OLS?

7 Regression with indicator variables

One of the most useful concepts in applied economics is the *indicator variable*, which signals the presence or absence of a characteristic. Indicator variables are also known as *binary* or *Boolean* variables and are well known to econometricians as *dummy variables* (although the meaning of that latter term is shrouded in the mists of time). Here we consider how to use indicator variables

- to evaluate the effects of qualitative factors;
- in models that mix quantitative and qualitative factors;
- in seasonal adjustment; and
- to evaluate structural stability and test for structural change.

7.1 Testing for significance of a qualitative factor

Economic data come in three varieties: quantitative (or cardinal), ordinal (or ordered), and qualitative.[1] In chapter 3, I described the first category as *continuous* data to stress that their values are quantities on the real line that may conceptually take on any value. We also may work with *ordinal* or ordered data. They are distinguished from cardinal measurements in that an ordinal measure can express only inequality of two items and not the magnitude of their difference; for example, a Likert scale of "How good a job has the president done? 5 = great, 4 = good, 3 = fair , 2 = poor, 1 = very poor" will generate ordered numeric responses. A response of 5 beats 4, which in turn beats 3 for voter satisfaction. But we cannot state that a respondent of 5 is five times more likely to support the president than a voter responding 1, nor 25% more likely than a respondent of 4, and so on. The numbers can be taken only as *ordered*. They could be any five ordered points on the real line (or the set of integers). The implication: if data are actually ordinal rather than cardinal, we should not mistake them for cardinal measures and should not use them as a response variable or as a regressor in a linear regression model.

In contrast, we often encounter economic data that are purely *qualitative*, lacking any obvious ordering. If these data are coded as string variables, such as `M` and `F` for survey respondents' genders, we are not likely to mistake them for quantitative values. We hope that few researchers would contemplate using five-digit ZIP codes (U.S. postal codes) in a quantitative setting. But where a quality may be coded numerically, there

1. I discuss censored data in chapter 10.

is the potential to misuse this qualitative factor as quantitative. This misuse of course is nonsensical: as described in section 2.2.4, we can `encode` a two-letter U.S. state code (AK, AL, AZ, ..., WY) into a set of integers 1, ..., 50 for ease of manipulation, but we should never take those numeric values as quantitative measures.

How should we evaluate the effects of purely qualitative measures? Since the answer to this question will apply largely to ordinal measures as well, it may be taken to cover all nonquantitative economic and financial data. To test the hypothesis that a qualitative factor has an effect on a response variable, we must convert the qualitative factor into a set of *indicator variables*, or dummy variables. Following the discussion in section 4.5.3, we then conduct a *joint test* on their coefficients. If the hypothesis to be tested includes one qualitative factor, the estimation problem may be described as a one-way ANOVA. Economic researchers consider that ANOVA models may be expressed as linear regressions on an appropriate set of indicator variables.[2]

The equivalence of one-way ANOVA and linear regression on a set of indicator variables that correspond to one qualitative factor generalizes to multiple qualitative factors. If two qualitative factors (e.g., race and sex) are hypothesized to affect income, an economic researcher would regress income on two appropriate sets of indicator variables, each representing one of the qualitative factors. If we include one or many qualitative factors in a model, we will estimate a linear regression on several indicator (dummy) variables.

7.1.1 Regression with one qualitative measure

Consider measures of the six New England states' per capita disposable personal income (`dpipc`) for 1981–2000 as presented in section 6.2.2. Does the state of residence explain a significant proportion of the variation in `dpipc` over these two decades? We calculate the average `dpipc` (in thousands of dollars) over the two decades by using `mean` (see [R] **mean**):

2. Stata's `anova` command has a `regress` option that presents the results of ANOVA models in a regression framework.

```
. use http://www.stata-press.com/data/imeus/NEdata, clear
. mean dpipc, over(state)
Mean estimation                    Number of obs    =      120

          CT: state = CT
          MA: state = MA
          ME: state = ME
          NH: state = NH
          RI: state = RI
          VT: state = VT
```

Over	Mean	Std. Err.	[95% Conf.	Interval]
dpipc				
CT	22.32587	1.413766	19.52647	25.12527
MA	19.77681	1.298507	17.20564	22.34798
ME	15.17391	.9571251	13.27871	17.06911
NH	18.66835	1.193137	16.30582	21.03088
RI	17.26529	1.045117	15.19586	19.33473
VT	15.73786	1.020159	13.71784	17.75788

States' average `dpipc` in 2000 varies considerably between Connecticut ($22,326) and Maine ($15,174). But are these differences statistically significant? Let us test this hypothesis with `regress`. We first must create the appropriate indicator variables. One way to do this (which I prefer to using `xi`) is, as described in section 2.2.4, to use `tabulate` and its `generate()` option to produce the desired variables. The following command generates six indicator variables, but we recognize that these six indicator variables must be *mutually exclusive and exhaustive* (MEE). Each observation must belong to one and only one state. Also the mean of an indicator variable is the fraction or proportion of the sample satisfying that characteristic. Those means must sum to 1.0 across any complete set of indicator variables.

If `tabulate` generates a set of indicator variables $\mathbf{D}_{N\times g}$, where there are G groups (here, six), then $\mathbf{D}\boldsymbol{\iota} = \boldsymbol{\iota}$, where $\boldsymbol{\iota}$ is the units vector. If we sum the indicator variables across the g categories, we must produce an N-vector of ones. For that reason, we must drop one of the indicator variables when running a regression to avoid perfect collinearity with the constant term. We fit the regression model, dropping the first indicator variable (that for CT):

```
. tabulate state, generate(NE)
```

state	Freq.	Percent	Cum.
CT	20	16.67	16.67
MA	20	16.67	33.33
ME	20	16.67	50.00
NH	20	16.67	66.67
RI	20	16.67	83.33
VT	20	16.67	100.00
Total	120	100.00	

```
. regress dpipc NE2-NE6
      Source |       SS       df       MS              Number of obs =     120
-------------+------------------------------           F(  5,   114) =    5.27
       Model |  716.218512     5  143.243702           Prob > F      =  0.0002
    Residual |  3099.85511   114  27.1917115           R-squared     =  0.1877
-------------+------------------------------           Adj R-squared =  0.1521
       Total |  3816.07362   119  32.0678456           Root MSE      =  5.2146

------------------------------------------------------------------------------
       dpipc |      Coef.   Std. Err.      t    P>|t|     [95% Conf. Interval]
-------------+----------------------------------------------------------------
         NE2 |  -2.549057   1.648991    -1.55   0.125    -5.815695    .7175814
         NE3 |  -7.151959   1.648991    -4.34   0.000     -10.4186    -3.88532
         NE4 |   -3.65752   1.648991    -2.22   0.029    -6.924158   -.3908815
         NE5 |  -5.060575   1.648991    -3.07   0.003    -8.327214   -1.793937
         NE6 |  -6.588007   1.648991    -4.00   0.000    -9.854646   -3.321369
       _cons |   22.32587   1.166013    19.15   0.000     20.01601    24.63573
------------------------------------------------------------------------------
```

This regression produces estimates of a constant term and five coefficients. We have excluded the first state (CT), so the constant term is the mean of CT values over time, identical to the `means` output above. The coefficients reported by `regress` represent the differences between each state's mean `dpipc` and that of CT.[3] The state means shown in the `mean` output above are six points on the real line. Are their differences statistically significant? It does not matter how we measure those differences, whether from the VT mean value of 15.7 or from the CT mean value of 22.3. Although we must exclude one state's indicator variable from the regression, the choice of the excluded class is arbitrary and will not affect the statistical judgments.

The test for relevance of the qualitative factor `state` is merely the ANOVA F statistic for this regression. The ANOVA F, as section 4.3.2 describes, tests the null hypothesis that all slope coefficients are jointly zero. In this context, that is equivalent to testing that all six state means of `dpipc` equal a common μ. The strong rejection of that hypothesis from the ANOVA F statistic implies that the New England states have significantly different levels of per capita disposable personal income.

Another transformation of indicator variables to produce *centered indicators* is often useful. If we create new indicators $d_i^* = d_i - d_g$, where d_g is the indicator for the excluded class, we can use the $(g-1)$ d_i^* variables in the model rather than the original d_i variables. As discussed above, the coefficients on the original d_i variables are contrasts with the excluded class. The d_i^* variables, which are trinary (taking on values of $-1, 0, 1$) will be contrasts with the grand mean. The constant term in the regression on d_i^* will be the grand mean, and the individual d_i^* coefficients are contrasts with that mean. To illustrate,

3. For instance, $22.32587 - 2.549057 = 19.77681$, the mean estimate for MA given above.

```
. forvalues i=1/5 {
  2.    generate NE_`i' = NE`i'-NE6
  3.    }

. regress dpipc NE_*

      Source |       SS       df       MS              Number of obs =     120
-------------+------------------------------           F(  5,   114) =    5.27
       Model |  716.218512     5  143.243702           Prob > F      =  0.0002
    Residual |  3099.85511   114  27.1917115           R-squared     =  0.1877
-------------+------------------------------           Adj R-squared =  0.1521
       Total |  3816.07362   119  32.0678456           Root MSE      =  5.2146

------------------------------------------------------------------------------
       dpipc |      Coef.   Std. Err.      t    P>|t|     [95% Conf. Interval]
-------------+----------------------------------------------------------------
        NE_1 |   4.167853   1.064419     3.92   0.000     2.059247    6.276459
        NE_2 |   1.618796   1.064419     1.52   0.131      -.48981    3.727402
        NE_3 |  -2.984106   1.064419    -2.80   0.006    -5.092712   -.8754996
        NE_4 |   .5103331   1.064419     0.48   0.633    -1.598273    2.618939
        NE_5 |  -.8927223   1.064419    -0.84   0.403    -3.001328    1.215884
       _cons |   18.15802   .4760227    38.15   0.000     17.21502    19.10101
------------------------------------------------------------------------------
```

This algebraically equivalent model has the same explanatory power in terms of its ANOVA F statistic and R^2 as the model including five indicator variables. For example, $4.168 + 18.158 = 22.326$, the mean income in CT. Below we use `lincom` to compute the coefficient on the excluded class as minus the sum of the coefficients on the included classes.

```
. lincom -(NE_1+NE_2+NE_3+NE_4+NE_5)
 ( 1)  - NE_1 - NE_2 - NE_3 - NE_4 - NE_5 = 0

------------------------------------------------------------------------------
       dpipc |      Coef.   Std. Err.      t    P>|t|     [95% Conf. Interval]
-------------+----------------------------------------------------------------
         (1) |  -2.420154   1.064419    -2.27   0.025     -4.52876   -.3115483
------------------------------------------------------------------------------
```

7.1.2 Regression with two qualitative measures

We can use two sets of indicator variables to evaluate the effects of two qualitative factors on a response variable. Take for example the Stata manual dataset `nlsw88`, an extract of the U.S. National Longitudinal Survey (NLSW) for employed women in 1988. We restrict the sample of 2,246 working women to a subsample for which data on hourly `wage`, `race`, and an indicator of `union` status are available. This step reduces the sample to 1,878 workers. We also have data on a measure of job `tenure` in years.

```
. use http://www.stata-press.com/data/imeus/nlsw88, clear
(NLSW, 1988 extract)

. keep if !missing(wage + race + union)
(368 observations deleted)

. generate lwage = log(wage)
```

```
. summarize wage race union tenure, sep(0)
    Variable |       Obs        Mean    Std. Dev.       Min        Max
-------------+--------------------------------------------------------
        wage |      1878    7.565423    4.168369   1.151368   39.23074
        race |      1878    1.292332    .4822417          1          3
       union |      1878    .2454739    .4304825          0          1
      tenure |      1868    6.571065    5.640675          0   25.91667
```

We model `lwage`, the log of the reported wage, as the response variable. The variable `race` is coded 1, 2, or 3 for `white`, `black`, or `other`. We want to determine whether the variance in (log) wages is significantly related to the factors `race` and `union`. We cannot fit a regression model with two complete sets of dummies, so we will exclude one dummy from each group.[4] The regression estimates show the following:

```
. tabulate race, generate(R)
       race |      Freq.     Percent        Cum.
------------+-----------------------------------
      white |      1,353       72.04       72.04
      black |        501       26.68       98.72
      other |         24        1.28      100.00
------------+-----------------------------------
      Total |      1,878      100.00
. regress lwage R1 R2 union
      Source |       SS       df       MS              Number of obs =    1878
-------------+------------------------------           F(  3,  1874) =   38.73
       Model |  29.3349228     3  9.77830761           Prob > F      =  0.0000
    Residual |  473.119209  1874  .252464893           R-squared     =  0.0584
-------------+------------------------------           Adj R-squared =  0.0569
       Total |  502.454132  1877  .267690001           Root MSE      =  .50246

------------------------------------------------------------------------------
       lwage |      Coef.   Std. Err.      t    P>|t|     [95% Conf. Interval]
-------------+----------------------------------------------------------------
          R1 |  -.0349326   .1035125    -0.34   0.736    -.2379444    .1680793
          R2 |  -.2133924   .1049954    -2.03   0.042    -.4193126   -.0074721
       union |    .239083   .0270353     8.84   0.000     .1860606    .2921054
       _cons |   1.913178   .1029591    18.58   0.000     1.711252    2.115105
------------------------------------------------------------------------------

. test R1 R2  //         joint test for the effect of race
 ( 1)  R1 = 0
 ( 2)  R2 = 0

       F(  2,  1874) =   23.25
            Prob > F =    0.0000
```

A test for the significance of the qualitative factor `race` is the joint test for the coefficients of `R1, R2` equaling zero. When taking `other` as the excluded class for `race` we do not find that β_{R1} (the coefficient for `white`) differs from zero. But this coefficient is the contrast between the mean of `lwage` for `other` and the mean for `white`. The mean for `R2` (`black`), on the other hand, is distinguishable from that for `other`. These coefficients,

4. We could include one complete set of dummies in an equation without a constant term, but I do not recommend that approach. The absence of a constant term alters the meaning of many summary statistics.

taken together, reflect the effects of `race` on `lwage`. Those regressors should be kept or removed *as a group*. In particular, we should not use the t statistics for individual indicator variables to make inferences beyond noting, as above, the differences between group means. The magnitudes of those coefficients and their t statistics depend on the choice of excluded class, which is arbitrary.

The model of two qualitative factors illustrated here is a special case in that it assumes that the effects of the two qualitative factors are independent and strictly additive. That is, if you are black, your (log) wage is expected to be 0.213 lower than that of the `other` race category,[5] whereas if you are a union member, it is predicted to be 0.239 higher. What would this regression model predict that a black union member would earn, relative to the excluded class (a nonunion member of `other` race)? It would predict merely the sum of those two effects, or +0.026, since the union effect is slightly stronger than the black effect. We have a 3×2 two-way table of `race` and `union` categories. We can fill in the six cells of that table from the four coefficients estimated in the regression. For that approach to be feasible, we must assume independence of the qualitative effects so that the joint effect (reflected by a cell within the table) is the sum of the marginal effects. The effect of being black and a union member is taken to be the sum of the effects of being black, independent of union status, and that of being a union member, independent of `race`.

Interaction effects

Although sometimes this independence of qualitative factors is plausible, often it is not an appropriate assumption. Consider variations of the unemployment rate across age and race. Teenagers have a hard time landing a job because they lack labor market experience, so teenage unemployment rates are high relative to those of prime-aged workers. Likewise, minority participants generally have higher unemployment rates, whether due to discrimination or other factors such as the quality of their education. These two effects may not be merely additive. Perhaps being a minority teenager involves two strikes against you when seeking employment. If so, the effects of being both minority and a teenager are greater than the sum of their individual contributions. This reasoning implies that we should allow for *interaction effects* in evaluating these qualitative factors, which will allow their effects to be correlated, and requires that we estimate all six elements in the 3×2 table from the last regression example.

In regression, interactions involve products of indicator variables. Dummy variables may be treated as algebraic or Boolean. Adding indicator variables is equivalent to the Boolean "or" operator (`|`), denoting the union of two sets, whereas multiplying two indicator variables is equivalent to the Boolean "and" operator (`&`), denoting the intersection of sets. We may use either syntax in Stata's `generate` statements, remembering that we need to handle missing values properly.

5. This prediction translates into roughly 21%, using the rough approximation that $\log(1 + x) \simeq x$, although this approximation should really be used only for single-digit x.

How can we include a `race*union` interaction in the last regression? Since we need two `race` dummies to represent the three classes and one `union` dummy to reflect that factor, we need two interaction terms in the model: the interaction of each included `race` dummy with the `union` dummy. In the model

$$\texttt{lwage}_i = \beta_1 + \beta_2\texttt{R1}_i + \beta_3\texttt{R2}_i + \beta_4\texttt{union}_i + \beta_5(\texttt{R1}_i \times \texttt{union}_i) + \beta_6(\texttt{R2}_i \times \texttt{union}_i) + u_i$$

the mean log wage for those in race `R1` (`white`) is $\beta_1 + \beta_2$ for nonunion members, but $\beta_1 + \beta_2 + \beta_4 + \beta_5$ for union members. Fitting this model yields the following:

```
. generate R1u = R1*union

. generate R2u = R2*union

. regress lwage R1 R2 union R1u R2u

      Source |       SS       df       MS              Number of obs =    1878
-------------+------------------------------           F(  5,  1872) =   26.63
       Model |  33.3636017     5  6.67272035           Prob > F      =  0.0000
    Residual |   469.09053  1872  .250582548           R-squared     =  0.0664
-------------+------------------------------           Adj R-squared =  0.0639
       Total |  502.454132  1877  .267690001           Root MSE      =  .50058

------------------------------------------------------------------------------
       lwage |      Coef.   Std. Err.      t    P>|t|     [95% Conf. Interval]
-------------+----------------------------------------------------------------
          R1 |  -.1818955   .1260945    -1.44   0.149    -.4291962    .0654051
          R2 |  -.4152863   .1279741    -3.25   0.001    -.6662731   -.1642995
       union |  -.2375316   .2167585    -1.10   0.273    -.6626452     .187582
         R1u |   .4232627   .2192086     1.93   0.054    -.0066561    .8531816
         R2u |   .6193578   .2221704     2.79   0.005     .1836302    1.055085
       _cons |    2.07205   .1251456    16.56   0.000      1.82661    2.317489
------------------------------------------------------------------------------

. test R1u R2u     //    joint test for the interaction effect of race*union

 ( 1)  R1u = 0
 ( 2)  R2u = 0

       F(  2,  1872) =    8.04
            Prob > F =    0.0003
```

The joint test of the two interaction coefficients `R1u` and `R2u` rejects the null hypothesis of independence of the qualitative factors `race` and `union` at all conventional levels. Because the interaction terms are jointly significant, it would be a misspecification to fit the earlier regression rather than this expanded form. In regression, we can easily consider the model with and without interactions by merely fitting the model with interactions and performing the joint test that all interaction coefficients are equal to zero.

7.2 Regression with qualitative and quantitative factors

Earlier, we fitted several regression models in which all the regressors are indicator variables. In economic research, we often want to combine quantitative and qualitative information in a regression model by including both continuous and indicator regressors.

Returning to the `nlsw88` dataset, we might model the log(wage) for qualitative factors `race` and `union`, as well as a quantitative factor `tenure`, the number of years worked in the current job. Estimation of that regression yields

```
. regress lwage R1 R2 union tenure
      Source |       SS       df       MS              Number of obs =    1868
-------------+------------------------------           F(  4,  1863) =   85.88
       Model |  77.1526731     4  19.2881683           Prob > F      =  0.0000
    Residual |  418.434693  1863  .224602626           R-squared     =  0.1557
-------------+------------------------------           Adj R-squared =  0.1539
       Total |  495.587366  1867  .265445831           Root MSE      =  .47392

------------------------------------------------------------------------------
       lwage |      Coef.   Std. Err.      t    P>|t|     [95% Conf. Interval]
-------------+----------------------------------------------------------------
          R1 |   -.070349   .0976711    -0.72   0.471    -.2619053    .1212073
          R2 |  -.2612185   .0991154    -2.64   0.008    -.4556074   -.0668297
       union |   .1871116   .0257654     7.26   0.000     .1365794    .2376438
      tenure |   .0289352   .0019646    14.73   0.000     .0250823    .0327882
       _cons |   1.777386   .0975549    18.22   0.000     1.586058    1.968715
------------------------------------------------------------------------------

. test R1 R2  // joint test for the effect of race
 ( 1)  R1 = 0
 ( 2)  R2 = 0
       F(  2,  1863) =   29.98
            Prob > F =    0.0000
```

These results illustrate that this analysis-of-covariance model accounts for considerably more of the variation in `lwage` than does its counterpart based on only qualitative factors.[6] How might we interpret $\widehat{\beta}_{\texttt{tenure}}$? Using the standard approximation that $\log(1+x) \simeq x$,[7] we see that a given worker with 1 more year on her current job can expect to earn about 2.89% more (roughly, the semielasticity of `wage` with respect to `tenure`). How do we interpret the constant term? It is the mean log wage for a nonunion worker of `other` race with zero years of job tenure. Here that is a plausible category, since you might have less than 1 year's tenure in your current job. In other cases—for instance, where age is used as a regressor in a labor market study—the constant term may not correspond to any observable cohort.

The predictions of this model generate a series of parallel lines in {log(wage), tenure} space: a total of six lines, corresponding to the six possible combinations of `race` and `union`, with their intercepts computed from their coefficients and the constant term. We can separately test that those lines are distinct with respect to a qualitative factor: for instance, following the regression above, we jointly tested `R1` and `R2` for significance. If that test could not reject its null that each of those coefficients is zero, we would conclude that the {log(wage), tenure} profiles do not differ according to the qualitative factor `race`, and the six profiles would collapse to two.

6. I earlier noted that the form of this model with interaction terms was to be preferred; for pedagogical reasons, we return to the simpler form of the model.

7. See section 4.3.4.

Testing for slope differences

The model we have fitted is parsimonious and successful, given that it considers one quantitative factor. But are the true {log(wage), tenure} profiles parallel? Say that the unionized sector achieves larger annual wage increments by using its organized bargaining power. Might we expect two otherwise identical workers—one union, one nonunion—to have different profiles, with the unionized worker's profile steeper? To test that hypothesis, I return to the notion of an *interaction effect*, but here we interact a continuous measure (`tenure`) with the indicator variable `union`:

```
. quietly generate uTen = union*tenure
. regress lwage R1 R2 union tenure uTen
      Source |       SS       df       MS              Number of obs =    1868
-------------+------------------------------           F(  5,  1862) =   69.27
       Model |   77.726069     5  15.5452138           Prob > F      =  0.0000
    Residual |  417.861297  1862  .224415304           R-squared     =  0.1568
-------------+------------------------------           Adj R-squared =  0.1546
       Total |  495.587366  1867  .265445831           Root MSE      =  .47372

------------------------------------------------------------------------------
       lwage |      Coef.   Std. Err.      t    P>|t|     [95% Conf. Interval]
-------------+----------------------------------------------------------------
          R1 |  -.0715443   .0976332    -0.73   0.464    -.2630264    .1199377
          R2 |  -.2638742   .0990879    -2.66   0.008    -.4582093   -.0695391
       union |   .2380442   .0409706     5.81   0.000      .157691    .3183975
      tenure |   .0309616   .0023374    13.25   0.000     .0263774    .0355458
        uTen |  -.0068913   .0043112    -1.60   0.110    -.0153467     .001564
       _cons |   1.766484   .0977525    18.07   0.000     1.574768      1.9582
------------------------------------------------------------------------------
```

The tenure effect is now measured as $\partial \texttt{lwage}/\partial \texttt{tenure} = \widehat{\beta}_{\texttt{tenure}}$ for nonunion members, but $(\widehat{\beta}_{\texttt{tenure}} + \widehat{\beta}_{\texttt{uTen}})$ for union members. The difference between those values is the estimated coefficient $\widehat{\beta}_{\texttt{uTen}}$, which is not significantly different from zero at the 10% level, but negative. Counter to our intuition, the data cannot reject the hypothesis that the slopes of the union and nonunion profiles are equal.

But what about the profiles for `race`? It is often claimed that minority hires are not treated equally over time, for instance, that promotions and larger increments go to whites rather than to blacks or Hispanics. We interact the race categories with `tenure`, in effect allowing the slopes of the {log(wage), tenure} profiles to differ by race:

```
. quietly generate R1ten = R1*tenure
. quietly generate R2ten = R2*tenure
. regress lwage R1 R2 union tenure R1ten R2ten
      Source |       SS       df       MS              Number of obs =    1868
-------------+------------------------------           F(  6,  1861) =   57.26
       Model |  77.2369283     6  12.8728214           Prob > F      =  0.0000
    Residual |  418.350438  1861  .224798731           R-squared     =  0.1558
-------------+------------------------------           Adj R-squared =  0.1531
       Total |  495.587366  1867  .265445831           Root MSE      =  .47413
```

```
       lwage        Coef.   Std. Err.      t    P>|t|     [95% Conf. Interval]

          R1     -.082753       .1395    -0.59   0.553    -.3563459    .1908398
          R2     -.291495    .1422361    -2.05   0.041     -.570454   -.012536
       union     .1876079    .0257915     7.27   0.000     .1370246    .2381912
      tenure     .0257611    .0186309     1.38   0.167    -.0107785    .0623007
       R1ten     .0024973    .0187646     0.13   0.894    -.0343045    .0392991
       R2ten     .0050825     .018999     0.27   0.789     -.032179    .0423441
       _cons     1.794018    .1382089    12.98   0.000     1.522957    2.065078

. test R1ten R2ten
 ( 1)  R1ten = 0
 ( 2)  R2ten = 0
       F(  2,  1861) =    0.19
            Prob > F =    0.8291
```

We cannot reject the null hypothesis that both interaction coefficients are zero, implying that we do not have evidence against the hypothesis that one slope over categories of `race` suffices to express the effect of `tenure` on the wage. There does not seem to be evidence of *statistical discrimination* in wage increments, in the sense that the growth rates of female workers' wages do not appear to be race related.[8]

This last regression estimates five {log(wage), tenure} profiles, where the profiles for union members and nonunion members have equal slopes for a given race (with intercepts 0.188 higher for union members). We could fully interact `tenure` with both qualitative factors and estimate six {log(wage), tenure} profiles with different slopes:

```
. regress lwage R1 R2 union tenure uTen R1ten R2ten
      Source         SS       df       MS              Number of obs =    1868
                                                       F(  7,  1860) =   49.48
       Model    77.8008722     7  11.1144103           Prob > F      =  0.0000
    Residual    417.786494  1860  .224616394           R-squared     =  0.1570
                                                       Adj R-squared =  0.1538
       Total    495.587366  1867  .265445831           Root MSE      =  .47394

       lwage        Coef.   Std. Err.      t    P>|t|     [95% Conf. Interval]

          R1    -.0697096    .1396861    -0.50   0.618    -.3436676    .2042485
          R2    -.2795277    .1423788    -1.96   0.050    -.5587668   -.0002886
       union      .238244    .0410597     5.80   0.000     .1577161    .3187718
      tenure     .0304528    .0188572     1.61   0.106    -.0065308    .0674364
        uTen    -.0068628    .0043311    -1.58   0.113    -.0153572    .0016316
       R1ten    -.0001912    .0188335    -0.01   0.992    -.0371283    .0367459
       R2ten     .0023429    .0190698     0.12   0.902    -.0350576    .0397433
       _cons      1.76904    .1390492    12.72   0.000     1.496331    2.041749
```

8. We could certainly use these findings to argue that black women with a given job tenure earn lower wages than do white women or those of other races, but that outcome could be related to other factors: the workers' ages, levels of education, employment location, and so forth.

```
. test uTen R1ten R2ten
 ( 1)  uTen = 0
 ( 2)  R1ten = 0
 ( 3)  R2ten = 0
       F(  3,  1860) =    0.96
            Prob > F =    0.4098
```

The joint test conducted here considers the null of one slope for all six categories versus six separate slopes. That null is not rejected by the data, so one slope will suffice.

Before leaving this topic, consider a simpler model in which we consider only the single indicator variable `union` and one quantitative measure, `tenure`. Compare the equation

$$\texttt{lwage}_i = \beta_1 + \beta_2\texttt{union}_i + \beta_3\texttt{tenure}_i + \beta_4(\texttt{union}_i \times \texttt{tenure}_i) + u_i \tag{7.1}$$

with the equations

$$\begin{aligned} \texttt{lwage}_i &= \gamma_1 + \gamma_2\texttt{tenure}_i + v_i,\ \ i \neq \texttt{union} \\ \texttt{lwage}_i &= \delta_1 + \delta_2\texttt{tenure}_i + \omega_i,\ \ i = \texttt{union} \end{aligned} \tag{7.2}$$

That is, we estimate separate equations from the nonunion and union cohorts. The point estimates of β from (7.2) are identical to those that may be computed from (7.1), but their standard errors will differ since the former are computed from smaller samples. Furthermore, when the two equations are estimated separately, each has its own σ^2 estimate. In estimating (7.1), we assume that u is homoskedastic over union and nonunion workers, but that may not be an appropriate assumption. From a behavioral standpoint, collective bargaining may reduce the *volatility* of wages (e.g., by ruling out merit increments in favor of across-the-board raises), regardless of the effects of collective bargaining on the *level* of wages. Estimating these equations for the `nlsw88` data illustrates these points. First, I present the regression over the full sample:

```
. regress lwage union tenure uTen
      Source |       SS       df       MS              Number of obs =    1868
-------------+------------------------------           F(  3,  1864) =   92.25
       Model |  64.0664855     3  21.3554952           Prob > F      =  0.0000
    Residual |   431.52088  1864  .231502618           R-squared     =  0.1293
-------------+------------------------------           Adj R-squared =  0.1279
       Total |  495.587366  1867  .265445831           Root MSE      =  .48115

------------------------------------------------------------------------------
       lwage |      Coef.   Std. Err.      t    P>|t|     [95% Conf. Interval]
-------------+----------------------------------------------------------------
       union |   .2144586   .0414898     5.17   0.000     .1330872      .29583
      tenure |   .0298926   .0023694    12.62   0.000     .0252456    .0345395
        uTen |  -.0056219   .0043756    -1.28   0.199    -.0142035    .0029597
       _cons |   1.655054   .0193938    85.34   0.000     1.617018     1.69309
------------------------------------------------------------------------------
```

The t test for `uTen` indicates that the effects of `tenure` do not differ significantly across the classifications. We now fit the model over the union and nonunion subsamples:

```
. regress lwage tenure if !union

      Source |       SS       df       MS              Number of obs =    1408
-------------+------------------------------           F(  1,  1406) =  148.43
       Model |  36.8472972     1  36.8472972           Prob > F      =  0.0000
    Residual |  349.032053  1406  .248244703           R-squared     =  0.0955
-------------+------------------------------           Adj R-squared =  0.0948
       Total |   385.87935  1407  .274256823           Root MSE      =  .49824

------------------------------------------------------------------------------
       lwage |      Coef.   Std. Err.      t    P>|t|     [95% Conf. Interval]
-------------+----------------------------------------------------------------
      tenure |   .0298926   .0024536    12.18   0.000     .0250795    .0347056
       _cons |   1.655054   .0200828    82.41   0.000     1.615659     1.69445
------------------------------------------------------------------------------

. predict double unw if e(sample), res
(470 missing values generated)

. regress lwage tenure if union

      Source |       SS       df       MS              Number of obs =     460
-------------+------------------------------           F(  1,   458) =   55.95
       Model |  10.0775663     1  10.0775663           Prob > F      =  0.0000
    Residual |  82.4888278   458  .180106611           R-squared     =  0.1089
-------------+------------------------------           Adj R-squared =  0.1069
       Total |  92.5663941   459  .201669704           Root MSE      =  .42439

------------------------------------------------------------------------------
       lwage |      Coef.   Std. Err.      t    P>|t|     [95% Conf. Interval]
-------------+----------------------------------------------------------------
      tenure |   .0242707   .0032447     7.48   0.000     .0178944    .0306469
       _cons |   1.869513   .0323515    57.79   0.000     1.805937    1.933088
------------------------------------------------------------------------------

. predict double nunw if e(sample), res
(1418 missing values generated)
```

The Root MSE values are different for the two subsamples and could be tested for equality as described in section 6.2.2's treatment of groupwise heteroskedasticity:[9]

```
. generate double allres = nunw
(1418 missing values generated)

. replace allres = unw if unw<.
(1408 real changes made)

. sdtest allres, by(union)

Variance ratio test
------------------------------------------------------------------------------
   Group |     Obs        Mean    Std. Err.   Std. Dev.   [95% Conf. Interval]
---------+--------------------------------------------------------------------
nonunion |    1408    5.19e-17    .0132735    .4980645   -.0260379    .0260379
   union |     460    6.47e-17    .0197657    .4239271   -.0388425    .0388425
---------+--------------------------------------------------------------------
combined |    1868    5.50e-17    .0111235    .4807605   -.0218157    .0218157
------------------------------------------------------------------------------
    ratio = sd(nonunion) / sd(union)                              f =   1.3803
Ho: ratio = 1                                  degrees of freedom = 1407, 459

    Ha: ratio < 1               Ha: ratio != 1                 Ha: ratio > 1
  Pr(F < f) = 1.0000         2*Pr(F > f) = 0.0000           Pr(F > f) = 0.0000
```

9. We could instead use `egen double allres = rowtotal(nunw unw)`, but we would then have to use `replace allres=. if nunw ==. & unw==.` to deal with observations missing from both subsamples. Those observations would otherwise be coded as zeros.

We conclude that contrary to our prior results, nonunion workers have a significantly smaller variance of their disturbance process than union members. We should either correct for the heteroskedasticity across this classification or use robust standard errors to make inferences from a model containing both union and nonunion workers. To illustrate the latter point:

```
. regress lwage union tenure uTen, robust
Linear regression                                    Number of obs =     1868
                                                     F(  3,  1864) =   109.84
                                                     Prob > F      =   0.0000
                                                     R-squared     =   0.1293
                                                     Root MSE      =   .48115

------------------------------------------------------------------------------
             |               Robust
       lwage |      Coef.   Std. Err.      t    P>|t|     [95% Conf. Interval]
-------------+----------------------------------------------------------------
       union |   .2144586   .0407254     5.27   0.000     .1345864    .2943308
      tenure |   .0298926   .0023964    12.47   0.000     .0251928    .0345924
        uTen |  -.0056219   .0038631    -1.46   0.146    -.0131984    .0019546
       _cons |   1.655054   .0210893    78.48   0.000     1.613693    1.696415
------------------------------------------------------------------------------
```

Although robust standard errors increase the t statistic for `uTen`, the coefficient is not significantly different from zero at any conventional level of significance. We conclude that an interaction of `tenure` and `union` is not required for proper specification of the model.

7.3 Seasonal adjustment with indicator variables

Economic data with a time-series dimension often must be *seasonally adjusted.* For instance, monthly sales data for a set of retail firms will have significant variations around the holidays, and quarterly tax collections for municipalities located in a tourist area will fluctuate widely between the tourist season and off-season. A common method of seasonal adjustment involves modeling the seasonal factor in the time series as being either additive or multiplicative. An *additive* seasonal factor increases (decreases) the variable by the same *dollar amount* every January (or first quarter), with the amount denominated in units of the variable. In contrast, a *multiplicative* seasonal factor increases (decreases) the variable by the same *percentage* every January (or first quarter).

The primary concern here is that some economic data are made available in seasonally adjusted (SA) form. For flow series such as personal income, this concept is often indicated as seasonally adjusted at an annual rate (SAAR). Other economic data that may be used in a model of household or firm behavior are denoted as not seasonally adjusted (NSA). The two types of data should not be mixed in the same model: for instance, an NSA response variable versus a set of regressors, each of which is SA. Such a regression will contain seasonality in its residuals and will fail any test for independence of the errors that considers AR(4) models (for quarterly data) or AR(12) models (for monthly data). If we recognize that there are seasonal components in one or more data series, we should use some method of seasonal adjustment unless all series in the model are NSA.

Deseasonalization with either the additive or multiplicative form of the seasonal model requires that a set of *seasonal dummies* be created by defining the elements of the set with statements like

```
. generate mseas1 = (month(dofm(datevar)) == 1)
. generate qseas1 = (quarter(dofq(datevar)) == 1)
```

for data that have been identified as monthly or quarterly data to Stata, respectively, by `tsset datevar`. The variable `mseas1` will be 1 in January and 0 in other months; `qseas1` will be 1 in the first quarter of each year and 0 otherwise. The `month()` and `quarter()` functions, as well as the more arcane `dofm()` and `dofq`, are described in [D] **functions** under the headings *Date functions* and *Time-series functions*. The set of seasonal dummies is easily constructed with a `forvalues` loop, as shown in the example below.

To remove an additive seasonal factor from the data, we regress the series on a constant term and all but one of the seasonal dummies

```
. regress sales mseas*
. regress taxrev qseas*
```

for monthly or quarterly data, respectively. After the regression, we use `predict` with the `residuals` option to produce the deseasonalized series. Naturally, this series will have a mean of zero, since it comes from a regression with a constant term; usually it is "rebenched" to the original series' mean, as I illustrate below. We use the `turksales` dataset, which contains quarterly turkey sales data for 1990q1–1994q4, as described by `summarize`:

```
. use http://www.stata-press.com/data/imeus/turksales, clear
. summarize sales
    Variable |       Obs        Mean    Std. Dev.       Min        Max
-------------+--------------------------------------------------------
       sales |        40    105.6178    4.056961   97.84603   112.9617
```

We first find the mean of the quarterly `sales` series and generate three quarterly dummy variables:

```
. summarize sales, meanonly
. local mu = r(mean)
. forvalues i=1/3 {
  2.          generate qseas`i' = (quarter(dofq(t)) == `i')
  3. }
```

We then run the regression to evaluate the importance of seasonal factors:

```
. regress sales qseas*

      Source |       SS       df       MS              Number of obs =      40
-------------+------------------------------           F(  3,    36) =    4.03
       Model |  161.370376     3  53.7901254           Prob > F      =  0.0143
    Residual |   480.52796    36  13.3479989           R-squared     =  0.2514
-------------+------------------------------           Adj R-squared =  0.1890
       Total |  641.898336    39  16.4589317           Root MSE      =  3.6535

------------------------------------------------------------------------------
       sales |      Coef.   Std. Err.      t    P>|t|     [95% Conf. Interval]
-------------+----------------------------------------------------------------
      qseas1 |  -5.232047   1.633891    -3.20   0.003    -8.545731   -1.918362
      qseas2 |  -2.842753   1.633891    -1.74   0.090    -6.156437    .4709317
      qseas3 |  -.8969368   1.633891    -0.55   0.586    -4.210621    2.416748
       _cons |   107.8608   1.155335    93.36   0.000     105.5177    110.2039
------------------------------------------------------------------------------
```

The ANOVA F statistic from the regression indicates that seasonal factors explain much of the variation in `sales`. To generate the deseasonalized series, we use `predict` to recover the residuals and add the original mean of the series to them:

```
. predict double salesSA, residual
. replace salesSA = salesSA + `mu'
(40 real changes made)
```

We can now compare the two series:

```
. summarize sales salesSA

    Variable |       Obs        Mean    Std. Dev.       Min        Max
-------------+--------------------------------------------------------
       sales |        40    105.6178    4.056961   97.84603   112.9617
     salesSA |        40    105.6178    3.510161   97.49429   111.9563

. label var salesSA "sales, seasonally adjusted"
. tsline sales salesSA, lpattern(solid dash)
```

The deseasonalized series has a smaller standard deviation than the original as the fraction of the variation because the seasonality has been removed. This effect is apparent in the graph of the original series and the smoother deseasonalized series in figure 7.1.

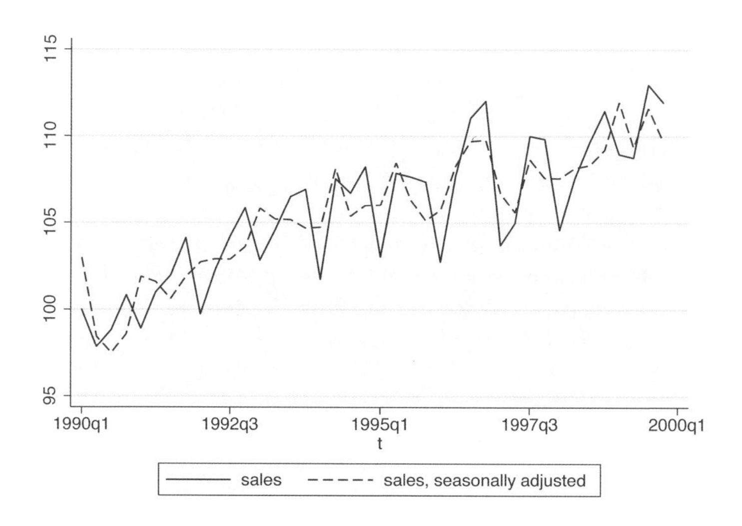

Figure 7.1: Seasonal adjustment of time series

We may also want to remove the trend component from a series. To remove a linear trend, we merely regress the series on a time trend. For a multiplicative (geometric, or constant growth rate) trend, we regress the logarithm of the series on the time trend. In either case, the residuals from that regression represent the detrended series.[10] We may remove both the trend and seasonal components from the series in the same regression, as illustrated here:

```
. regress sales qseas* t

      Source |       SS       df       MS              Number of obs =      40
-------------+------------------------------           F(  4,    35) =   54.23
       Model |  552.710487     4  138.177622           Prob > F      =  0.0000
    Residual |  89.1878487    35  2.54822425           R-squared     =  0.8611
-------------+------------------------------           Adj R-squared =  0.8452
       Total |  641.898336    39  16.4589317           Root MSE      =  1.5963

------------------------------------------------------------------------------
       sales |      Coef.   Std. Err.      t    P>|t|     [95% Conf. Interval]
-------------+----------------------------------------------------------------
      qseas1 |  -4.415311   .7169299    -6.16   0.000    -5.870756   -2.959866
      qseas2 |  -2.298262   .7152449    -3.21   0.003    -3.750287    -.846238
      qseas3 |  -.6246916   .7142321    -0.87   0.388     -2.07466    .8252766
           t |   .2722452   .0219686    12.39   0.000     .2276466    .3168438
       _cons |   69.47421   3.138432    22.14   0.000     63.10285    75.84556
------------------------------------------------------------------------------

. test qseas1 qseas2 qseas3

 ( 1)  qseas1 = 0
 ( 2)  qseas2 = 0
 ( 3)  qseas3 = 0

       F(  3,    35) =   15.17
            Prob > F =    0.0000
```

10. For more detail, see Davidson and MacKinnon (2004, 72–73).

```
. predict double salesSADT, residual
. replace salesSADT = salesSADT + `mu'
(40 real changes made)
. label var salesSADT "sales, detrended and SA"
. tsline sales salesSADT, lpattern(solid dash) yline(`mu')
```

The trend t is highly significant in these data. A joint F test for the seasonal factors shows that they are also significant beyond a trend term. The detrended and deseasonalized series, rebenched to the mean of the original series (shown by the horizontal line), is displayed in figure 7.2.

Figure 7.2: Seasonal adjustment and detrending of time series

Several other methods of seasonal adjustment and detrending for time-series data are implemented in Stata under the heading tssmooth; see in particular [TS] **tssmooth shwinters**. As Davidson and MacKinnon (2004, 584–585) point out, the seasonal adjustment methods used by government statistics bureaus can be approximated by a *linear filter*, or τ-term moving average. In this context, tssmooth ma or the egen function filter() available in the egenmore package from ssc may be helpful.

If you are interested in filtering time-series data to identify business cycles, see the author's bking (Baxter–King bandpass filter) and hprescott (Hodrick–Prescott filter) routines, both available from the SSC archive (see [R] **ssc**).

7.4 Testing for structural stability and structural change

Indicator variables are used to test for structural stability in a regression function in which we specify a priori the location of the possible structural breakpoints. In (7.1) and (7.2), we found that the intercept of the regression differed significantly between union and nonunion cohorts but that one slope parameter for `tenure` was adequate. In further testing, we found that the σ_u^2 differed significantly between these two cohorts in the sample. If we doubt structural stability—for instance, an industry-level regression over a set of natural resource–intensive and manufacturing industries—we may use indicator variables to identify groups within the sample and test whether the intercept and slope parameters are stable over these groups. In household data, a function predicting food expenditures might not be stable over families with different numbers of children. Merely including the number of children as a regressor might not be adequate if this relationship is nonlinear in the number of mouths to feed.

Structural instability over cohorts of the sample need not be confined to shifts in the intercept of the relationship. A structural shift may not be present in the intercept, but it may be an important factor for one or more slope parameters. If we question structural stability, we should formulate a general model in which all regressors (including the constant term) are interacted with cohort indicators and test down where coefficients appear to be stable across cohorts.

Section 6.2.2 considers the possibility of heteroskedasticity over groups or cohorts in the data that may have been pooled. Beyond the possibility that σ_u^2 may differ across groups, we should be concerned with the stability of the regression function's coefficients over the groups. Whereas groupwise heteroskedasticity may be readily diagnosed and corrected, improperly specifying the regression function to be constant over groups of the sample will be far more damaging, rendering regression estimates biased and inconsistent. For instance, if those firms who are subject to liquidity constraints (because of poor credit history or inadequate collateral) behave differently from firms that have ready access to financial markets, combining both sets of firms in the same regression will yield a regression function that is a mix of the two groups' dissimilar behavior. Such a regression is unlikely to provide reasonable predictions for firms in *either* group. Placing the two groups in the same regression, with indicator variables used to allow for potential differences in structure between their coefficient vectors, is more sensible. That approach will allow those differences to be estimated and tested for significance.

7.4.1 Constraints of continuity and differentiability

It is easy to determine that the regression function should be allowed to exhibit various structural breaks. Tests may show that a representative worker's earnings–tenure profile should be allowed to have different slopes over different ranges of job tenure. You could accomplish this configuration by using a polynomial in tenure, but doing so may introduce unacceptable behavior (for instance, with `tenure` and `tenure`^{2}, there must be some tenure at which the profile turns downward, predicting that wages will fall with each additional year on the job). If we use the interaction terms with no further constraints on the regression function, that piecewise linear function exhibits discontinuities over the groups identified by the interaction terms (e.g., the age categories in the sample). I illustrate, returning to the NLSW dataset and defining four job tenure categories: fewer than 2 years, 2–7 years, 7–12 years, and more than 12 years:

```
. use http://www.stata-press.com/data/imeus/nlsw88, clear
(NLSW, 1988 extract)
. generate lwage = log(wage)
. generate Ten2 = tenure<=2
. generate Ten7 = !Ten2 & tenure<=7
. generate Ten12 = !Ten2 & !Ten7 & tenure<=12
. generate Ten25 = !Ten2 & !Ten7 & !Ten12 & tenure<.
```

We now generate interactions of `tenure` with each of the tenure categories, run the regression on the categories and interaction terms,[11] and generate predicted values:

```
. generate tTen2 = tenure*Ten2
(15 missing values generated)
. generate tTen7 = tenure*Ten7
(15 missing values generated)
. generate tTen12 = tenure*Ten12
(15 missing values generated)
. generate tTen25 = tenure*Ten25
(15 missing values generated)
. regress lwage Ten* tTen*, nocons hascons
      Source |       SS       df       MS              Number of obs =    2231
-------------+------------------------------           F(  7,  2223) =   37.12
       Model |  76.6387069     7  10.9483867           Prob > F      =  0.0000
    Residual |  655.578361  2223  .294907045           R-squared     =  0.1047
-------------+------------------------------           Adj R-squared =  0.1018
       Total |  732.217068  2230  .328348461           Root MSE      =  .54305

------------------------------------------------------------------------------
       lwage |      Coef.   Std. Err.      t    P>|t|     [95% Conf. Interval]
-------------+----------------------------------------------------------------
        Ten2 |    1.55662   .0383259    40.62   0.000     1.481462    1.631778
        Ten7 |   1.708728    .060084    28.44   0.000     1.590901    1.826554
       Ten12 |   1.870808   .1877798     9.96   0.000     1.502566     2.23905
       Ten25 |   1.751961   .1691799    10.36   0.000     1.420194    2.083728
       tTen2 |   .0897426   .0331563     2.71   0.007     .0247221    .1547631
       tTen7 |   .0434089   .0140739     3.08   0.002     .0158095    .0710083
      tTen12 |   .0154208    .019786     0.78   0.436    -.0233801    .0542218
      tTen25 |   .0238014   .0102917     2.31   0.021     .0036191    .0439837
------------------------------------------------------------------------------
. predict double lwagehat
(option xb assumed; fitted values)
(15 missing values generated)
. label var lwagehat "Predicted log(wage)"
. sort tenure
```

11. We exclude the constant term so that all four tenure dummies can be included. The option `hascons` indicates to Stata that we have the equivalent of a constant term in the four tenure dummies `Ten2-Ten25`.

The predicted values for each segment of the wage–tenure profile can now be graphed:

```
. twoway (line lwagehat tenure if tenure<=2)
> (line lwagehat tenure if tenure>2 & tenure<=7)
> (line lwagehat tenure if tenure>7 & tenure<=12)
> (line lwagehat tenure if tenure>12 & tenure<.), legend(off)
```

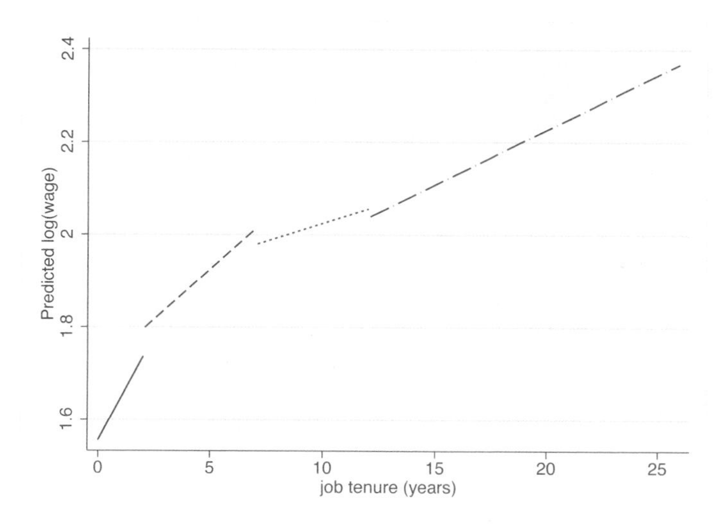

Figure 7.3: Piecewise wage–tenure profile

As we see in figure 7.3, this piecewise function allows for a different slope and intercept for each of the four ranges of job tenure, but it is not continuous. For instance, the estimates predict that at the point of 2 years' tenure, the average worker's log wage will abruptly jump from 1.73 per hour to 1.80 per hour and then decline from 2.01 per hour to 1.98 per hour at the point of 7 years' tenure.

We may want to allow such a profile to be flexible over different ranges of job tenure but force the resulting function to be *piecewise continuous* by using a *linear spline*: a mathematical function that enforces continuity between the adjacent segments. Spline functions are characterized by their *degree*. A linear spline is degree 1, a quadratic spline is degree 2, and so on. A linear spline will be continuous but not differentiable at the *knot points*: those points on the profile that define segments of the function. A quadratic spline is continuous and once differentiable. Since the function has constant first derivatives on both sides of the knot, there will be no kinks in the curve. Likewise, a cubic spline will be continuous and twice differentiable, and so on.

I illustrate using a linear spline to generate a piecewise continuous earnings–tenure profile. Stata's `mkspline` (see [R] **mkspline**) command automates this process for linear splines. Higher-order splines must be defined algebraically or by using a user-written routine. We can use the **mkspline** command to generate a spline with knots placed at specified points or a spline with equally spaced knots.[12] Here we use the former syntax:

mkspline *newvar*$_1$ *#1* [*newvar*$_2$ *#2* [...]] *newvar*$_k$ = *oldvar* [*if*] [*in*]

where k *newvars* are specified to define a linear spline of *varname* with $(k-1)$ knots, placed at the values *#1*, *#2*, ..., *#*$(k-1)$ of the splined variable. The resulting set of *newvarname* variables may then be used as regressors.

In the piecewise regression above, we estimated four slopes and four intercepts for a total of eight regression parameters. Fitting this model as a linear spline places constraints on the parameters. At each of the three knot points (2, 7, and 12 years) along the tenure axis, $\gamma + \delta$ **tenure** must be equal from the left and right. Simple algebra shows that each of the three knot points imposes one constraint on the parameter vector. The piecewise linear regression using a linear spline will have five parameters rather than eight:

```
. mkspline sTen2 2 sTen7 7 sTen12 12 sTen25 = tenure
. regress lwage sTen*
      Source |       SS       df       MS              Number of obs =    2231
-------------+------------------------------           F(  4,   2226) =   64.55
       Model |  76.1035947     4  19.0258987           Prob > F      =  0.0000
    Residual |  656.113473  2226  .294749988           R-squared     =  0.1039
-------------+------------------------------           Adj R-squared =  0.1023
       Total |  732.217068  2230  .328348461           Root MSE      =  .54291

------------------------------------------------------------------------------
       lwage |      Coef.   Std. Err.      t    P>|t|     [95% Conf. Interval]
-------------+----------------------------------------------------------------
       sTen2 |   .1173168   .0248619     4.72   0.000     .0685619    .1660716
       sTen7 |   .0471177    .009448     4.99   0.000       .02859    .0656455
      sTen12 |   .0055041   .0111226     0.49   0.621    -.0163076    .0273158
      sTen25 |   .0237767   .0083618     2.84   0.005      .007379    .0401744
       _cons |   1.539985   .0359605    42.82   0.000     1.469465    1.610505
------------------------------------------------------------------------------

. predict double lwageSpline
(option xb assumed; fitted values)
(15 missing values generated)
. label var lwageSpline "Predicted log(wage), splined"
. twoway line lwageSpline tenure
```

The result of the piecewise linear estimation, displayed in figure 7.4, is a continuous earnings–tenure profile with kinks at the three knot points. From an economic standpoint, the continuity is highly desirable. The model's earnings predictions for tenures of 1.9, 2.0, and 2.1 years will now be smooth, without implausible jumps at the knot points.

12. The alternative syntax can also place knots at equally spaced percentiles of the variable with the **pctile** option.

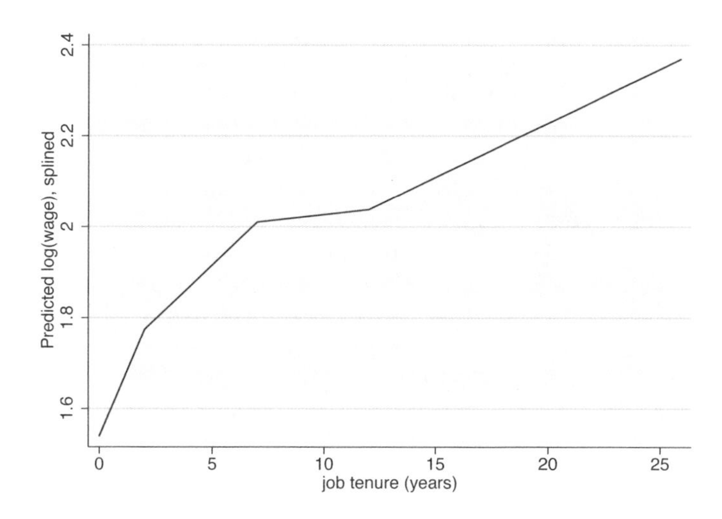

Figure 7.4: Piecewise linear wage–tenure profile

7.4.2 Structural change in a time-series model

With time-series data, a concern for structural stability is usually termed a test for *structural change*. We can allow for different slopes or intercepts for different periods in a time-series regression (e.g., allowing for a household consumption function to shift downward during wartime). Just as in a cross-sectional context, we should consider that both intercept and slope parameters may differ over various periods. Older econometrics texts often discuss this difference in terms of a *Chow test* and provide formulas that manipulate error sums of squares from regressions run over different periods to generate a test statistic. This step is not necessary since the Chow test is nothing more than the F test that all *regime dummy* coefficients are jointly zero. For example,

$$y_t = \beta_1 + \beta_2 x_{2t} + \beta_3 x_{3t} + \beta_4 gw_t + \beta_5(x_{2t} \times gw_t) + \beta_6(x_{3t} \times gw_t) + u_t$$

where $gw_t = 1$ during calendar quarters of the Gulf War. The joint test $\beta_4 = \beta_5 = \beta_6 = 0$ would test that this regression function is stable during the two regimes. We may also consider the intermediate cases: for instance, the coefficient on x_2 may be stable over peacetime and wartime, but the coefficient on x_3 (or the intercept) may not. We can easily handle more than two regimes by merely adding regime dummies for each regime and their interactions with the other regressors. We should also be concerned about the realistic possibility that the σ_u^2 has changed over regimes. We may deal with this possibility by computing robust standard errors for the regression with regime dummies, but we might want to estimate the differing variances for each regime, as this is a sort of groupwise heteroskedasticity where the groups are time-series regimes.

Sometimes a regime may be too short to set up the fully interacted model since it requires that the regression model be fitted over the observations of that regime. Since the model above contains three parameters per regime, it cannot be estimated over a regime with 4 or fewer observations. This problem often arises at the end of a time series. We may want to test the hypothesis that the last T_2 observations were generated by the same regime as the previous T_1 observations. Then we construct an F test by estimating the regression over all $T = T_1 + T_2$ observations and then estimating it again over the first T_1 observations. The sum of squared residuals ($\Sigma \widehat{u}_t^2$) for the full sample will exceed that from the first T_1 observations unless the regression fits perfectly over the additional T_2 data points. If the fit is very poor over the additional T_2 data points, we can reject the null of model stability over $[T_1, T_2]$. This Chow predictive F test has T_2 degrees of freedom in the numerator:

$$F(T_2, T_1 - k) = \frac{(\widehat{\mathbf{u}}_T'\widehat{\mathbf{u}}_T - \widehat{\mathbf{u}}_{T1}'\widehat{\mathbf{u}}_{T1})/T_2}{(\widehat{\mathbf{u}}_{T1}'\widehat{\mathbf{u}}_{T1})/(T_1 - k)}$$

where $\widehat{\mathbf{u}}_T$ is the residual vector from the full sample. Following a regression, the error sum of squares may be accessed as `e(rss)` (see [P] **ereturn**).

These dummy variable methods are useful when the timing of one or more structural breaks is known a priori from the economic history of the period. However, we often are not sure whether (and if so, when) a relationship may have undergone a structural shift. This uncertainty is particularly problematic when a change may be a gradual process rather than an abrupt and discernible break. Several tests have been devised to evaluate the likelihood that a change has taken place, and if so, when that break may have occurred. Those techniques are beyond the scope of this text. See Bai and Perron (2003).

Exercises

1. Using the dataset of section 7.1.2, test that `race` explains much of the variation in `lwage`.
2. Consider the model used in section 7.2 to search for evidence of statistical discrimination. Test a model that includes interactions of the factors `race` and `tenure`.
3. Consider the model used in section 7.3 to seasonally adjust turkey sales data. Fit a multiplicative seasonal model to these data. Is an additive seasonal factor or a multiplicative seasonal factor preferred?
4. Consider the model used in section 7.3 to seasonally adjust turkey sales data. Apply Holt–Winters seasonal smoothing (`tssmooth shwinters`), and compare the resulting series to that produced by seasonal adjustment with indicator variables.
5. Consider the model used in section 7.4.1. Use the alternate syntax of `mkspline` to generate three equally placed knots and estimate the equation. Repeat the exercise, using the `pctile` option. How sensitive are the results to the choice of linear spline technique?

8 Instrumental-variables estimators

8.1 Introduction

The zero-conditional-mean assumption presented in section 4.2 must hold for us to use linear regression. There are three common instances where this assumption may be violated in economic research: endogeneity (simultaneous determination of response variable and regressors), omitted-variable bias, and errors in variables (measurement error in the regressors). Although these problems arise for different reasons in microeconomic models, the solution to each is the same econometric tool: the *instrumental-variables* (IV) estimator, described in this chapter. The most common problem, endogeneity, is presented in the next section. The other two problems are discussed in chapter appendices. The following sections discuss the IV and two-stage least-squares (2SLS) estimators, identification and tests of overidentifying restrictions, and the generalization to generalized method-of-moments (GMM) estimators. The last three sections of the chapter consider testing for heteroskedasticity in the IV context, testing the relevance of instruments, and testing for endogeneity.

A variable is *endogenous* if it is correlated with the disturbance. In the model

$$y = \beta_1 x_1 + \beta_2 x_2 + \cdots + \beta_k x_k + u$$

x_j is endogenous if $\text{Cov}[x_j, u] \neq 0$. x_j is exogenous if $\text{Cov}[x_j, u] = 0$. The OLS estimator will be consistent only if $\text{Cov}[x_j, u] = 0$, $j = 1, 2, \ldots, k$. This zero-covariance assumption and our convention that x_1 is a constant imply that $E[u] = 0$. Following Wooldridge (2002, 2006), we use the zero-conditional-mean assumption

$$E[u|x_1, x_2, \ldots, x_k] = 0$$

which is sufficient for the zero-covariance condition.

Although the rest of this chapter uses economic intuition to determine when a variable is likely to be endogenous in an empirical study, it is the above definition of endogeneity that matters for empirical work.

8.2 Endogeneity in economic relationships

Economists often model behavior as simultaneous-equations systems in which economically endogenous variables are determined by each other and some additional economically exogenous variables. The simultaneity gives rise to empirical models with variables that do not satisfy the zero-conditional-mean assumption. Consider the textbook supply-and-demand paradigm. We commonly write

$$q^d = \beta_1 + \beta_2 p + \beta_3 inc \tag{8.1}$$

to indicate that the quantity demanded of a good (q^d) depends on its price (p) and the level of purchasers' income (inc). When $\beta_1 > 0$, $\beta_2 < 0$, and $\beta_3 > 0$, the demand curve in $[p, q]$ space slopes downward, and for any given price the quantity demanded will rise for a higher level of purchasers' income.

If this equation reflected an individual's demand function, we might argue that the individual is a price taker who pays the posted price if she chooses to purchase the good and has a fixed income at her disposal on shopping day. But we often lack *microdata*, or household-level data, for the estimation of this relationship for a given good. Rather, we have data generated by the market for the good. The observations on p and q are equilibrium prices and quantities in successive trading periods.

If we append an error term, u, to (8.1) and estimate OLS from these $[p, q]$ pairs, the estimates will be inconsistent. It does not matter whether the model is specified as above with q as the response variable or in inverse form with p as the response variable. In either case, the regressor is endogenous. Simple algebra shows that the regressor must be correlated with the error term, violating the zero-conditional-mean assumption. In (8.1), a shock to the demand curve must alter both the equilibrium price and quantity in the market. By definition, the shock u is correlated with p.

How can we use these market data to estimate a demand curve for the product? We must specify an *instrument* for p that is uncorrelated with u but highly correlated with p. In an economic model, this is termed the *identification problem*: what will allow us to *identify* or trace out the demand curve? Consider the other side of the market. Any factor in the supply function that does not appear in the demand function will be a valid instrument. If we are modeling the demand for an agricultural commodity, a factor like rainfall or temperature would suffice. Those factors are determined outside the economic model but may have an important effect on the yield of the commodity and thus the quantity that the grower will bring to market. In the economic model, these factors will appear in the *reduced-form* equations for both q and p: the algebraic solution to the simultaneous system.

To derive consistent estimates of (8.1), we must find an IV that satisfies two properties: the instrument z must be uncorrelated with u but must be highly correlated with p.[1] A variable that meets those two conditions is an IV or *instrument* for p that deals with the correlation of p and the error term. Because we cannot observe u, we cannot directly test the assumption of zero correlation between z and u, which is known as an orthogonality assumption. We will see that in the presence of multiple instruments, such a test can be constructed. But we can readily test the second assumption and should always do so by regressing the included regressor p on the instrument z:

1. The meaning of "highly correlated" is the subject of section 8.10.

$$p_i = \pi_1 + \pi_2 z_i + \zeta_i \tag{8.2}$$

If we fail to reject the null hypothesis H_0: $\pi_2 = 0$, we conclude that z is not a valid instrument. Unfortunately, rejecting the null of irrelevance is not sufficient to imply that the instrument is not "weak", as discussed in section 8.10.[2] There is no unique choice of an instrument here. We discuss below how we can construct an instrument if more than one is available.

If we decide that we have a valid instrument, how can we use it? Return to (8.1), and write it in matrix form in terms of $\mathbf{y}$ and $\mathbf{X}$

$$\mathbf{y} = \mathbf{X}\boldsymbol{\beta} + \mathbf{u}$$

where $\boldsymbol{\beta}$ is the vector of coefficients $(\beta_1, \beta_2, \beta_3)'$ and $\mathbf{X}$ is $N \times k$. Define a matrix $\mathbf{Z}$ of the same dimension as $\mathbf{X}$ in which the endogenous regressor—p in our example above—is replaced by z. Then

$$\mathbf{Z}'\mathbf{y} = \mathbf{Z}'\mathbf{X}\boldsymbol{\beta} + \mathbf{Z}'\mathbf{u}$$

The assumption that $\mathbf{Z}$ is unrelated to $\mathbf{u}$ implies that $1/N(\mathbf{Z}'\mathbf{u})$ goes to zero in probability as N becomes large. Thus we may define the estimator $\widehat{\boldsymbol{\beta}}_{\text{IV}}$ from

$$\begin{aligned} \mathbf{Z}'\mathbf{y} &= \mathbf{Z}'\mathbf{X}\,\widehat{\boldsymbol{\beta}}_{\text{IV}} \\ \widehat{\boldsymbol{\beta}}_{\text{IV}} &= (\mathbf{Z}'\mathbf{X})^{-1}\mathbf{Z}'\mathbf{y} \end{aligned} \tag{8.3}$$

We may also use the zero-conditional-mean assumption to define a *method-of-moments* estimator of the IV model. In the linear regression model presented in section 4.2.1, the zero-conditional-mean assumption held for each of the k variables in $\mathbf{X}$, giving rise to a set of k moment conditions. In the IV model, we cannot assume that each $\mathbf{X}$ satisfies the zero-conditional-mean assumption: an endogenous x does not. But we can define a matrix $\mathbf{Z}$ as above in which each endogenous regressor will be replaced by its instrument, yielding a method-of-moments estimator for β:

$$\begin{aligned} \mathbf{Z}'\mathbf{u} &= 0 \\ \mathbf{Z}'(\mathbf{y} - \mathbf{X}\boldsymbol{\beta}) &= 0 \end{aligned} \tag{8.4}$$

We may then substitute calculated moments from our sample of data into the expression and replace the unknown coefficients $\boldsymbol{\beta}$ with estimated values $\widehat{\boldsymbol{\beta}}$ in (8.4) to derive

$$\begin{aligned} \mathbf{Z}'\mathbf{y} - \mathbf{Z}'\mathbf{X}\widehat{\boldsymbol{\beta}}_{\text{IV}} &= 0 \\ \widehat{\boldsymbol{\beta}}_{\text{IV}} &= (\mathbf{Z}'\mathbf{X})^{-1}\mathbf{Z}'\mathbf{y} \end{aligned}$$

The IV estimator has an interesting special case. If the zero-conditional-mean assumption holds, each explanatory variable can serve as its own instrument, $\mathbf{X} = \mathbf{Z}$, and

2. Bound, Jaeger, and Baker (1995) proposed the rule of thumb that this F statistic must be at least 10. In more recent work, table 1 of Stock, Wright, and Yogo (2002) provides critical values that depend on the number of instruments.

the IV estimator reduces to the OLS estimator. Thus OLS is a special case of IV that is appropriate when the zero-conditional-mean assumption is satisfied. When that assumption cannot be made, the IV estimator is consistent and has a large-sample normal distribution as long as the two key assumptions about the instrument's properties are satisfied. However, the IV estimator is not an unbiased estimator, and in small samples its bias may be substantial.

8.3 2SLS

Consider the case where we have one endogenous regressor and more than one potential instrument. In (8.1), we might have two candidate instruments: z_1 and z_2. We could apply the IV estimator of (8.3) with z_1 entering $\mathbf{z}$, and generate an estimate of $\widehat{\boldsymbol{\beta}}_{\mathrm{IV}}$. If we repeated the process with z_2 entering $\mathbf{z}$, we would generate another $\widehat{\boldsymbol{\beta}}_{\mathrm{IV}}$ estimate, and those two estimates would differ.

Obtaining the simple IV estimator of (8.3) for each candidate instrument raises the question of how we could combine them. An alternative approach, 2SLS, combines multiple instruments into one optimal instrument, which can then be used in the simple IV estimator. This optimal combination, conceptually, involves running a regression. Consider the auxiliary regression of (8.2), which we use to check that a candidate $\mathbf{z}$ is reasonably well correlated with the regressor that it is instrumenting. Merely extend that regression model,

$$p_i = \pi_1 + \pi_2 z_{i1} + \pi_3 z_{i2} + \omega_i$$

and generate the instrument as the predicted values of this equation: $\widehat{p}$. Given the mechanics of least squares, $\widehat{p}$ is an optimal linear combination of the information in z_1 and z_2. We may then estimate the parameters of (8.3), using the IV estimator with $\widehat{p}$ as a column of $\mathbf{Z}$.

2SLS is nothing more than the IV estimator with a decision rule that reduces the number of instruments to the exact number needed to estimate the equation and fill in the $\mathbf{Z}$ matrix. To clarify the mechanics, define matrix $\mathbf{Z}$ of dimension $N \times \ell$, $\ell \geq k$, of instruments. Then the first-stage regressions define the instruments as

$$\widehat{\mathbf{X}} = \mathbf{Z}(\mathbf{Z}'\mathbf{Z})^{-1}\mathbf{Z}'\mathbf{X} \tag{8.5}$$

Denote the projection matrix $\mathbf{Z}(\mathbf{Z}'\mathbf{Z})^{-1}\mathbf{Z}'$ as $\mathbf{P}_Z$. Then from (8.3),

$$\begin{aligned}
\widehat{\boldsymbol{\beta}}_{\mathrm{2SLS}} &= (\widehat{\mathbf{X}}'\mathbf{X})^{-1}\widehat{\mathbf{X}}'\mathbf{y} \\
&= \{\mathbf{X}'\mathbf{Z}(\mathbf{Z}'\mathbf{Z})^{-1}\mathbf{Z}'\mathbf{X}\}^{-1}\{\mathbf{X}'\mathbf{Z}(\mathbf{Z}'\mathbf{Z})^{-1}\mathbf{Z}'\mathbf{y}\} \\
&= (\mathbf{X}'\mathbf{P}_Z\mathbf{X})^{-1}\mathbf{X}'\mathbf{P}_Z\mathbf{y}
\end{aligned} \tag{8.6}$$

where the "two-stage" estimator can be calculated in one computation using the data on $\mathbf{X}$, $\mathbf{Z}$, and $\mathbf{y}$. When $\ell = k$, 2SLS reduces to IV, so the 2SLS formulas presented below also cover the IV estimator.

Assuming i.i.d. disturbances, a consistent large-sample estimator of the VCE of the 2SLS estimator is

$$\text{Var}[\widehat{\boldsymbol{\beta}}_{\text{2SLS}}] = \widehat{\sigma}^2\{\mathbf{X}'\mathbf{Z}(\mathbf{Z}'\mathbf{Z})^{-1}\mathbf{Z}'\mathbf{X}\}^{-1} = \widehat{\sigma}^2(\mathbf{X}'\mathbf{P}_Z\mathbf{X})^{-1} \tag{8.7}$$

where $\widehat{\sigma}^2$ is computed as

$$\widehat{\sigma}^2 = \frac{\widehat{\mathbf{u}}'\widehat{\mathbf{u}}}{N}$$

calculated from the 2SLS residuals

$$\widehat{\mathbf{u}} = \mathbf{y} - \mathbf{X}\widehat{\boldsymbol{\beta}}_{\text{2SLS}}$$

defined by the *original* regressors and the estimated 2SLS coefficients.[3]

The point of using the 2SLS estimator is the consistent estimation of $\widehat{\boldsymbol{\beta}}_{\text{2SLS}}$ in a model containing response variable $\mathbf{y}$ and regressors $\mathbf{X}$, some of which are correlated with the disturbance process $\mathbf{u}$. The predictions of that model involve the original regressors $\mathbf{X}$, not the instruments $\widehat{\mathbf{X}}$. Although from a pedagogical standpoint we speak of 2SLS as a sequence of first-stage and second-stage regressions, we should never perform those two steps by hand. If we did so, we would generate predicted values $\{\widehat{\mathbf{X}}\}$ from first-stage regressions of endogenous regressors on instruments and then run the second-stage OLS regression using those predicted values. Why should we avoid this? Because the second stage will yield the incorrect residuals,

$$\widehat{\mathbf{u}}_i = \mathbf{y}_i - \widehat{\mathbf{X}}\widehat{\boldsymbol{\beta}}_{\text{2SLS}} \tag{8.8}$$

rather than the correct residuals,

$$\widehat{\mathbf{u}}_i = \mathbf{y}_i - \mathbf{X}\widehat{\boldsymbol{\beta}}_{\text{2SLS}}$$

which would be calculated by `predict` after a 2SLS estimation. Statistics computed from the incorrect residuals, such as an estimate of σ^2 and the estimated standard error for each $\widehat{\boldsymbol{\beta}}_{\text{2SLS}}$ in (8.7), will be inconsistent since the $\widehat{X}$ variables are not the true explanatory variables (see Davidson and MacKinnon 2004, 324). Using Stata's 2SLS command `ivreg` avoids these problems, as I now discuss.

8.4 The ivreg command

The `ivreg` command has the following partial syntax:

`ivreg` *depvar* [*varlist1*] `(`*varlist2* `=` *instlist*`)` [*if*] [*in*] [`,` *options*]

where *depvar* is the response variable, *varlist2* contains the endogenous regressors, *instlist* contains the excluded instruments, and the optional *varlist1* contains any exogenous regressors included in the equation. In our example from the demand for an agricultural commodity in (8.1), we could specify

3. Some packages, including Stata's `ivreg`, include a degrees-of-freedom correction to the estimate of $\widehat{\sigma}^2$ by replacing N with $N-k$. This correction is unnecessary since the estimate of $\widehat{\sigma}^2$ would not be unbiased anyway (Greene 2000, 373).

```
. ivreg q inc (p = rainfall temperature)
```

to indicate that `q` is to be regressed on `inc` and `p` with `rainfall` and `temperature` as excluded instruments. Stata reports that the instruments used in estimation include `inc rainfall temperature`, considering that `inc` is serving as its own instrument. Just as with `regress`, a constant term is included in the equation by default. If a constant appears in the equation, it also implicitly appears in the instrument list used to specify $\mathbf{Z}$, the matrix of instruments in the first-stage regression. The first-stage regression (one for each endogenous regressor) may be displayed with the `first` option.

In a situation with multiple endogenous regressors such as

```
. ivreg y x2 (x3 x4 = za zb zc zd)
```

novice users of instrumental variables often ask, "How do I tell Stata that I want to use `za, zb` as instruments for `x3`, and `zc, zd` as instruments for `x4`?" You cannot, but not because of any limitation of Stata's `ivreg` command. The theory of 2SLS estimation does not allow such designations. *All* instruments—included and excluded—must be used as regressors in *all* first-stage regressions. Here both `x3` and `x4` are regressed on `z`: `x2 za zb zc zd` and a constant term to form the $\widehat{\mathbf{X}}$ matrix.

We noted above that summary statistics such as `Root MSE` should be calculated from the appropriate residuals using the original regressors in $\mathbf{X}$. If we compare the `Root MSE` from `ivreg` and the `Root MSE` from `regress` on the same model, the former will inevitably be larger. It appears that taking account of the endogeneity of one or more regressors has cost us something in goodness of fit: least squares is least squares. The minimum sum of squared errors from a model including $\{y\ \mathbf{x}\}$ is by definition that computed by `regress`. The 2SLS estimator calculated by `ivreg` is a least-squares estimator, but the criterion minimized involves the improper residuals of (8.8). The 2SLS method is fitting $\mathbf{y}$ to $\widehat{\mathbf{X}}$ by least squares to generate consistent estimates $\widehat{\boldsymbol{\beta}}_{2SLS}$, thereby minimizing sum of squared errors with respect to $\widehat{\mathbf{X}}$. As long as $\widehat{\mathbf{X}} \neq \mathbf{X}$, those $\widehat{\boldsymbol{\beta}}_{2SLS}$ estimates cannot also minimize the sum of squared errors calculated by [R] **regress**.

Before I present an example of `ivreg`, we must define *identification* of a structural equation.

8.5 Identification and tests of overidentifying restrictions

The parameters in an equation are said to be identified when we have sufficient valid instruments so that the 2SLS estimator produces unique estimates. In econometrics, we say that an equation is identified, if the parameters in that equation are identified.[4] Equation (8.6) shows that $\widehat{\boldsymbol{\beta}}_{2SLS}$ is unique only if $(\mathbf{Z}'\mathbf{Z})$ is a nonsingular $\ell \times \ell$ matrix and $(\mathbf{Z}'\mathbf{X})$ has full rank k. As long as the instruments are linearly independent, $(\mathbf{Z}'\mathbf{Z})$ will be a nonsingular $\ell \times \ell$ matrix, so this requirement is usually taken for granted. That

4. This terminology comes from literature on estimating the structural parameters in systems of simultaneous equations.

$(\mathbf{Z}'\mathbf{X})$ be of rank k is known as the *rank condition*. That $\ell \geq k$ is known as the *order condition*. Because the exogenous regressors in $\mathbf{X}$ serve as their own instruments, the order condition is often stated as requiring that there be at least as many instruments as endogenous variables. The order condition is necessary, but not sufficient, for the rank condition to hold.

If the rank condition fails, the equation is said to be *underidentified*, and no econometric procedure can produce consistent estimates. If the rank of $(\mathbf{Z}'\mathbf{X})$ is k, the equation is said to be *exactly identified*. If the rank of $(\mathbf{Z}'\mathbf{X}) > k$, the equation is said to be *overidentified*.

The rank condition requires only that there be enough correlation between the instruments and the endogenous variables to guarantee that we can compute unique parameter estimates. For the large-sample approximations to be useful, we need much higher correlations between the instruments and the regressors than the minimal level required by the rank condition. Instruments that satisfy the rank condition but are not sufficiently correlated with the endogenous variables for the large-sample approximations to be useful are known as *weak instruments*. We discuss weak instruments in section 8.10.

The parameters of exactly identified equations can be estimated by IV. The parameters of overidentified equations can be estimated by IV, after combining the instruments as in 2SLS. Although overidentification might sound like a nuisance to be avoided, it is actually preferable to working with an exactly identified equation. Overidentifying restrictions produce more efficient estimates in large samples. Furthermore, recall that the first essential property of an instrument is statistical independence from the disturbance process. Although we cannot test the validity of that assumption directly, we can assess the adequacy of instruments in an overidentified context with a *test of overidentifying restrictions*.

In such a test, the residuals from a 2SLS regression are regressed on all exogenous variables: both included exogenous regressors and excluded instruments. Under the null hypothesis that all instruments are uncorrelated with u, an LM statistic of the $N \times R^2$ form has a large-sample $\chi^2(r)$ distribution, where r is the number of *overidentifying restrictions*: the number of excess instruments. If we reject this hypothesis, we cast doubt on the suitability of the instrument set. One or more of the instruments do not appear to be uncorrelated with the disturbance process. This Sargan (1958) or Basmann (1960) test is available in Stata as the `overid` command (Baum, Schaffer, and Stillman 2003). This command can be installed from `ssc` for use after estimation with `ivreg`. I present an example of its use below.

8.6 Computing IV estimates

I illustrate how to use `ivreg` with a regression from Griliches (1976), a classic study of the wages of a sample of 758 young men.[5] Griliches models their wages as a function of several continuous factors: `s`, `expr`, and `tenure` (years of schooling, experience, and job tenure, respectively); `rns`, an indicator for residency in the South; `smsa`, an indicator for urban versus rural; and a set of year dummies since the data are a set of pooled cross sections. The endogenous regressor is `iq`, the worker's IQ score, which is considered as a potentially mismeasured version of ability. Here we do not consider that `wage` and `iq` are simultaneously determined, but rather that `iq` cannot be assumed independent of the error term: the same correlation that arises in the context of an endogenous regressor in a structural equation.[6] The IQ score is instrumented with four factors excluded from the equation: `med`, the mother's level of education; `kww`, the score on another standardized test; `age`, the worker's age; and `mrt`, an indicator of marital status. I present the descriptive statistics with `summarize` and then fit the IV model.

```
. use http://www.stata-press.com/data/imeus/griliches, clear
(Wages of Very Young Men, Zvi Griliches, J.Pol.Ec. 1976)

. summarize lw s expr tenure rns smsa iq med kww age mrt, sep(0)

    Variable |       Obs        Mean    Std. Dev.       Min        Max
-------------+--------------------------------------------------------
          lw |       758    5.686739    .4289494      4.605      7.051
           s |       758    13.40501    2.231828          9         18
        expr |       758    1.735429    2.105542          0     11.444
      tenure |       758    1.831135     1.67363          0         10
         rns |       758    .2691293    .4438001          0          1
        smsa |       758    .7044855     .456575          0          1
          iq |       758    103.8562    13.61867         54        145
         med |       758    10.91029     2.74112          0         18
         kww |       758    36.57388    7.302247         12         56
         age |       758    21.83509    2.981756         16         30
         mrt |       758    .5145119    .5001194          0          1
```

We use the `first` option for `ivreg` to evaluate the degree of correlation between these four factors and the endogenous regressor `iq`:

5. These data were later used by Blackburn and Neumark (1992). I am grateful to Professor Fumio Hayashi for his permission to use the version of the Blackburn–Neumark data circulated as `grilic` with his econometrics textbook (Hayashi 2000).

6. Measurement error problems are discussed in appendix B to this chapter.

```
. ivreg lw s expr tenure rns smsa _I* (iq=med kww age mrt), first
First-stage regressions
-----------------------

      Source |       SS       df       MS              Number of obs =     758
-------------+------------------------------           F( 15,   742) =   25.03
       Model |  47176.4676    15  3145.09784           Prob > F      =  0.0000
    Residual |  93222.8583   742  125.637275           R-squared     =  0.3360
-------------+------------------------------           Adj R-squared =  0.3226
       Total |  140399.326   757  185.468066           Root MSE      =  11.209

------------------------------------------------------------------------------
          iq |      Coef.   Std. Err.      t    P>|t|     [95% Conf. Interval]
-------------+----------------------------------------------------------------
           s |   2.497742   .2858159     8.74   0.000     1.936638    3.058846
        expr |   -.033548   .2534458    -0.13   0.895    -.5311042    .4640082
      tenure |   .6158215   .2731146     2.25   0.024     .0796522    1.151991
         rns |  -2.610221   .9499731    -2.75   0.006    -4.475177   -.7452663
        smsa |   .0260481   .9222585     0.03   0.977    -1.784499    1.836595
   _Iyear_67 |   .9254935   1.655969     0.56   0.576    -2.325449    4.176436
   _Iyear_68 |   .4706951   1.574561     0.30   0.765    -2.620429     3.56182
   _Iyear_69 |   2.164635   1.521387     1.42   0.155    -.8221007     5.15137
   _Iyear_70 |   5.734786   1.696033     3.38   0.001     2.405191    9.064381
   _Iyear_71 |   5.180639   1.562156     3.32   0.001     2.113866    8.247411
   _Iyear_73 |   4.526686    1.48294     3.05   0.002     1.615429    7.437943
         med |   .2877745   .1622338     1.77   0.077    -.0307176    .6062665
         kww |   .4581116   .0699323     6.55   0.000     .3208229    .5954003
         age |  -.8809144   .2232535    -3.95   0.000    -1.319198   -.4426307
         mrt |   -.584791    .946056    -0.62   0.537    -2.442056    1.272474
       _cons |   67.20449   4.107281    16.36   0.000     59.14121    75.26776
------------------------------------------------------------------------------

Instrumental variables (2SLS) regression
      Source |       SS       df       MS              Number of obs =     758
-------------+------------------------------           F( 12,   745) =   45.91
       Model |  59.2679161    12  4.93899301           Prob > F      =  0.0000
    Residual |  80.0182337   745  .107407025           R-squared     =  0.4255
-------------+------------------------------           Adj R-squared =  0.4163
       Total |   139.28615   757  .183997556           Root MSE      =  .32773

------------------------------------------------------------------------------
          lw |      Coef.   Std. Err.      t    P>|t|     [95% Conf. Interval]
-------------+----------------------------------------------------------------
          iq |   .0001747   .0039374     0.04   0.965    -.0075551    .0079044
           s |   .0691759    .013049     5.30   0.000     .0435587    .0947931
        expr |    .029866    .006697     4.46   0.000     .0167189    .0430132
      tenure |   .0432738   .0076934     5.62   0.000     .0281705     .058377
         rns |  -.1035897   .0297371    -3.48   0.001    -.1619682   -.0452111
        smsa |   .1351148   .0268889     5.02   0.000     .0823277    .1879019
   _Iyear_67 |   -.052598   .0481067    -1.09   0.275    -.1470388    .0418428
   _Iyear_68 |   .0794686   .0451078     1.76   0.079     -.009085    .1680222
   _Iyear_69 |   .2108962   .0443153     4.76   0.000     .1238984    .2978939
   _Iyear_70 |   .2386338   .0514161     4.64   0.000     .1376962    .3395714
   _Iyear_71 |   .2284609   .0441236     5.18   0.000     .1418396    .3150823
   _Iyear_73 |   .3258944   .0410718     7.93   0.000     .2452642    .4065247
       _cons |    4.39955   .2708771    16.24   0.000     3.867777    4.931323
------------------------------------------------------------------------------
Instrumented:  iq
Instruments:   s expr tenure rns smsa _Iyear_67 _Iyear_68
               _Iyear_69 _Iyear_70 _Iyear_71 _Iyear_73 med
               kww age mrt
------------------------------------------------------------------------------
```

The first-stage regression results suggest that three of the four excluded instruments are highly correlated with `iq`. The exception is `mrt`, the indicator of marital status. However, the endogenous regressor `iq` has an IV coefficient that cannot be distinguished from zero. Conditioning on the other factors included in the equation, `iq` does not seem to play an important role in determining the wage. The other coefficient estimates agree with the predictions of theory and empirical findings.

Are the instruments for `iq` appropriately uncorrelated with the disturbance process? To answer that, we compute the test for overidentifying restrictions:

```
. overid
Tests of overidentifying restrictions:
Sargan N*R-sq test        87.655  Chi-sq(3)     P-value = 0.0000
Basmann test              97.025  Chi-sq(3)     P-value = 0.0000
```

The above test signals a strong rejection of the null hypothesis that the instruments are uncorrelated with the error term and suggests that we should not be satisfied with this specification of the equation. We return to this example in the next section.

In the following sections, I present several topics related to the IV estimator and a generalization of that estimator. These capabilities are not provided by Stata's `ivreg` but are available in the extension of that routine known as `ivreg2` (Baum, Schaffer, and Stillman 2003, 2005).[7]

8.7 ivreg2 and GMM estimation

In defining the simple IV estimator and the 2SLS estimator, we assumed the presence of i.i.d. errors. As for linear regression, when the errors do not satisfy the i.i.d. assumption, the simple IV and 2SLS estimators produce consistent but inefficient estimates whose large-sample VCE must be estimated by a robust method. In another parallel to the linear regression case, there is a more general estimator based on the GMM that will produce consistent and efficient estimates in the presence of non-i.i.d. errors. Here I describe and illustrate this more general estimation technique.

The equation of interest is

$$\mathbf{y} = \mathbf{X}\boldsymbol{\beta} + \mathbf{u}, \qquad E[\mathbf{u}\mathbf{u}'|\mathbf{X}] = \boldsymbol{\Omega}$$

The matrix of regressors $\mathbf{X}$ is $N \times k$, where N is the number of observations. The error term $\mathbf{u}$ is distributed with mean zero, and its covariance matrix $\boldsymbol{\Omega}$ is $N \times N$. We consider four cases for $\boldsymbol{\Omega}$: homoskedasticity, conditional heteroskedasticity, clustering, and the combination of heteroskedasticity and autocorrelation. The last three cases correspond to those described in section 6.1.

7. I am deeply indebted to collaborators Mark E. Schaffer and Steven Stillman for their efforts in crafting the software in the `ivreg2` suite of programs and its description in our cited article. Much of this section is adapted from that article and subsequent joint work.

Some of the regressors are endogenous, so $E[\mathbf{x}u] \neq \mathbf{0}$. We partition the set of regressors into $\{\mathbf{x}_1\ \mathbf{x}_2\}$ with the k_1 regressors $\mathbf{x}_1$ considered endogenous and the $(k - k_1)$ remaining regressors $\mathbf{x}_2$ assumed to be exogenous.

The matrix of instrumental variables $\mathbf{Z}$ is $N \times \ell$. These variables are assumed to be exogenous: $E[\mathbf{z}u] = \mathbf{0}$. We partition the instruments into $\{\mathbf{z}_1\ \mathbf{z}_2\}$ where the ℓ_1 instruments $\mathbf{z}_1$ are excluded instruments and the remaining $(\ell - \ell_1)$ instruments $\mathbf{z}_2 \equiv \mathbf{x}_2$ are the included instruments or exogenous regressors.

8.7.1 The GMM estimator

The standard IV and 2SLS estimators are special cases of the GMM estimator. As with the simple IV case discussed in section 8.2, the assumption that the instruments $\mathbf{z}$ are exogenous can be expressed as a set of *moment conditions* $E[\mathbf{z}u] = \mathbf{0}$. The ℓ instruments give us a set of ℓ moments:

$$g_i(\boldsymbol{\beta}) = \mathbf{Z}_i' u_i = \mathbf{Z}_i'(y_i - \mathbf{x}_i\boldsymbol{\beta})$$

where g_i is $\ell \times 1$.[8] Just as in method-of-moments estimators of linear regression and simple IV, each of the ℓ moment equations corresponds to a sample moment. We write these ℓ sample moments as

$$\overline{g}(\boldsymbol{\beta}) = \frac{1}{N}\sum_{i=1}^{N} g_i(\boldsymbol{\beta}) = \frac{1}{N}\sum_{i=1}^{N} \mathbf{z}_i'(y_i - \mathbf{x}_i\boldsymbol{\beta}) = \frac{1}{N}\mathbf{Z}'\mathbf{u}$$

The intuition behind GMM is to choose an estimator for $\boldsymbol{\beta}$ that solves $\overline{g}(\widehat{\boldsymbol{\beta}}_{\text{GMM}}) = 0$.

If the equation to be estimated is *exactly identified* ($\ell = k$), we have just as many moment conditions as we do unknowns. We can exactly solve the ℓ moment conditions for the k coefficients in $\widehat{\boldsymbol{\beta}}_{\text{GMM}}$. Here there is a unique $\widehat{\boldsymbol{\beta}}_{\text{GMM}}$ that solves $\overline{g}(\widehat{\boldsymbol{\beta}}_{\text{GMM}}) = 0$. This GMM estimator is identical to the standard IV estimator of (8.3).

If the equation is *overidentified*, $\ell > k$, we have more equations than we do unknowns. We will not be able to find a k-vector $\widehat{\boldsymbol{\beta}}_{\text{GMM}}$ that will set all ℓ sample moment conditions to zero. We want to choose $\widehat{\boldsymbol{\beta}}_{\text{GMM}}$ so that the elements of $\overline{g}(\widehat{\boldsymbol{\beta}}_{\text{GMM}})$ are as close to zero as possible. We could do so by minimizing $\overline{g}(\widehat{\boldsymbol{\beta}}_{\text{GMM}})'\overline{g}(\widehat{\boldsymbol{\beta}}_{\text{GMM}})$, but this method offers no way to produce more efficient estimates when the errors are not i.i.d. For this reason, the GMM estimator chooses the $\widehat{\boldsymbol{\beta}}_{\text{GMM}}$ that minimizes

$$J(\widehat{\boldsymbol{\beta}}_{\text{GMM}}) = N\ \overline{g}(\widehat{\boldsymbol{\beta}}_{\text{GMM}})'\mathbf{W}\overline{g}(\widehat{\boldsymbol{\beta}}_{\text{GMM}}) \tag{8.9}$$

where $\mathbf{W}$ is an $\ell \times \ell$ *weighting matrix* that accounts for the correlations among the $\overline{g}(\widehat{\boldsymbol{\beta}}_{\text{GMM}})$ when the errors are not i.i.d.

8. Because these conditions imply that $(\mathbf{Z}, \mathbf{u})$ will be uncorrelated, they are often termed *orthogonality conditions* in the literature.

A GMM estimator for $\boldsymbol{\beta}$ is the $\widehat{\boldsymbol{\beta}}$ that minimizes $J(\widehat{\boldsymbol{\beta}}_{\text{GMM}})$. Deriving and solving the k first-order conditions

$$\frac{\partial J(\widehat{\boldsymbol{\beta}})}{\partial \widehat{\boldsymbol{\beta}}} = \mathbf{0}$$

yields the GMM estimator of an overidentified equation:

$$\widehat{\beta}_{\text{GMM}} = (\mathbf{X}'\mathbf{Z}\mathbf{W}\mathbf{Z}'\mathbf{X})^{-1}\mathbf{X}'\mathbf{Z}\mathbf{W}\mathbf{Z}'\mathbf{y} \tag{8.10}$$

The results of the minimization—and hence the GMM estimator—will be identical for all weighting matrices $\mathbf{W}$ that differ by a constant of proportionality. We use knowledge of this fact below. However, there are as many GMM estimators as there are choices of weighting matrix $\mathbf{W}$. For an exactly identified equation, $\mathbf{W} = \mathbf{I}_N$. The weighting matrix only plays a role in the presence of overidentifying restrictions.

The *optimal* weighting matrix is that which produces the most efficient estimate.[9] Hansen (1982) showed that this process involves choosing $\mathbf{W} = \mathbf{S}^{-1}$, where $\mathbf{S}$ is the covariance matrix of the moment conditions g:

$$\mathbf{S} = E[\mathbf{Z}'\mathbf{u}\mathbf{u}'\mathbf{Z}] = E[\mathbf{Z}'\boldsymbol{\Omega}\mathbf{Z}] \tag{8.11}$$

where $\mathbf{S}$ is an $\ell \times \ell$ matrix. Substitute this matrix into (8.10) to obtain the efficient GMM estimator:

$$\widehat{\beta}_{\text{EGMM}} = (\mathbf{X}'\mathbf{Z}\mathbf{S}^{-1}\mathbf{Z}'\mathbf{X})^{-1}\mathbf{X}'\mathbf{Z}\mathbf{S}^{-1}\mathbf{Z}'\mathbf{y}$$

Note the generality (the G of GMM) of this approach. We have made no assumptions about $\boldsymbol{\Omega}$, the covariance matrix of the disturbance process.[10] But the efficient GMM estimator is not a feasible estimator since the matrix $\mathbf{S}$ is not known. To implement the estimator, we need to estimate $\mathbf{S}$, so we must make some assumptions about $\boldsymbol{\Omega}$, as we discuss next.

Assume that we have developed a consistent estimator of $\mathbf{S}$, denoted $\widehat{\mathbf{S}}$. Generally such an estimator will involve the 2SLS residuals. Then we may use that estimator to define the feasible efficient two-step GMM estimator (FEGMM) implemented in `ivreg2` when the `gmm` option is used.[11] In the first step, we use standard 2SLS estimation to generate parameter estimates and residuals. In the second step, we use an assumption about the structure of $\boldsymbol{\Omega}$ to produce $\widehat{\mathbf{S}}$ from those residuals and define the FEGMM:

$$\widehat{\beta}_{\text{FEGMM}} = (\mathbf{X}'\mathbf{Z}\widehat{\mathbf{S}}^{-1}\mathbf{Z}'\mathbf{X})^{-1}\mathbf{X}'\mathbf{Z}\widehat{\mathbf{S}}^{-1}\mathbf{Z}'\mathbf{y}$$

8.7.2 GMM in a homoskedastic context

If we assume that $\boldsymbol{\Omega} = \sigma^2 I_N$, the optimal weighting matrix implied by (8.11) will be proportional to the identity matrix. Since no weighting is involved in calculating

9. Efficiency is described in section 4.2.3.

10. Aside from some conditions that guarantee $\frac{1}{\sqrt{N}}\mathbf{Z}'u$ is a vector of well-behaved random variables.

11. This estimator goes under various names: two-stage instrumental variables (2SIV), White (1982); two-step two-stage least squares, Cumby, Huizinga, and Obstfeld (1983); heteroskedastic two-stage least squares (H2SLS), Davidson and MacKinnon (1993, 599).

(8.10), the GMM estimator is merely the standard IV estimator in point and interval form. The IV estimator of (8.6) and (8.7) is the FEGMM estimator under conditional homoskedasticity of $\mathbf{\Omega}$.

8.7.3 GMM and heteroskedasticity-consistent standard errors

One of the most commonly encountered problems in economic data is heteroskedasticity of unknown form, as described in section 6.2. We need a heteroskedasticity-consistent estimator of $\mathbf{S}$. Such an $\widehat{\mathbf{S}}$ is available by using the standard sandwich approach to robust covariance estimation described in section 6.1.2. Define the 2SLS residuals as $\widehat{u}_i$ and the ith row of the instrument matrix as $\mathbf{Z}_i$. Then a consistent estimator of $\mathbf{S}$ is given by

$$\widehat{\mathbf{S}} = \frac{1}{N}\sum_{i=1}^{N} \widehat{u}_i^2 \mathbf{Z}_i' \mathbf{Z}_i$$

The residuals can come from any consistent estimator of $\boldsymbol{\beta}$ because efficiency of the parameter estimates used to compute the $\widehat{u}_i$ is not required. In practice, 2SLS residuals are almost always used. For more details, see Davidson and MacKinnon (1993, 607–610).

If the regression equation is exactly identified with $\ell = k$, the results from `ivreg2, gmm` will be identical to those of `ivreg2, robust` or from `ivreg` with the `robust` option. For overidentified models, the GMM approach makes more efficient use of the information in the ℓ moment conditions than the standard 2SLS approach that reduces them to k instruments in $\widehat{\mathbf{X}}$. The 2SLS estimator can be considered a GMM estimator with a suboptimal weighting matrix when the errors are not i.i.d.

To compare GMM with 2SLS, we reestimate the wage equation displayed earlier by using the `gmm` option. This step automatically generates heteroskedasticity-robust standard errors. By default, `ivreg2` reports large-sample `z` statistics for the coefficients.

(*Continued on next page*)

```
. ivreg2 lw s expr tenure rns smsa _I* (iq=med kww age mrt), gmm
GMM estimation
--------------

                                                      Number of obs =      758
                                                      F( 12,    745) =    49.67
                                                      Prob > F      =   0.0000
Total (centered) SS     =  139.2861498                Centered R2   =   0.4166
Total (uncentered) SS   =  24652.24662                Uncentered R2 =   0.9967
Residual SS             =  81.26217887                Root MSE      =    .3274

------------------------------------------------------------------------------
             |               Robust
          lw |      Coef.   Std. Err.      z    P>|z|     [95% Conf. Interval]
-------------+----------------------------------------------------------------
          iq |  -.0014014   .0041131    -0.34   0.733     -.009463    .0066602
           s |   .0768355   .0131859     5.83   0.000     .0509915    .1026794
        expr |   .0312339   .0066931     4.67   0.000     .0181157    .0443522
      tenure |   .0489998   .0073437     6.67   0.000     .0346064    .0633931
         rns |  -.1006811   .0295887    -3.40   0.001    -.1586738   -.0426884
        smsa |   .1335973   .0263245     5.08   0.000     .0820021    .1851925
   _Iyear_67 |  -.0210135   .0455433    -0.46   0.645    -.1102768    .0682498
   _Iyear_68 |   .0890993    .042702     2.09   0.037     .0054049    .1727937
   _Iyear_69 |   .2072484   .0407995     5.08   0.000     .1272828     .287214
   _Iyear_70 |   .2338308   .0528512     4.42   0.000     .1302445    .3374172
   _Iyear_71 |   .2345525   .0425661     5.51   0.000     .1511244    .3179805
   _Iyear_73 |   .3360267   .0404103     8.32   0.000     .2568239    .4152295
       _cons |   4.436784   .2899504    15.30   0.000     3.868492    5.005077
------------------------------------------------------------------------------
Anderson canon. corr. LR statistic (identification/IV relevance test):  54.338
                                                   Chi-sq(4) P-val =    0.0000
------------------------------------------------------------------------------
Hansen J statistic (overidentification test of all instruments):        74.165
                                                   Chi-sq(3) P-val =    0.0000
------------------------------------------------------------------------------
Instrumented:         iq
Included instruments: s expr tenure rns smsa _Iyear_67 _Iyear_68 _Iyear_69
                      _Iyear_70 _Iyear_71 _Iyear_73
Excluded instruments: med kww age mrt
------------------------------------------------------------------------------
```

We see that the endogenous regressor `iq` still does not play a role in the equation. The Hansen J statistic displayed by `ivreg2` is the GMM equivalent of the Sargan test produced by `overid` above. The independence of the instruments and the disturbance process is called into question by this strong rejection of the J test null hypothesis.

8.7.4 GMM and clustering

When the disturbances have within-cluster correlation, `ivreg2` can compute the cluster–robust estimator of the VCE, and it can optionally use the cluster–robust estimator of $\widehat{\mathbf{S}}$ to produce more efficient parameter estimates when the model is overidentified. A consistent estimate of $\mathbf{S}$ in the presence of within-cluster correlated disturbances is

$$\widehat{\mathbf{S}} = \sum_{j=1}^{M} \widehat{\mathbf{u}}_j' \widehat{\mathbf{u}}_j$$

where

$$\widehat{\mathbf{u}}_j = (y_j - \mathbf{x}_j\widehat{\boldsymbol{\beta}})\mathbf{X}'\mathbf{Z}(\mathbf{Z}'\mathbf{Z})^{-1}\mathbf{z}_j$$

and y_j is the jth observation on y, $\mathbf{x}_j$ is the jth row of $\mathbf{X}$, and $\mathbf{z}_j$ is the jth row of $\mathbf{Z}$. The fact that we are summing over the M clusters instead of the N observations leads to the intuition that we essentially do not have N observations: we have M, where M is the number of clusters. That number, M, must exceed ℓ if we are to estimate the equation, since $M - \ell$ is the effective degrees of freedom of the cluster estimator. If we are using a sizable number of instruments, this constraint may bind.[12]

Specifying the `cluster()` option will cause `ivreg2` to compute the cluster–robust estimator of the VCE. If the equation is overidentified, adding the `gmm` option will cause `ivreg2` to use the cluster–robust estimate of $\mathbf{S}$ to compute more efficient parameter estimates.

8.7.5 GMM and HAC standard errors

When the disturbances are conditionally heteroskedastic and autocorrelated, we can compute HAC estimates of the VCE and if the equation is overidentified, we can optionally use an HAC estimate of $\mathbf{S}$ to compute more efficient parameter estimates. The `ivreg2` routine will compute Newey–West estimates of the VCE using the Bartlett-kernel weighting when the `robust` and `bw()` options are specified. When there are no endogenous regressors, the results will be the same as those computed by `newey`. If some of the regressors are endogenous, then specifying the `robust` and `bw()` options will cause `ivreg2` to compute an HAC estimator of the VCE. If the equation is overidentified and the `robust` and `gmm` options are specified, the resulting GMM estimates will be more efficient than those produced by 2SLS.

The number specified in the `bw()` (bandwidth) option should be greater than that specified in the `lag()` option in `newey`. The `ivreg2` routine lets us choose several alternative kernel estimators (see the `kernel()` option) as described in the online help for that command.[13]

To illustrate, we estimate a Phillips curve relationship with annual time-series data for the United States, 1948–1996. The descriptive statistics for consumer price inflation (`cinf`) and the unemployment rate (`unem`) are as follows:

```
. use http://www.stata-press.com/data/imeus/phillips, clear
. summarize cinf unem if cinf < .

    Variable |       Obs        Mean    Std. Dev.       Min        Max
-------------+--------------------------------------------------------
        cinf |        48     -.10625    2.566926       -9.3        6.6
        unem |        48     5.78125    1.553261        2.9        9.7
```

12. Official `ivreg` is more forgiving and will complain only if $M < k$. On the other hand, `ivreg2` insists that $M > \ell$.

13. If we do not doubt the homoskedasticity assumption but want to deal with autocorrelation of unknown form, we should use the AC correction without the H correction for arbitrary heteroskedasticity. `ivreg2` allows us to select H, AC, or HAC VCEs by combining the `robust` (or `gmm`), `bw()`, and `kernel()` options.

A Phillips curve is the relation between price or wage inflation and the unemployment rate. In Phillips' model, these variables should have a negative relationship with lower unemployment leading to inflationary pressures. Since both variables are determined within the macroeconomic environment, we cannot consider either as exogenous.

Using these data, we regress the rate of consumer price inflation on the unemployment rate. To deal with simultaneity, we instrument the unemployment rate with its second and third lags. Specifying `bw(3)`, `gmm`, and `robust` causes `ivreg2` to compute the efficient GMM estimates.

```
. ivreg2 cinf (unem = l(2/3).unem), bw(3) gmm robust
GMM estimation
---------------
Heteroskedasticity and autocorrelation-consistent statistics
  kernel=Bartlett; bandwidth=3
  time variable (t):  year
                                                      Number of obs =        46
                                                      F(  1,     44) =      0.39
                                                      Prob > F      =    0.5371
Total (centered) SS     =  217.4271745                Centered R2   =   -0.1266
Total (uncentered) SS   =  217.4900005                Uncentered R2 =   -0.1262
Residual SS             =  244.9459113                Root MSE      =     2.308

------------------------------------------------------------------------------
             |               Robust
        cinf |      Coef.   Std. Err.      z    P>|z|     [95% Conf. Interval]
-------------+----------------------------------------------------------------
        unem |   .1949334   .3064662     0.64   0.525    -.4057292     .795596
       _cons |  -1.144072   1.686995    -0.68   0.498    -4.450522    2.162378
------------------------------------------------------------------------------
Anderson canon. corr. LR statistic (identification/IV relevance test):  13.545
                                                   Chi-sq(2) P-val =    0.0011
------------------------------------------------------------------------------
Hansen J statistic (overidentification test of all instruments):         0.589
                                                   Chi-sq(1) P-val =    0.4426
------------------------------------------------------------------------------
Instrumented:          unem
Excluded instruments:  L2.unem L3.unem
------------------------------------------------------------------------------
```

The hypothesized relationship is not borne out by these estimates, as many researchers have found. The original relationship over the period ending in the late 1960s broke down badly in the presence of 1970s supply shocks and high inflation. To focus on the IV technique, we see that the Hansen J test statistic indicates that the instruments are appropriately uncorrelated with the disturbance process. If the first and second lags of `unem` are used, the J test rejects its null with a p-value of 0.02. The first lag of `unem` appears to be inappropriate as an instrument in this specification.

8.8 Testing overidentifying restrictions in GMM

Just as for 2SLS (see section 8.5), the validity of the overidentifying restrictions imposed on a GMM estimator can be tested. The test, which can and should be performed as a standard diagnostic in any overidentified model,[14] has a null hypothesis of correct model specification and valid overidentifying restrictions. A rejection calls either or both of those hypotheses into question.

With GMM, the overidentifying restrictions may be tested by the commonly used J statistic of Hansen (1982).[15] This statistic is merely the value of the GMM objective function (8.9), evaluated at the efficient GMM estimator $\widehat{\boldsymbol{\beta}}_{\text{EGMM}}$. Under the null,

$$J(\widehat{\boldsymbol{\beta}}_{\text{EGMM}}) = N\ \overline{g}(\widehat{\boldsymbol{\beta}}_{\text{EGMM}})'\widehat{\mathbf{S}}^{-1}\overline{g}(\widehat{\boldsymbol{\beta}}_{\text{EGMM}}) \overset{A}{\sim} \chi^2_{\ell-k}$$

where the matrix $\widehat{\mathbf{S}}$ is estimated using the two-step methods described above.

The J statistic is asymptotically distributed as χ^2 with degrees of freedom equal to the number of overidentifying restrictions $\ell - k$ rather than the total number of moment conditions, ℓ. In effect, k degrees of freedom are spent in estimating the coefficients β. Hansen's J is the most common diagnostic used in GMM estimation to evaluate the suitability of the model. A rejection of the null hypothesis implies that the instruments do not satisfy the required orthogonality conditions—either because they are not truly exogenous or because they are being incorrectly excluded from the regression. The J statistic is calculated and displayed by `ivreg2` when the `gmm` or `robust` options is specified.[16]

8.8.1 Testing a subset of the overidentifying restrictions in GMM

The Hansen–Sargan tests for overidentification presented above evaluate the entire set of overidentifying restrictions. In a model containing a very large set of excluded instruments, such a test may have little power. Another common problem arises when you have suspicions about the validity of a subset of instruments and want to test them.

In these contexts, you can use a *difference-in-Sargan* test.[17] The C test allows us to test a subset of the original set of orthogonality conditions. The statistic is computed as the difference between two J statistics. The first is computed from the fully efficient regression using the entire set of overidentifying restrictions. The second is that of the inefficient but consistent regression using a smaller set of restrictions in which a specified set of instruments are removed from the instrument list. For excluded instruments, this

14. Thus Davidson and MacKinnon (1993, 236): "Tests of overidentifying restrictions should be calculated routinely whenever one computes IV estimates." Sargan's own view, cited in Godfrey (1988, 145), was that regression analysis without testing the orthogonality assumptions is a "pious fraud".

15. For conditional homoskedasticity (see section 8.7.2), this statistic is numerically identical to the Sargan test statistic discussed above.

16. Despite the importance of testing the overidentifying restrictions, the J test is known to overreject the null hypothesis in certain circumstances. Using the "continuous updating" GMM estimator discussed in the help file for `ivreg2` may produce rejection rates that are closer to the level of the test. See Hayashi (2000, 218) for more information.

17. See Hayashi (2000, 218–221 and 232–234) or Ruud (2000, chap. 22), for comprehensive presentations. The test is known under other names as well; e.g., Ruud (2000) calls it the distance difference statistic, and Hayashi (2000) follows Eichenbaum, Hansen, and Singleton (1988) and dubs it the C statistic. I use the latter term.

step is equivalent to dropping them from the instrument list. For included instruments, the C test places them in the list of included endogenous variables, treating them as endogenous regressors. The order condition must still be satisfied for this form of the equation. Under the null hypothesis that the specified variables are proper instruments, the difference-in-Sargan C test statistic is distributed χ^2 with degrees of freedom equal to the loss of overidentifying restrictions or the number of suspect instruments being tested.[18]

Specifying `orthog(`*instlist*`)` with the suspect instruments causes `ivreg2` to compute the C test with *instlist* as the excluded instruments. The equation must still be identified with these instruments removed (or placed in the endogenous regressor list) to compute the C test. If the equation excluding suspect instruments is exactly identified, the J statistic for that equation will be zero and the C statistic will coincide with the statistic for the original equation. This property illustrates how the J test of overidentifying restrictions is an *omnibus* test for the failure of any of the instruments to satisfy the orthogonality conditions. At the same time, the test requires that the investigator believe the nonsuspect instruments to be valid (see Ruud 2000, 577).

Below we use the C statistic to test whether `s`, years of schooling, is a valid instrument in the wage equation estimated above by 2SLS and GMM. In those examples, the Sargan and J tests of overidentifying restrictions signaled a problem with the instruments used.

18. Although the C statistic can be calculated as the simple difference between the Hansen–Sargan statistics for two regressions, this procedure can generate a negative test statistic in finite samples. For 2SLS, this problem can be avoided and the C statistic guaranteed to be nonnegative if the estimate of the error variance $\widehat{\sigma}^2$ from the original 2SLS regression is used to calculate the Sargan statistic for the regression with nonsuspect instruments as well. The equivalent procedure in GMM is to use the $\widehat{\mathbf{S}}$ matrix from the original estimation to calculate both J statistics. More precisely, $\widehat{\mathbf{S}}$ from the original equation is used to form the first J statistic, and the submatrix of $\widehat{\mathbf{S}}$ with rows/columns corresponding to the reduced set of instruments is used to form the J statistic for the second equation (see Hayashi 2000, 220).

```
. ivreg2 lw s expr tenure rns smsa _I* (iq=med kww age mrt), gmm orthog(s)
GMM estimation
______________

                                                   Number of obs =      758
                                                   F( 12,    745) =   49.67
                                                   Prob > F      =   0.0000
Total (centered) SS     =  139.2861498             Centered R2   =   0.4166
Total (uncentered) SS   =  24652.24662             Uncentered R2 =   0.9967
Residual SS             =  81.26217887             Root MSE      =    .3274

------------------------------------------------------------------------------
             |               Robust
          lw |      Coef.   Std. Err.      z    P>|z|     [95% Conf. Interval]
-------------+----------------------------------------------------------------
          iq |  -.0014014   .0041131    -0.34   0.733     -.009463    .0066602
           s |   .0768355   .0131859     5.83   0.000     .0509915    .1026794
        expr |   .0312339   .0066931     4.67   0.000     .0181157    .0443522
      tenure |   .0489998   .0073437     6.67   0.000     .0346064    .0633931
         rns |  -.1006811   .0295887    -3.40   0.001    -.1586738   -.0426884
        smsa |   .1335973   .0263245     5.08   0.000     .0820021    .1851925
   _Iyear_67 |  -.0210135   .0455433    -0.46   0.645    -.1102768    .0682498
   _Iyear_68 |   .0890993    .042702     2.09   0.037     .0054049    .1727937
   _Iyear_69 |   .2072484   .0407995     5.08   0.000     .1272828     .287214
   _Iyear_70 |   .2338308   .0528512     4.42   0.000     .1302445    .3374172
   _Iyear_71 |   .2345525   .0425661     5.51   0.000     .1511244    .3179805
   _Iyear_73 |   .3360267   .0404103     8.32   0.000     .2568239    .4152295
       _cons |   4.436784   .2899504    15.30   0.000     3.868492    5.005077
------------------------------------------------------------------------------
Anderson canon. corr. LR statistic (identification/IV relevance test):  54.338
                                                   Chi-sq(4) P-val =    0.0000
------------------------------------------------------------------------------
Hansen J statistic (overidentification test of all instruments):        74.165
                                                   Chi-sq(3) P-val =    0.0000
-orthog- option:
Hansen J statistic (eqn. excluding suspect orthog. conditions):         15.997
                                                   Chi-sq(2) P-val =    0.0003
C statistic (exogeneity/orthogonality of suspect instruments):          58.168
                                                   Chi-sq(1) P-val =    0.0000
Instruments tested:   s
------------------------------------------------------------------------------
Instrumented:         iq
Included instruments: s expr tenure rns smsa _Iyear_67 _Iyear_68 _Iyear_69
                      _Iyear_70 _Iyear_71 _Iyear_73
Excluded instruments: med kww age mrt
------------------------------------------------------------------------------
```

The C test rejects its null, indicating that the suspect instrument, `s`, fails the test for overidentifying restrictions. The significant J statistic of 15.997 for the equation excluding suspect instruments implies that treating s as endogenous still results in an unsatisfactory equation. The remaining instruments do not appear to be independent of the error distribution.

Now we use the `orthog()` option to test whether a subset of the excluded instruments are appropriately exogenous. We include age and the marital status indicator (`age` and `mrt`) in the option's *varlist*. The equation estimated without suspect instruments merely drops those instruments from the list of excluded instruments:

```
. ivreg2 lw s expr tenure rns smsa _I* (iq=med kww age mrt), gmm orthog(age mrt)
GMM estimation
--------------
                                                      Number of obs =       758
                                                      F( 12,    745) =    49.67
                                                      Prob > F       =   0.0000
Total (centered) SS     =  139.2861498                Centered R2    =   0.4166
Total (uncentered) SS   =  24652.24662                Uncentered R2  =   0.9967
Residual SS             =  81.26217887                Root MSE       =    .3274

------------------------------------------------------------------------------
             |               Robust
          lw |      Coef.   Std. Err.      z    P>|z|     [95% Conf. Interval]
-------------+----------------------------------------------------------------
          iq |  -.0014014   .0041131    -0.34   0.733     -.009463    .0066602
           s |   .0768355   .0131859     5.83   0.000     .0509915    .1026794
        expr |   .0312339   .0066931     4.67   0.000     .0181157    .0443522
      tenure |   .0489998   .0073437     6.67   0.000     .0346064    .0633931
         rns |  -.1006811   .0295887    -3.40   0.001    -.1586738   -.0426884
        smsa |   .1335973   .0263245     5.08   0.000     .0820021    .1851925
   _Iyear_67 |  -.0210135   .0455433    -0.46   0.645    -.1102768    .0682498
   _Iyear_68 |   .0890993    .042702     2.09   0.037     .0054049    .1727937
   _Iyear_69 |   .2072484   .0407995     5.08   0.000     .1272828     .287214
   _Iyear_70 |   .2338308   .0528512     4.42   0.000     .1302445    .3374172
   _Iyear_71 |   .2345525   .0425661     5.51   0.000     .1511244    .3179805
   _Iyear_73 |   .3360267   .0404103     8.32   0.000     .2568239    .4152295
       _cons |   4.436784   .2899504    15.30   0.000     3.868492    5.005077
------------------------------------------------------------------------------
Anderson canon. corr. LR statistic (identification/IV relevance test):  54.338
                                                   Chi-sq(4) P-val =    0.0000
------------------------------------------------------------------------------
Hansen J statistic (overidentification test of all instruments):        74.165
                                                   Chi-sq(3) P-val =    0.0000
-orthog- option:
Hansen J statistic (eqn. excluding suspect orthog. conditions):          1.176
                                                   Chi-sq(1) P-val =    0.2782
C statistic (exogeneity/orthogonality of suspect instruments):          72.989
                                                   Chi-sq(2) P-val =    0.0000
Instruments tested:   age mrt
------------------------------------------------------------------------------
Instrumented:         iq
Included instruments: s expr tenure rns smsa _Iyear_67 _Iyear_68 _Iyear_69
                      _Iyear_70 _Iyear_71 _Iyear_73
Excluded instruments: med kww age mrt
------------------------------------------------------------------------------
```

The equation estimated without suspect instruments, free of the two additional orthogonality conditions on `age` and `mrt`, has an insignificant J statistic, whereas the C statistic for those two instruments is highly significant. These two instruments do not appear valid in this context. To evaluate whether we have found a more appropriate specification, we reestimate the equation with the reduced instrument list:

```
. ivreg2 lw s expr tenure rns smsa _I* (iq=med kww), gmm
GMM estimation
---------------

                                                    Number of obs =       758
                                                    F( 12,    745) =    30.77
                                                    Prob > F       =   0.0000
Total (centered) SS      =  139.2861498             Centered R2    =   0.1030
Total (uncentered) SS    =  24652.24662             Uncentered R2  =   0.9949
Residual SS              =  124.9413508             Root MSE       =     .406

------------------------------------------------------------------------------
             |               Robust
          lw |      Coef.   Std. Err.      z    P>|z|     [95% Conf. Interval]
-------------+----------------------------------------------------------------
          iq |   .0240417   .0060961     3.94   0.000     .0120936    .0359899
           s |   .0009181   .0194208     0.05   0.962    -.0371459     .038982
        expr |   .0393333   .0088012     4.47   0.000     .0220833    .0565834
      tenure |   .0324916   .0091223     3.56   0.000     .0146122     .050371
         rns |  -.0326157   .0376679    -0.87   0.387    -.1064433     .041212
        smsa |    .114463   .0330718     3.46   0.001     .0496434    .1792825
   _Iyear_67 |  -.0694178   .0568781    -1.22   0.222    -.1808968    .0420613
   _Iyear_68 |   .0891834   .0585629     1.52   0.128    -.0255977    .2039645
   _Iyear_69 |   .1780712   .0532308     3.35   0.001     .0737407    .2824016
   _Iyear_70 |    .139594   .0677261     2.06   0.039     .0068533    .2723346
   _Iyear_71 |   .1730151   .0521623     3.32   0.001      .070779    .2752512
   _Iyear_73 |    .300759   .0490919     6.13   0.000     .2045407    .3969772
       _cons |   2.859113   .4083706     7.00   0.000     2.058721    3.659504
------------------------------------------------------------------------------
Anderson canon. corr. LR statistic (identification/IV relevance test):  35.828
                                                   Chi-sq(2) P-val =   0.0000
------------------------------------------------------------------------------
Hansen J statistic (overidentification test of all instruments):         0.781
                                                   Chi-sq(1) P-val =   0.3768
------------------------------------------------------------------------------
Instrumented:         iq
Included instruments: s expr tenure rns smsa _Iyear_67 _Iyear_68 _Iyear_69
                      _Iyear_70 _Iyear_71 _Iyear_73
Excluded instruments: med kww
------------------------------------------------------------------------------
```

In these results, we find that in line with theory, `iq` appears as a significant regressor for the first time and the equation's J statistic is satisfactory. The regressor `s`, which appeared in an earlier test to be inappropriately considered exogenous, plays no role in this form of the estimated equation.[19]

8.9 Testing for heteroskedasticity in the IV context

This section discusses the Pagan and Hall (1983) test for heteroskedasticity in 2SLS models and the `ivhettest` command (Baum, Schaffer, and Stillman 2003), which im-

19. You might wonder why the J statistic for this equation is not equal to that of the equation lacking the suspect instruments in the C test above. As explained in an earlier footnote, a positive C statistic is guaranteed by computing both J statistics using the estimated error variance of the full equation. Those two error variances differ—the equation above has a larger `Root MSE`—so that the J statistics differ as well.

plements this test in Stata. The idea behind the test—similar to that of the Breusch–Pagan (Breusch and Pagan 1979) and White tests for heteroskedasticity discussed in section 6.2.1—is that if any of the exogenous variables can predict the squared residuals, the errors are conditionally heteroskedastic.[20] Under the null of conditional homoskedasticity in the 2SLS regression, the Pagan–Hall statistic is distributed as χ^2_p, irrespective of the presence of heteroskedasticity elsewhere in the system.[21]

The **ivhettest** command follows the abbreviated syntax:

ivhettest [*varlist*] [, *options*]

where the optional *varlist* specifies the exogenous variables to be used to model the squared errors. Common choices for those variables include the following:

1. The levels only of the instruments Z (excluding the constant). This choice is available in **ivhettest** by specifying the **ivlev** option, which is the default option.
2. The levels and squares of the instruments Z, available as the **ivsq** option.
3. The levels, squares, and cross products of the instruments Z (excluding the constant), as in the White (1980) test: available as the **ivcp** option.
4. The fitted value of the response variable.[22] This choice is available in **ivhettest** by specifying the **fitlev** option.
5. The fitted value of the response variable and its square, available as the **fitsq** option.
6. A user-defined set of variables may also be provided.

The tradeoff in the choice of variables to be used is that a smaller set of variables will conserve degrees of freedom, at the cost of being unable to detect heteroskedasticity in certain directions.

The Pagan–Hall statistic has not been widely used, perhaps because it is not a standard feature of most regression packages.[23] However, from an analytical standpoint, it is clearly superior to the techniques more commonly used since it is robust to the presence of heteroskedasticity elsewhere in a system of simultaneous equations and to non–normally distributed disturbances.[24]

20. The Breusch–Pagan and White tests for heteroskedasticity (Breusch and Pagan 1979) discussed in section 6.2.1 can be applied in 2SLS models, but Pagan and Hall (1983) point out that they will be valid only if heteroskedasticity is present in that equation and *nowhere else in the system*. The other structural equations in the system corresponding to the endogenous regressors must also be homoskedastic even though they are not being explicitly estimated.

21. A more general form of this test was separately proposed by White (1982).

22. This fitted value is not the usual fitted value of the response variable, $X\widehat{\boldsymbol{\beta}}_{\mathrm{IV}}$. It is, rather, $\widehat{X}\widehat{\boldsymbol{\beta}}_{\mathrm{IV}}$, i.e., the prediction based on the IV estimator $\widehat{\boldsymbol{\beta}}_{\mathrm{IV}}$, the exogenous regressors Z_2, and the fitted values of the endogenous regressors $\widehat{X}_1$.

23. Although we discuss its use in 2SLS, **ivhettest** may also be used after **regress**.

24. White's general test (White 1980), or its generalization by Koenker (1981), also relaxes the assumption of normality underlying the Breusch–Pagan test.

We compute several of the tests for heteroskedasticity appropriate in the IV context with `ivhettest` from the last regression reported above. The default setting uses the levels of the instruments as associated variables. Results from the `fitsq` option are also displayed.

```
. ivhettest, all
IV heteroskedasticity test(s) using levels of IVs only
Ho: Disturbance is homoskedastic
    Pagan-Hall general test statistic     :   8.645  Chi-sq(13) P-value = 0.7992
    Pagan-Hall test w/assumed normality :   9.539  Chi-sq(13) P-value = 0.7311
    White/Koenker nR2 test statistic      :  13.923  Chi-sq(13) P-value = 0.3793
    Breusch-Pagan/Godfrey/Cook-Weisberg :  15.929  Chi-sq(13) P-value = 0.2530
. ivhettest, fitsq all
IV heteroskedasticity test(s) using fitted value (X-hat*beta-hat) & its square
Ho: Disturbance is homoskedastic
    Pagan-Hall general test statistic     :   0.677  Chi-sq(2) P-value = 0.7127
    Pagan-Hall test w/assumed normality :   0.771  Chi-sq(2) P-value = 0.6799
    White/Koenker nR2 test statistic      :   0.697  Chi-sq(2) P-value = 0.7056
    Breusch-Pagan/Godfrey/Cook-Weisberg :   0.798  Chi-sq(2) P-value = 0.6710
```

None of the tests signal any problem of heteroskedasticity in the estimated equation's disturbance process.

8.10 Testing the relevance of instruments

As discussed above, an instrumental variable must not be correlated with the equation's disturbance process and it must be highly correlated with the included endogenous regressors. We may test the latter condition by examining the fit of the first-stage regressions. The first-stage regressions are reduced-form regressions of the endogenous regressors, $\mathbf{x}_1$, on the full set of instruments, $\mathbf{z}$. The relevant test statistics here relate to the explanatory power of the excluded instruments, $\mathbf{z}_1$, in these regressions. A statistic commonly used, as recommended by Bound, Jaeger, and Baker (1995), is the R^2 of the first-stage regression with the included instruments partialled out.[25] This test may be expressed as the F test of the joint significance of the $\mathbf{z}_1$ instruments in the first-stage regression. But the distribution of this F statistic is nonstandard.[26] Also, for models with multiple endogenous variables, these indicators may not be sufficiently informative.

To grasp the pitfalls facing empirical researchers here, consider the following simple example. You have a model with two endogenous regressors and two excluded instruments. One of the two excluded instruments is highly correlated with each of the two endogenous regressors, but the other excluded instrument is just noise. Your model is basically *underidentified*. You have one valid instrument but two endogenous regressors. The Bound, Jaeger, and Baker F statistics and partial R^2 measures from the two

25. More precisely, this is the squared partial correlation between the excluded instruments $\mathbf{z}_1$ and the endogenous regressor in question. It is defined as $(\text{RSS}_{\mathbf{z}_2} - \text{RSS}_{\mathbf{z}})/\text{TSS}$, where $\text{RSS}_{\mathbf{z}_2}$ is the residual sum of squares in the regression of the endogenous regressor on $\mathbf{z}_2$ and $\text{RSS}_{\mathbf{z}}$ is the RSS when the full set of instruments is used.

26. See Bound, Jaeger, and Baker (1995) and Stock, Wright, and Yogo (2002).

first-stage regressions will not reveal this weakness. Indeed, the F statistics will be statistically significant, and without investigation you may not realize that the model cannot be estimated in this form. To deal with this problem of instrument irrelevance, either more relevant instruments are needed or one of the endogenous regressors must be dropped from the model. The statistics proposed by Bound, Jaeger, and Baker can diagnose instrument relevance only in the presence of one endogenous regressor. When multiple endogenous regressors are used, other statistics are required.

One such statistic has been proposed by Shea (1997): a partial R^2 measure that takes the intercorrelations among the instruments into account.[27] For a model containing one endogenous regressor, the two R^2 measures are equivalent. The distribution of Shea's partial R^2 statistic has not been derived, but it may be interpreted like any R^2. As a rule of thumb, if an estimated equation yields a large value of the standard (Bound, Jaeger, and Baker 1995) partial R^2 and a small value of the Shea measure, you should conclude that the instruments lack sufficient relevance to explain all the endogenous regressors. Your model may be essentially underidentified. The Bound, Jaeger, and Baker measures and the Shea partial R^2 statistic are provided by the `first` or `ffirst` options of the `ivreg2` command.

A more general approach to the problem of instrument relevance was proposed by Anderson (1984) and discussed in Hall, Rudebusch, and Wilcox (1996).[28] Anderson's approach considers the *canonical correlations* of the $\mathbf{X}$ and $\mathbf{Z}$ matrices. These measures, r_i, $i = 1, \dots, k$ represent the correlations between linear combinations of the k columns of $\mathbf{X}$ and linear combinations of the ℓ columns of $\mathbf{Z}$.[29] If an equation to be estimated by instrumental variables is identified from a numerical standpoint, all k of the canonical correlations must be significantly different from zero. Anderson's likelihood-ratio test has the null hypothesis that the smallest canonical correlation is zero and assumes that the regressors are distributed multivariate normal. Under the null, the test statistic is distributed χ^2 with $(\ell - k + 1)$ degrees of freedom, so that it may be calculated even for an exactly identified equation. A failure to reject the null hypothesis calls the identification status of the estimated equation into question. The Anderson statistic is displayed in `ivreg2`'s standard output.

The canonical correlations between $\mathbf{X}$ and $\mathbf{Z}$ may also be used to test a set of instruments for *redundancy* following Hall and Peixe (2000). In an overidentified context with $\ell \geq k$, if some of the instruments are redundant then the large-sample efficiency of the estimation is not improved by including them. The test statistic is a likelihood-ratio statistic based on the canonical correlations with and without the instruments being

27. The Shea partial R^2 statistic may be easily computed according to the simplification presented in Godfrey (1999), who demonstrates that Shea's statistic for endogenous regressor i may be expressed as $R_p^2 = (\nu_{i,i,\mathrm{OLS}})/(\nu_{i,i,\mathrm{IV}}) \left\{(1 - R_{\mathrm{IV}}^2)/(1 - R_{\mathrm{OLS}}^2)\right\}$, where $\nu_{i,i}$ is the estimated asymptotic variance of the coefficient.

28. Hall, Rudebusch, and Wilcox state that the test is closely related to the minimum-eigenvalue test statistic proposed by Cragg and Donald (1993). This test is displayed with the `first` or `ffirst` option of `ivreg2`: see the following example.

29. The squared canonical correlations may be calculated as the eigenvalues of $(\mathbf{X}'\mathbf{X})^{-1}(\mathbf{X}'\mathbf{Z})(\mathbf{Z}'\mathbf{Z})^{-1}(\mathbf{Z}'\mathbf{X})$; see Hall, Rudebusch, and Wilcox (1996, 287).

tested. Under the null hypothesis that the specified instruments are redundant, the statistic is distributed as χ^2 with degrees of freedom equal to the number of endogenous regressors times the number of instruments being tested. Like the Anderson test, the redundancy test assumes that the regressors are distributed multivariate normal. This test is available in `ivreg2` with the `redundant()` option.

I illustrate the weak-instruments problem with a variation on the log wage equation using only `age` and `mrt` as instruments.

```
. ivreg2 lw s expr tenure rns smsa _I* (iq = age mrt), ffirst redundant(mrt)
Summary results for first-stage regressions
-------------------------------------------

                Shea
Variable     | Partial R2    |     Partial R2    F(  2,    744)     P-value
iq           |    0.0073     |       0.0073           2.72          0.0665

Underidentification tests:
                                                 Chi-sq(2)        P-value
Anderson canon. corr. likelihood ratio stat.          5.52         0.0632
Cragg-Donald N*minEval stat.                          5.54         0.0626
Ho: matrix of reduced form coefficients has rank=K-1 (underidentified)
Ha: matrix has rank>=K (identified)

Weak identification statistics:
Cragg-Donald (N-L)*minEval/L2 F-stat        2.72

Anderson-Rubin test of joint significance of
endogenous regressors B1 in main equation, Ho:B1=0
  F(2,744)=       43.83     P-val=0.0000
  Chi-sq(2)=      89.31     P-val=0.0000

Number of observations N            =        758
Number of regressors   K            =         13
Number of instruments  L            =         14
Number of excluded instruments L2   =          2
```

(*Continued on next page*)

```
Instrumental variables (2SLS) regression

                                                      Number of obs =      758
                                                      F( 12,    745) =     3.95
                                                      Prob > F       =   0.0000
Total (centered) SS     =  139.2861498                Centered R2    =  -6.4195
Total (uncentered) SS   =  24652.24662                Uncentered R2  =   0.9581
Residual SS             =  1033.432656                Root MSE       =    1.168

------------------------------------------------------------------------------
          lw |      Coef.   Std. Err.      z    P>|z|     [95% Conf. Interval]
-------------+----------------------------------------------------------------
          iq |  -.0948902   .0433073    -2.19   0.028    -.1797708   -.0100095
           s |   .3397121    .125526     2.71   0.007     .0936856    .5857386
        expr |   -.006604    .028572    -0.23   0.817     -.062604    .0493961
      tenure |   .0848854   .0327558     2.59   0.010     .0206852    .1490856
         rns |  -.3769393   .1584438    -2.38   0.017    -.6874834   -.0663952
        smsa |   .2181191   .1022612     2.13   0.033     .0176908    .4185474
   _Iyear_67 |   .0077748   .1733579     0.04   0.964    -.3320005    .3475501
   _Iyear_68 |   .0377993   .1617101     0.23   0.815    -.2791466    .3547452
   _Iyear_69 |   .3347027   .1666592     2.01   0.045     .0080568    .6613487
   _Iyear_70 |   .6286425   .2486186     2.53   0.011      .141359    1.115926
   _Iyear_71 |   .4446099    .182733     2.43   0.015     .0864599    .8027599
   _Iyear_73 |    .439027   .1542401     2.85   0.004      .136722    .7413321
       _cons |   10.55096   2.821406     3.74   0.000      5.02111    16.08082
------------------------------------------------------------------------------
Anderson canon. corr. LR statistic (identification/IV relevance test):   5.522
                                                   Chi-sq(2) P-val =    0.0632
-redundant- option:
LR IV redundancy test (redundancy of specified instruments):             0.002
                                                   Chi-sq(1) P-val =    0.9685
Instruments tested:   mrt
------------------------------------------------------------------------------
Sargan statistic (overidentification test of all instruments):           1.393
                                                   Chi-sq(1) P-val =    0.2379
------------------------------------------------------------------------------
Instrumented:         iq
Included instruments: s expr tenure rns smsa _Iyear_67 _Iyear_68 _Iyear_69
                      _Iyear_70 _Iyear_71 _Iyear_73
Excluded instruments: age mrt
------------------------------------------------------------------------------
```

In the first-stage regression results, Shea's partial R^2 statistic is very small for this equation, and the Cragg–Donald statistic marginally rejects its null hypothesis of underidentification. The Anderson canonical correlation statistic fails to reject its null hypothesis at the 5% level, suggesting that although we have more instruments than coefficients the instruments may be inadequate to identify the equation. The `redundant(mrt)` option indicates that `mrt` provides no useful information to identify the equation. This equation may be only exactly identified.

The consequence of excluded instruments with little explanatory power is increased bias in the estimated IV coefficients (Hahn and Hausman 2002b) and worsening of the large-sample approximations to the finite-sample distributions. If these instruments' explanatory power in the first-stage regression is nil, the model is in effect unidentified with respect to that endogenous variable. Here the large-sample bias of the IV estimator is the same as that of the OLS estimator, IV becomes inconsistent, and nothing is gained from instrumenting (Hahn and Hausman 2002b). What is surprising is that, as Staiger and Stock (1997) and others have shown, the weak-instrument problem can arise even when the first-stage tests are significant at conventional levels (5% or 1%) and the researcher is using a large sample. One rule of thumb is that for one endogenous regressor, an F statistic less than 10 is cause for concern (Staiger and Stock 1997, 557). The magnitude of large-sample bias of the IV estimator increases with the number of instruments (Hahn and Hausman 2002b). Given that, one recommendation when faced with a weak-instrument problem is to be parsimonious in the choice of instruments. For further discussion, see Staiger and Stock (1997); Hahn and Hausman (2002a,b); Stock, Wright, and Yogo (2002); Chao and Swanson (2005); and references therein.

8.11 Durbin–Wu–Hausman tests for endogeneity in IV estimation

There may well be reason to suspect a failure of the zero-conditional-mean assumption presented in section 4.2 in many regression models. Turning to IV or efficient GMM estimation for the sake of consistency must be balanced against the inevitable loss of efficiency. As Wooldridge states, "[there is an] important cost of performing IV estimation when $\mathbf{x}$ and u are uncorrelated: the asymptotic variance of the IV estimator is *always* larger, and sometimes *much* larger, than the asymptotic variance of the OLS estimator" (Wooldridge 2006, 516; emphasis added). This loss of efficiency is a price worth paying if the OLS estimator is biased and inconsistent. A test of the appropriateness of OLS and the necessity to resort to IV or GMM methods would be useful.[30] The intuition for such a test may also be couched in the number of orthogonality conditions available. Can all or some of the included endogenous regressors be appropriately treated as exogenous? If so, these restrictions can be added to the set of moment conditions, and more efficient estimation will be possible.

Many econometrics texts discuss the issue of OLS versus IV in the context of the Durbin–Wu–Hausman (DWH) tests. These tests involve fitting the model by both OLS and IV approaches and comparing the resulting coefficient vectors. In the Hausman form of the test, a quadratic form in the differences between the two coefficient vectors scaled by the precision matrix gives rise to a test statistic for the null hypothesis that the OLS estimator is consistent and fully efficient.

Denote by $\widehat{\boldsymbol{\beta}}^c$ the estimator that is consistent under both the null and the alternative hypotheses, and by $\widehat{\boldsymbol{\beta}}^e$ the estimator that is fully efficient under the null but inconsistent if the null is not true. The Hausman (1978) specification test takes the quadratic form

$$H = (\widehat{\boldsymbol{\beta}}_c - \widehat{\boldsymbol{\beta}}_e)'\mathbf{D}^-(\widehat{\boldsymbol{\beta}}_c - \widehat{\boldsymbol{\beta}}_e)$$

30. As discussed in Baum, Schaffer, and Stillman (2003, 11), GMM may have poor small-sample properties. If the zero-conditional-mean assumption cannot be refuted, we should use linear regression rather than IV or GMM, especially in small samples.

where

$$\mathbf{D} = \text{Var}[\widehat{\boldsymbol{\beta}}_c] - \text{Var}[\widehat{\boldsymbol{\beta}}_e]$$

$\text{Var}[\widehat{\boldsymbol{\beta}}]$ denotes a consistent estimate of the asymptotic variance of $\boldsymbol{\beta}$, and the operator $^-$ denotes a generalized inverse.

A Hausman statistic for a test of endogeneity in an IV regression is formed by choosing OLS as the efficient estimator $\widehat{\boldsymbol{\beta}}_e$ and IV as the inefficient but consistent estimator $\widehat{\boldsymbol{\beta}}_c$. The test statistic is distributed as χ^2 with k_1 degrees of freedom: the number of regressors being tested for endogeneity. The test is perhaps best interpreted not as a test for the endogeneity or exogeneity of regressors per se but rather as a test of the consequence of using different estimation methods on the same equation. Under the null hypothesis that OLS is an appropriate estimation technique, only efficiency should be lost by turning to IV. The point estimates should be qualitatively unaffected.

There are many ways to conduct a DWH endogeneity test in Stata for the standard IV case with conditional homoskedasticity. Three equivalent ways of obtaining the Durbin component of the DWH statistic in Stata are

1. Fit the less efficient but consistent model using IV, followed by the command `estimates store iv` (where `iv` is a name of your choice that is attached to this set of estimates; see the discussion of stored estimates in section 4.4). Then fit the fully efficient model with `regress` (or with `ivreg` if only a subset of regressors is being tested for endogeneity), followed by `hausman iv ., constant sigmamore`.[31]
2. Fit the fully efficient model using `ivreg2` and specify the regressors to be tested in the `orthog()` option.
3. Fit the less efficient but consistent model using `ivreg` and use `ivendog` to conduct an endogeneity test. The `ivendog` command takes as its argument a *varlist* consisting of the subset of regressors to be tested for endogeneity. If the *varlist* is empty, the full set of endogenous regressors is tested.

The last two methods are more convenient than the first because the test can be done in one step. Furthermore, the `hausman` command will often generate a negative χ^2 statistic, rendering the test infeasible. Stata's documentation describes this result as a small-sample problem in which the variance of the difference of the coefficient vectors is not necessarily positive definite in finite samples.[32] The different commands implement distinct versions of the tests, which although asymptotically equivalent can lead to different inference from finite samples.

---

31. You should disregard the note produced by `hausman` regarding the rank of the differenced matrix. As the documentation of the `sigmamore` option indicates, this is the proper setting for a test of exogeneity comparing linear regression and IV estimates.

32. The description of `hausman` suggests that a generalized Hausman test can be performed by `suest`. However, this command does not support the `ivreg` estimator.

I first illustrate using the **hausman** command for the wage equation:

```
. quietly ivreg2 lw s expr tenure rns smsa _I* (iq=med kww), small
. estimates store iv
. quietly regress lw s expr tenure rns smsa _I* iq
. hausman iv ., constant sigmamore
Note: the rank of the differenced variance matrix (1) does not equal the number
        of coefficients being tested (13); be sure this is what you expect, or
        there may be problems computing the test.  Examine the output of your
        estimators for anything unexpected and possibly consider scaling your
        variables so that the coefficients are on a similar scale.

                 ---- Coefficients ----
             |      (b)          (B)            (b-B)     sqrt(diag(V_b-V_B))
             |      iv            .          Difference          S.E.
-------------+----------------------------------------------------------------
          iq |    .0243202     .0027121        .021608          .0046882
           s |    .0004625     .0619548      -.0614923          .0133417
        expr |     .039129     .0308395       .0082896          .0017985
      tenure |    .0327048     .0421631      -.0094582          .0020521
         rns |   -.0341617    -.0962935       .0621318          .0134804
        smsa |    .1140326     .1328993      -.0188667          .0040934
   _Iyear_67 |   -.0679321    -.0542095      -.0137226          .0029773
   _Iyear_68 |    .0900522     .0805808       .0094714           .002055
   _Iyear_69 |    .1794505     .2075915       -.028141          .0061056
   _Iyear_70 |    .1395755     .2282237      -.0886482          .0192335
   _Iyear_71 |    .1735613     .2226915      -.0491302          .0106595
   _Iyear_73 |    .2971599     .3228747      -.0257148          .0055792
       _cons |    2.837153     4.235357      -1.398204          .3033612
------------------------------------------------------------------------------
                           b = consistent under Ho and Ha; obtained from ivreg2
            B = inconsistent under Ha, efficient under Ho; obtained from regress

    Test:  Ho:  difference in coefficients not systematic

                  chi2(1) = (b-B)'[(V_b-V_B)^(-1)](b-B)
                          =       21.24
                Prob>chi2 =      0.0000
                (V_b-V_B is not positive definite)
```

The comparison here is restricted to the point estimate and estimated standard error of the endogenous regressor, **iq**; the **hausman** test statistic rejects exogeneity of this variable. The command also warns of difficulties computing a positive-definite covariance matrix. The large χ^2 value indicates that estimation of the equation with **regress** yields inconsistent results.

I now illustrate the second method, using **ivreg2** and the **orthog()** option. You should notice the peculiar syntax of the parenthesized list in which no variable is identified as endogenous. This argument (and the equals sign) is still required to signal to Stata that **med kww** are to be considered as instruments in the unrestricted equation in which **iq** is considered endogenous. This treatment causes **ivreg2** to perform the reported estimation using linear regression and consider the alternative model to be IV.[33]

33. I use the **small** option to ensure that the χ^2 statistic takes on the same value in the second and third methods.

```
. ivreg2 lw s expr tenure rns smsa _I* iq (=med kww), orthog(iq) small
Ordinary Least Squares (OLS) regression
---------------------------------------

                                                  Number of obs =      758
                                                  F( 12,    745) =    46.86
                                                  Prob > F      =   0.0000
Total (centered) SS     =  139.2861498            Centered R2   =   0.4301
Total (uncentered) SS   =  24652.24662            Uncentered R2 =   0.9968
Residual SS             =  79.37338879            Root MSE      =    .3264

------------------------------------------------------------------------------
          lw |      Coef.   Std. Err.      t    P>|t|     [95% Conf. Interval]
-------------+----------------------------------------------------------------
           s |   .0619548   .0072786     8.51   0.000     .0476658    .0762438
        expr |   .0308395   .0065101     4.74   0.000     .0180592    .0436198
      tenure |   .0421631   .0074812     5.64   0.000     .0274763    .0568498
         rns |  -.0962935   .0275467    -3.50   0.001    -.1503719   -.0422151
        smsa |   .1328993   .0265758     5.00   0.000     .0807268    .1850717
   _Iyear_67 |  -.0542095   .0478522    -1.13   0.258    -.1481506    .0397317
   _Iyear_68 |   .0805808   .0448951     1.79   0.073    -.0075551    .1687168
   _Iyear_69 |   .2075915   .0438605     4.73   0.000     .1214867    .2936963
   _Iyear_70 |   .2282237   .0487994     4.68   0.000      .132423    .3240245
   _Iyear_71 |   .2226915   .0430952     5.17   0.000     .1380889     .307294
   _Iyear_73 |   .3228747   .0406574     7.94   0.000     .2430579    .4026915
          iq |   .0027121   .0010314     2.63   0.009     .0006873    .0047369
       _cons |   4.235357   .1133489    37.37   0.000     4.012836    4.457878
------------------------------------------------------------------------------
Sargan statistic (Lagrange multiplier test of excluded instruments):    22.659
                                                   Chi-sq(2) P-val =    0.0000
-orthog- option:
Sargan statistic (eqn. excluding suspect orthogonality conditions):      1.045
                                                   Chi-sq(1) P-val =    0.3067
C statistic (exogeneity/orthogonality of suspect instruments):          21.614
                                                   Chi-sq(1) P-val =    0.0000
Instruments tested:   iq
------------------------------------------------------------------------------
Included instruments: s expr tenure rns smsa _Iyear_67 _Iyear_68 _Iyear_69
                      _Iyear_70 _Iyear_71 _Iyear_73 iq
Excluded instruments: med kww
------------------------------------------------------------------------------
```

The second method's C test statistic from `ivendog` agrees qualitatively with that from `hausman`. I now illustrate the third method's use of `ivendog`:

```
. quietly ivreg lw s expr tenure rns smsa _I* (iq=med kww)
. ivendog
Tests of endogeneity of: iq
H0: Regressor is exogenous
    Wu-Hausman F test:                  21.83742  F(1,744)     P-value = 0.00000
    Durbin-Wu-Hausman chi-sq test:      21.61394  Chi-sq(1)    P-value = 0.00000
```

The test statistic is identical to that provided by the C statistic above. All forms of the test agree that estimation of this equation with linear regression yields inconsistent results. The regressor `iq` must be considered endogenous in the fitted model.

Exercises

1. Following the discussion in section 8.3, use the Griliches data in section 8.6 to estimate two-stage least squares "by hand". Compare the residuals and s^2 with those computed by `ivreg` on the same equation.

2. When we presented robust linear regression estimates, the estimated coefficients and summary statistics were unchanged; only the VCE was affected. Compare the estimates displayed in section 8.6 with those of section 8.7.3. Why do the coefficient estimates and summary statistics such as `R-squared` and `Root MSE` differ?

3. Using the Griliches data, estimate the equation

```
. ivreg2 lw s expr rns smsa (iq=med kww age mrt) if year==67, gmm
```

   What comments can you make about these estimates? Reestimate the equation, adding the `cluster(age)` option. What is the rationale for clustering by age? Evaluate this form of the equation versus that estimated without clustering. What are its problems?

4. Following the discussion in section 8.7.5, refit the Phillips curve model (a) without the `gmm` option and (b) without the `gmm` and `robust` options. How do these estimates—corresponding to 2SLS–HAC and 2SLS–AC—compare with the GMM–HAC estimates displayed in the text?

5. Refit the Phillips curve model using lags 1, 2, and 3 of `unem` as instruments for the unemployment rate. What do you find?

6. Does the Phillips curve require an IV estimator, or can it be consistently estimated with linear regression? Refit the model of section 8.7.5, using the `orthog()` option of `ivreg2` to decide whether linear regression is satisfactory using the DWH framework.

7. Does the Phillips curve exhibit heteroskedasticity in the time dimension? Refit the model of section 8.7.5 without the `robust` option, and use the options of `ivhettest` to test this hypothesis.

8.A Appendix: Omitted-variables bias

The OLS estimator cannot produce consistent estimates if the zero-conditional-mean assumption (4.2) is violated. I illustrate an alternative solution by considering the omitted-variables problem discussed above in section 5.2: an unobserved but relevant omitted explanatory factor. Consider the relationship among high schools' average Scholastic Aptitude Test (SAT) scores (`sat`),[34] expenditure per pupil (`spend`), and the poverty rate in each district (`poverty`):

$$\texttt{sat} = \beta_1 + \beta_2\texttt{expend} + \beta_3\texttt{poverty} + u_i \tag{8.12}$$

We cannot estimate this equation because we do not have access to poverty rates at the school-district level. However, that factor is thought to play an important role in educational attainment, proxying for the quality of the student's home environment. If we had a *proxy variable* available, we could substitute it for `poverty`, for example, the median income in the school district. Whether this strategy would succeed depends on how highly the proxy variable is correlated with the unobserved `poverty`. If no proxy is available, we might estimate the equation, ignoring `poverty`:

$$\log(\texttt{sat}_i) = \beta_1 + \beta_2\texttt{expend}_i + v_i$$

The disturbance process v_i in this equation is composed of $(\beta_3\texttt{poverty}_i + u_i)$. If `expend` and `poverty` are correlated—as they are likely to be—regression will yield biased and inconsistent estimates of β_1 and β_2 because the zero-conditional-mean assumption is violated.

To derive consistent estimates of this equation, we must find an IV, as discussed in section 8.2. Many potential variables could be uncorrelated with the unobservable factors influencing SAT performance (including `poverty`) and highly correlated with `expend`.[35] What might be an appropriate instrument for `expend`? Perhaps we could measure each school district's student–teacher ratio (`stratio`). This measure is likely to be (negatively) correlated with district expenditure. If states' education policy mandates that student–teacher ratios fall within certain bounds, `stratio` should not be correlated with district poverty rates.

8.B Appendix: Measurement error

I introduced the concept of measurement error in section 5.3 and now discuss its consequences. Measurement error could appear in the response variable. Say that the true relationship explains y^*, but we observe $y = y^* + \epsilon$, where ϵ is a mean-zero-error process. Then ϵ becomes a component of the regression error term, worsening the fit of the

34. The SAT is the most common standardized test taken by U.S. high school students for college entrance.

35. We are not searching for a proxy variable for `poverty`. If we had a good proxy for $\texttt{poverty}_i$, it would not make a satisfactory instrumental variable. Correlation with $\texttt{poverty}_i$ implies correlation with the composite error process v_i.

estimated equation. We assume that ϵ is not systematic in that it is not correlated with the independent variables x. Then measurement error does no real harm—it merely weakens the model without introducing bias in either point or interval estimates.[36]

On the other hand, measurement error in a regressor is a far more serious problem. Say that the true model is

$$y = \beta_1 + \beta_2 x_2^* + u$$

but that x_2^* is not observed: we observe $x_2 = x_2^* + \epsilon_2$. We assume that $E[\epsilon_2] = 0$. What should we assume about the relationship between ϵ_2 and x_2^*? First, let us assume that ϵ_2 is not correlated with the observed measure x_2: larger values of x_2 do not give rise to systematically larger or smaller errors of measurement, which we can write as $\text{Cov}[\epsilon_2, x_2] = 0$. But if so, $\text{Cov}[\epsilon_2, x_2^*] \neq 0$: that is, the error of measurement must be correlated with the true explanatory variable x_2^*. We can then write the estimated equation in which x_2^* is replaced with the observable x_2 as

$$y = \beta_1 + \beta_2 x_2 + (u - \beta_2 \epsilon_2) \tag{8.13}$$

Since both u and ϵ_2 have zero mean and, by assumption, are uncorrelated with x_2, the presence of measurement error merely inflates the error term. $\text{Var}\left[u - \beta_2 \epsilon_2\right] = \sigma_u^2 + \beta_2^2 \sigma_{\epsilon_2}^2$ given a zero correlation of u, ϵ. Measurement error in x_2^* does not damage the regression of y on x_2—it merely inflates the error variance, as does measurement error in the response variable.

However, this is not the case that is usually considered in applied econometrics as *errors in variables*. It is more reasonable to assume that the measurement error is uncorrelated with the true explanatory variable: $\text{Cov}[\epsilon_2, x_2^*] = 0$. For instance, we might assume that the discrepancy between reported income and actual income is not a function of actual income. If so, $\text{Cov}[\epsilon_2, x_2] = \text{Cov}\left[\epsilon_2, (x_2^* + \epsilon_2)\right] \neq 0$ by construction, and the regression of (8.13) will have a nonzero correlation between its explanatory variable x_2 and the composite error term. This result violates the zero-conditional-mean assumption of (4.2). The covariance of $(x_2, u - \beta_2\epsilon_2) = -\beta_2 \text{Cov}[\epsilon_2, x_2] = -\beta_2\, \sigma_{\epsilon_2}^2 \neq 0$, causing the OLS regression of y on x_2 to be biased and inconsistent. In this simple case of one explanatory variable measured with error, we can determine the nature of the bias because $\widehat{\beta}_2$ consistently estimates

$$\begin{aligned}
\widehat{\boldsymbol{\beta}}_2 &= \beta_2 + \frac{\text{Cov}\left[x_2, u - \beta_2 \epsilon_2\right]}{\text{Var}[x_2]} \\
&= \beta_2 \left(\frac{\sigma_{x_2}^2}{\sigma_{x_2}^2 + \sigma_{\epsilon_2}^2} \right)
\end{aligned}$$

This expression demonstrates that the OLS point estimate will be *attenuated*—biased toward zero even in large samples—because the bracketed expression of squared quantities must be a fraction. In the absence of measurement error, $\sigma_{\epsilon_2}^2 \to 0$, and the OLS coefficient becomes consistent and unbiased. As $\sigma_{\epsilon_2}^2$ increases relative to the variance

36. If the magnitude of the measurement error in y is correlated with one or more of the regressors in x, the point estimates will be biased.

in the (correctly measured) explanatory variable, the OLS estimate becomes more and more unreliable, shrinking toward zero.

We conclude that in a multiple regression equation in which one of the regressors is subject to measurement error, if the measurement error is uncorrelated with the true (correctly measured) explanatory variable, then the OLS estimates will be biased and inconsistent for all the regressors, not merely for the coefficient of the regressor measured with error. We cannot predict the direction of bias with multiple regressors. Realistically, more than one regressor in an economic model may be subject to measurement error. In a household survey, both reported income and reported wealth may be measured incorrectly. Since measurement error violates the zero-conditional-mean assumption in the same sense as simultaneity bias or omitted-variables bias, we can treat it similarly.

8.B.1 Solving errors-in-variables problems

We can use the IV estimator to deal with the errors-in-variables model discussed in section 8.B. To deal with measurement error in one or more regressors, we must be able to specify an instrument for the mismeasured x variable that satisfies the usual assumptions. The instrument must not be correlated with the disturbance process u but must be highly correlated with the mismeasured x. If we could find a second measurement of x—even one that is prone to measurement error—we could use it as an instrument, since it would presumably be well correlated with x itself. If it is generated by an independent measurement process, it will be uncorrelated with the original measurement error. For instance, we might have data from a household survey that inquired about each family's disposable income, consumption, and saving. The respondents' answers about their saving last year might well be mismeasured since it is much harder to track saving than, say, earned income. We could say the same for their estimates of how much they spent on various categories of consumption. But using income and consumption data, we could derive a second (mismeasured) estimate of saving, which we could use as an instrument to mitigate the problems of measurement error in the direct estimate.

9 Panel-data models

A panel dataset has multiple observations on the same economic units. For instance, we may have multiple observations on the same households or firms over time. In panel data, each element has two subscripts, the group identifier i and a within-group index denoted by t in econometrics, because it usually identifies time.

Given panel data, we can define several models that arise from the most general linear representation:

$$y_{it} = \sum_{k=1}^{k} x_{kit}\beta_{kit} + \epsilon_{it}, \; i = 1, \ldots, N, \; t = 1, \ldots, T \tag{9.1}$$

where N is the number of individuals and T is the number of periods.

In sections 9.1–9.3, I present methods designed for "large N, small T" panels in which there are many individuals and a few periods. These methods use the large number of individuals to construct the large-sample approximations. The small T puts limits on what can be estimated.

Assume a *balanced* panel in which there are T observations for each of the N individuals. Since this model contains $k \times N \times T$ regression coefficients, it cannot be estimated from $N \times T$ observations. We could ignore the nature of the panel data and apply pooled ordinary least squares, which would assume that $\beta = \beta_j \; \forall \; j, i, t$, but that model might be overly restrictive and can have a complicated error process (e.g., heteroskedasticity across panel units, serial correlation within panel units, and so forth). Thus the pooled OLS solution is not often considered to be practical.

One set of panel-data estimators allows for heterogeneity across panel units (and possibly across time) but confines that heterogeneity to the intercept terms of the relationship. I discuss these techniques, the *fixed-effects* (FE) and *random-effects* (RE) models, in the next section. They impose restrictions on the above model of $\beta_{jit} = \beta \; \forall i, t, \; j > 1$, thereby allowing only the constant to differ over i.

These estimation techniques can be extended to deal with endogenous regressors. The following section discusses several IV estimators that accommodate endogenous regressors. I then present the dynamic panel data (DPD) estimator, which is appropriate when lagged dependent variables are included in the set of regressors. The DPD estimator is applied to "large N, small T" panels, such as a few years of annual data on each of several hundred firms.

Section 9.4 discusses applying seemingly unrelated regression (SUR) estimators to "small N, large T" panels, in which there are a few individuals and many periods—for instance, financial variables of the 10 largest U.S. manufacturing firms, observed over the last 40 calendar quarters.

The last section of the chapter revisits the notion of moving-window estimation, demonstrating how to compute a moving-window regression for each unit of a panel.

9.1 FE and RE models

The structure represented in (9.1) may be restricted to allow for heterogeneity across units without the full generality (and infeasibility) that this equation implies. In particular, we might restrict the slope coefficients to be constant over both units and time and allow for an intercept coefficient that varies by unit or by time. For a given observation, an intercept varying over units results in the structure

$$y_{it} = \mathbf{x}_{it}\boldsymbol{\beta}_k + \mathbf{z}_i\boldsymbol{\delta} + u_i + \epsilon_{it} \tag{9.2}$$

where $\mathbf{x}_{it}$ is a $1 \times k$ vector of variables that vary over individual and time, $\boldsymbol{\beta}$ is the $k \times 1$ vector of coefficients on $\mathbf{x}$, $\mathbf{z}_i$ is a $1 \times p$ vector of time-invariant variables that vary only over individuals, $\boldsymbol{\delta}$ is the $p \times 1$ vector of coefficients on $\mathbf{z}$, u_i is the individual-level effect, and ϵ_{it} is the disturbance term.

The u_i are either correlated or uncorrelated with the regressors in $\mathbf{x}_{it}$ and $\mathbf{z}_i$. (The u_i are always assumed to be uncorrelated with ϵ_{it}.)

If the u_i are uncorrelated with the regressors, they are known as RE, but if the u_i are correlated with the regressors, they are known as FE. The origin of the term RE is clear: when u_i are uncorrelated with everything else in the model, the individual-level effects are simply parameterized as additional random disturbances. The sum $u_i + \epsilon_{it}$ is sometimes referred to as the composite-error term and the model is sometimes known as an error-components model. The origin of the term FE is more elusive. When the u_i are correlated with some of the regressors in the model, one estimation strategy is to treat them like parameters or FE. But simply including a parameter for every individual is not feasible, because it would imply an infinite number of parameters in our large-N, large-sample approximations. The solution is to remove the u_i from the estimation problem by a transformation that still identifies some of the coefficients of interest.

RE estimators use the assumptions that the u_i are uncorrelated with the regressors to identify the $\boldsymbol{\beta}$ and $\boldsymbol{\delta}$ coefficients. In the process of removing the u_i, FE estimators lose the ability to identify the $\boldsymbol{\delta}$ coefficients. An additional cost of using the FE estimator is that all inference is conditional on the u_i in the sample. In contrast, inference using RE estimators pertains to the population from which the RE were drawn.

We could treat a time-varying intercept term similarly, as either an FE (giving rise to an additional coefficient) or as a component of a composite-error term. We concentrate here on the one-way FE and RE models in which only the individual intercept is considered in the "large N, small T" context most commonly found in microeconomic research.[1]

9.1.1 One-way FE

The FE model modestly relaxes the assumption that the regression function is constant over time and space. A one-way FE model permits each cross-sectional unit to have its own constant term while the slope estimates ($\boldsymbol{\beta}$) are constrained across units, as is the σ_ϵ^2. This estimator is often termed the least-squares dummy variable (LSDV) model, since it is equivalent to including $N-1$ dummy variables in the OLS regression of y on $\mathbf{x}$ (including a units vector). However, the name LSDV is fraught with problems because it implies an infinite number of parameters in our estimator. A better way to understand the FE estimator is to see that removing panel-level averages from each side of (9.2) removes the FE from the model. Let $\overline{y}_i = (1/T)\sum_{t=1}^T y_{it}$, $\overline{\mathbf{x}}_i = (1/T)\sum_{t=1}^T \mathbf{x}_{it}$, and $\overline{\epsilon}_i = (1/T)\sum_{t=1}^T \epsilon_{it}$. Also note that $\mathbf{z}_i$ and u_i are panel-level averages. Then simple algebra on (9.2) implies

$$y_{it} - \overline{y}_i = (\mathbf{x}_{it} - \overline{\mathbf{x}}_i)\boldsymbol{\beta} + (\mathbf{z}_i - \mathbf{z}_i)\boldsymbol{\delta} + u_i - u_i + \epsilon_{it} - \overline{\epsilon}_i$$

which implies that

$$\widetilde{y}_{it} = (\widetilde{\mathbf{x}}_{it})\,\boldsymbol{\beta} + \widetilde{\epsilon}_{it} \tag{9.3}$$

Equation (9.3) implies that OLS on the within-transformed data will produce consistent estimates of $\boldsymbol{\beta}$. We call this estimator $\widehat{\boldsymbol{\beta}}_{\text{FE}}$. Equation (9.3) also shows that sweeping out the u_i also removes the $\boldsymbol{\delta}$. The large-sample estimator of the VCE of $\widehat{\boldsymbol{\beta}}_{\text{FE}}$ is just the standard OLS estimator of the VCE that has been adjusted for the degrees of freedom used up by the within transform

$$s^2\left(\sum_{i=1}^N \sum_{t=1}^T \widetilde{\mathbf{x}}_{it}\widetilde{\mathbf{x}}_{it}'\right)^{-1}$$

where $s^2 = \{1/(NT-N-k-1)\}\sum_{i=1}^N \sum_{t=1}^T \widehat{\widetilde{\epsilon}}_{it}^{\,2}$ and $\widehat{\widetilde{\epsilon}}_{it}$ are the residuals from the OLS regression of $\widetilde{y}_{it}$ on $\widetilde{\mathbf{x}}_{it}$.

This model will have explanatory power *only if* the individual's y above or below the individual's mean is significantly correlated with the individual's $\mathbf{x}$ values above or below the individual's vector of mean $\mathbf{x}$ values. For that reason, it is termed the *within estimator*, since it depends on the variation *within* the unit. It does not matter if some individuals have, e.g., very high y values and very high $\mathbf{x}$ values because it is only the within variation that will show up as explanatory power.[2] This outcome clearly implies that any characteristic that does not vary over time for each unit cannot be included

1. Stata's set of `xt` commands extends these panel-data models in a variety of ways. For more information, see [XT] **xt**.

2. This is the panel analogue to the notion that OLS on a cross-section does not seek to "explain" the mean of y, but only the variation around that mean.

in the model, for instance, an individual's gender or a firm's three-digit SIC (industry) code. The unit-specific intercept term absorbs all heterogeneity in y and $\mathbf{x}$ that is a function of the identity of the unit, and any variable constant over time for each unit will be perfectly collinear with the unit's indicator variable.

We can fit the one-way individual FE model with the Stata command `xtreg` by using the `fe` (FE) option. The command has a syntax similar to that of `regress`:

`xtreg` *depvar* [*indepvars*] `, fe` [*options*]

As with standard regression, options include `robust` and `cluster()`. The command output displays estimates of σ_u^2 (labeled `sigma_u`), σ_ϵ^2 (labeled `sigma_e`), and what Stata terms `rho`: the fraction of variance due to u_i. Stata fits a model in which the u_i of (9.2) are taken as deviations from one constant term, displayed as `_cons`. The empirical correlation between u_i and the fitted values is also displayed as `corr(u_i, Xb)`. The FE estimator does not require a balanced panel as long as there are at least 2 observations per unit.[3]

We wish to test whether the individual-specific heterogeneity of u_i is necessary: are there distinguishable intercept terms across units? `xtreg, fe` provides an F test of the null hypothesis that the constant terms are equal across units. A rejection of this null hypothesis indicates that pooled OLS would produce inconsistent estimates. The one-way FE model also assumes that the errors are not contemporaneously correlated across units of the panel. De Hoyos and Sarafidis (2006) describe some new tests for contemporaneous correlation, and their command `xtcsd` is available from SSC. Likewise, a departure from the assumed homoskedasticity of ϵ_{it} across units of the panel—that is, a form of groupwise heteroskedasticity as discussed in section 6.2.2—may be tested by an LM statistic (Greene 2003, 328), available as the author's `xttest3` routine from `ssc` (Baum 2001). `xttest3` will operate on unbalanced panels.

The example below uses 1982–1988 state-level data for 48 U.S. states on traffic fatality rates (deaths per 100,000). We model the highway fatality rates as a function of several common factors: `beertax`, the tax on a case of beer; `spircons`, a measure of spirits consumption; and two economic factors: the state unemployment rate (`unrate`) and state per capita personal income, in thousands (`perincK`). Descriptive statistics for these variables of the `traffic.dta` dataset are given below.

3. An alternative command for this model is `areg`, which used to provide options unavailable from `xtreg`. With Stata version 9 or better, there is no advantage to using `areg`.

```
. use http://www.stata-press.com/data/imeus/traffic, clear
. xtsum fatal beertax spircons unrate perincK state year
Variable              Mean   Std. Dev.        Min        Max    Observations
-----------------+--------------------------------------------+----------------
fatal    overall | 2.040444   .5701938     .82121    4.21784 |     N =     336
         between |            .5461407   1.110077   3.653197 |     n =      48
         within  |            .1794253    1.45556   2.962664 |     T =       7
                 |                                            |
beertax  overall |  .513256   .4778442   .0433109   2.720764 |     N =     336
         between |            .4789513   .0481679   2.440507 |     n =      48
         within  |            .0552203   .1415352   .7935126 |     T =       7
                 |                                            |
spircons overall |  1.75369   .6835745        .79        4.9 |     N =     336
         between |            .6734649   .8614286   4.388572 |     n =      48
         within  |             .147792   1.255119   2.265119 |     T =       7
                 |                                            |
unrate   overall | 7.346726   2.533405        2.4         18 |     N =     336
         between |            1.953377        4.1       13.2 |     n =      48
         within  |            1.634257   4.046726   12.14673 |     T =       7
                 |                                            |
perincK  overall | 13.88018   2.253046   9.513762   22.19345 |     N =     336
         between |            2.122712    9.95087   19.51582 |     n =      48
         within  |            .8068546   11.43261   16.55782 |     T =       7
                 |                                            |
state    overall |  30.1875   15.30985          1         56 |     N =     336
         between |            15.44883          1         56 |     n =      48
         within  |                   0    30.1875    30.1875 |     T =       7
                 |                                            |
year     overall |     1985   2.002983       1982       1988 |     N =     336
         between |                   0       1985       1985 |     n =      48
         within  |            2.002983       1982       1988 |     T =       7
```

The results for the panel identifier, `state`, and time variable, `year`, illustrate the importance of the additional information provided by `xtsum`. By construction, the panel identifier `state` does not vary within the panels; i.e., it is time invariant. `xtsum` informs us of this fact by reporting that the within standard deviation is zero. Any variable with a within standard deviation of zero will be dropped from the FE model. The coefficients on variables with small within standard deviations are not well identified. The above output indicates that the coefficient on `beertax` may not be as well identified as the others. Similarly, the between standard deviation of `year` is zero by construction.

(*Continued on next page*)

The results of the one-way FE model are

```
. xtreg fatal beertax spircons unrate perincK, fe
Fixed-effects (within) regression               Number of obs      =       336
Group variable (i): state                       Number of groups   =        48

R-sq:  within  = 0.3526                         Obs per group: min =         7
       between = 0.1146                                        avg =       7.0
       overall = 0.0863                                        max =         7

                                                F(4,284)           =     38.68
corr(u_i, Xb)  = -0.8804                        Prob > F           =    0.0000

------------------------------------------------------------------------------
       fatal |      Coef.   Std. Err.      t    P>|t|     [95% Conf. Interval]
-------------+----------------------------------------------------------------
     beertax |  -.4840728   .1625106    -2.98   0.003    -.8039508   -.1641948
    spircons |   .8169652   .0792118    10.31   0.000     .6610484    .9728819
      unrate |  -.0290499   .0090274    -3.22   0.001    -.0468191   -.0112808
     perincK |   .1047103   .0205986     5.08   0.000      .064165    .1452555
       _cons |   -.383783   .4201781    -0.91   0.362    -1.210841    .4432754
-------------+----------------------------------------------------------------
     sigma_u |  1.1181913
     sigma_e |  .15678965
         rho |  .98071823   (fraction of variance due to u_i)
------------------------------------------------------------------------------
F test that all u_i=0:     F(47, 284) =    59.77             Prob > F = 0.0000
```

All explanatory factors are highly significant, with the unemployment rate having a negative effect on the fatality rate (perhaps since those who are unemployed are income constrained and drive fewer miles) and having income a positive effect (as expected because driving is a normal good). The estimate of `rho` suggests that almost all the variation in `fatal` is related to interstate differences in fatality rates. The F test following the regression indicates that there are significant individual (state level) effects, implying that pooled OLS would be inappropriate.

9.1.2 Time effects and two-way FE

Stata lacks a command to automatically fit two-way FE models. If the number of periods is reasonably small, we can fit a two-way FE model by creating a set of time indicator variables and including all but one in the regression.[4] The joint test that all the coefficients on those indicator variables are zero will be a test of the significance of time FE. Just as the individual FE model requires regressors' variation over time within each unit, a time FE (implemented with a time indicator variable) requires regressors' variation over units within each period. Estimating an equation from individual or firm microdata implies that we cannot include a macrofactor such as the rate of GDP growth or price inflation in a model with a time FE because those factors do not vary across individuals. `xtsum` can be used to check that the between standard deviation is greater than zero.

4. In the context of a balanced panel, Hsiao (1986) proposes an algebraic solution involving "double demeaning", which allows estimation of a two-way FE model with no i or t indicator variables.

We consider the two-way FE model by adding time effects to the model of the previous example. The time effects are generated by `tabulate`'s `generate()` option and then transformed into centered indicators (as discussed in section 7.1.1) by subtracting the indicator for the excluded class from each of the other indicator variables. This transformation expresses the time effects as variations from the conditional mean of the sample rather than deviations from the excluded class (1988).

```
. quietly tabulate year, generate(yr)
. local j 0
. forvalues i=82/87 {
  2.         local ++j
  3.         rename yr`j' yr`i'
  4.         quietly replace yr`i' = yr`i' - yr7
  5.         }
. drop yr7
. xtreg fatal beertax spircons unrate perincK yr*, fe
Fixed-effects (within) regression               Number of obs      =       336
Group variable (i): state                       Number of groups   =        48

R-sq:  within  = 0.4528                         Obs per group: min =         7
       between = 0.1090                                        avg =       7.0
       overall = 0.0770                                        max =         7

                                                F(10,278)          =     23.00
corr(u_i, Xb)  = -0.8728                        Prob > F           =    0.0000

------------------------------------------------------------------------------
       fatal |      Coef.   Std. Err.      t    P>|t|     [95% Conf. Interval]
-------------+----------------------------------------------------------------
     beertax |  -.4347195   .1539564    -2.82   0.005    -.7377878   -.1316511
    spircons |    .805857   .1126425     7.15   0.000     .5841163    1.027598
      unrate |  -.0549084   .0103418    -5.31   0.000    -.0752666   -.0345502
     perincK |   .0882636   .0199988     4.41   0.000     .0488953    .1276319
        yr82 |   .1004321   .0355629     2.82   0.005     .0304253     .170439
        yr83 |   .0470609   .0321574     1.46   0.144    -.0162421    .1103638
        yr84 |  -.0645507   .0224667    -2.87   0.004    -.1087771   -.0203243
        yr85 |  -.0993055   .0198667    -5.00   0.000    -.1384139   -.0601971
        yr86 |   .0496288   .0232525     2.13   0.034     .0038554    .0954021
        yr87 |   .0003593   .0289315     0.01   0.990    -.0565933    .0573119
       _cons |   .0286246   .4183346     0.07   0.945    -.7948812    .8521305
-------------+----------------------------------------------------------------
     sigma_u |  1.0987683
     sigma_e |  .14570531
         rho |  .98271904   (fraction of variance due to u_i)
------------------------------------------------------------------------------
F test that all u_i=0:     F(47, 278) =    64.52             Prob > F = 0.0000
. test yr82 yr83 yr84 yr85 yr86 yr87
 ( 1)  yr82 = 0
 ( 2)  yr83 = 0
 ( 3)  yr84 = 0
 ( 4)  yr85 = 0
 ( 5)  yr86 = 0
 ( 6)  yr87 = 0

       F(  6,   278) =    8.48
            Prob > F =    0.0000
```

The four quantitative factors included in the one-way FE model retain their sign and significance in the two-way FE model. The time effects are jointly significant, suggesting that they should be included in a properly specified model. Otherwise, the model is qualitatively similar to the earlier model, with much variation explained by the individual FE.

9.1.3 The between estimator

Another estimator for a panel dataset is the *between estimator*, in which the group means of y are regressed on the group means of $\mathbf{x}$ in a regression of N observations. This estimator ignores all the individual-specific variation in y that is considered by the within estimator, replacing each observation for an individual with his or her mean behavior. The between estimator is the OLS estimator of $\boldsymbol{\beta}$ and $\boldsymbol{\delta}$ from the model

$$\overline{y}_i = \overline{\mathbf{x}}_i\boldsymbol{\beta} + \overline{\mathbf{z}}_i\boldsymbol{\delta} + u_i + \overline{\epsilon}_i \tag{9.4}$$

Equation (9.4) shows that if the u_i are correlated with any of the regressors in the model, the zero-conditional-mean assumption does not hold and the between estimator will produce inconsistent results.

This estimator is not widely used but has sometimes been applied where the time-series data for each individual are thought to be somewhat inaccurate or when they are assumed to contain random deviations from long-run means. If you assume that the inaccuracy has mean zero over time, a solution to this measurement error problem can be found by averaging the data over time and retaining only 1 observation per unit. We could do so explicitly with Stata's `collapse` command, which would generate a new dataset of that nature (see section 3.3). However, you need not form that dataset to use the between estimator because the command `xtreg` with the `be` (between) option will invoke it. Using the between estimator requires that $N > k$. Any macro factor that is constant over individuals cannot be included in the between estimator because its average will not differ by individual.

We can show that the pooled OLS estimator is a matrix-weighted average of the within and between estimators, with the weights defined by the relative precision of the two estimators. With panel data, we can identify whether the interesting sources of variation are in individuals' variation around their means or in those means themselves. The within estimator takes account of only the former, whereas the between estimator considers only the latter.

To show why we account for all the information present in the panel, we refit the first model above with the between estimator (the second model, containing year FE, is not appropriate, since the time dimension is suppressed by the between estimator). Interestingly, two of the factors that played an important role in the one- and two-way FE model, `beertax` and `unrate`, play no significant role in this regression on group (state) means.

```
. xtreg fatal beertax spircons unrate perincK, be
Between regression (regression on group means)  Number of obs      =       336
Group variable (i): state                       Number of groups   =        48
R-sq:  within  = 0.0479                         Obs per group: min =         7
       between = 0.4565                                        avg =       7.0
       overall = 0.2583                                        max =         7
                                                F(4,43)            =      9.03
sd(u_i + avg(e_i.))=  .4209489                  Prob > F           =    0.0000

       fatal       Coef.   Std. Err.      t    P>|t|     [95% Conf. Interval]

     beertax    .0740362   .1456333     0.51   0.614    -.2196614    .3677338
    spircons    .2997517   .1128135     2.66   0.011     .0722417    .5272618
      unrate    .0322333    .038005     0.85   0.401    -.0444111    .1088776
     perincK   -.1841747   .0422241    -4.36   0.000    -.2693277   -.0990218
       _cons    3.796343   .7502025     5.06   0.000     2.283415    5.309271
```

9.1.4 One-way RE

Rather than considering the individual-specific intercept as an FE of that unit, the RE model specifies the individual effect as a random draw that is uncorrelated with the regressors and the overall disturbance term

$$y_{it} = \mathbf{x}_{it}\boldsymbol{\beta} + \mathbf{z}_i\boldsymbol{\delta} + (u_i + \epsilon_{it}) \tag{9.5}$$

where $(u_i + \epsilon_{it})$ is a composite error term and the u_i are the individual effects. A crucial assumption of this model is that the u_i are uncorrelated with the regressors $\mathbf{x}_{it}$ and $\mathbf{z}_i$. This orthogonality assumption implies that the parameters can be consistently estimated by OLS and the between estimator, but neither of these estimators is efficient. The RE estimator uses the assumption that the u_i are uncorrelated with regressors to construct a more efficient estimator. If the regressors are correlated with the u_i, they are correlated with the composite error term and the RE estimator is inconsistent.

The RE model uses the orthogonality between the u_i and the regressors to greatly reduce the number of estimated parameters. In a large survey, with thousands of individuals, an RE model has $k+p$ coefficients and two variance parameters, whereas an FE model has $k-1+N$ coefficients and one variance parameter. The coefficients on time-invariant variables are identified in the RE model. Because the RE model identifies the population parameter that describes the individual-level heterogeneity, inference from the RE model pertains to the underlying population of individuals. In contrast, because the FE model cannot estimate the parameters that describe the individual-level heterogeneity, inference from the FE model is conditional on the FE in the sample. Therefore, the RE model is more efficient and allows a broader range of statistical inference. The key assumption that the u_i are uncorrelated with the regressors can and should be tested.

To implement the one-way RE formulation of (9.5), we assume that both u and ϵ are mean-zero processes, uncorrelated with the regressors; that they are each homoskedastic; that they are uncorrelated with each other; and that there is no correlation over individuals or time. For the T observations belonging to the ith unit of the panel, the composite error process

$$\eta_{it} = u_i + \epsilon_{it}$$

gives rise to the *error-components* model with conditional variance

$$E[\eta_{it}^2|\mathbf{x}^*] = \sigma_u^2 + \sigma_\epsilon^2$$

and conditional covariance within a unit of

$$E[\eta_{it}\eta_{is}|\mathbf{x}^*] = \sigma_u^2, \quad t \neq s$$

The covariance matrix of these T errors can then be written as

$$\Sigma = \sigma_\epsilon^2 I_T + \sigma_u^2 \iota_T \iota_T'$$

Since observations i and j are uncorrelated, the full covariance matrix of η across the sample is block diagonal in $\mathbf{\Sigma}$: $\mathbf{\Omega} = \mathbf{I}_n \otimes \mathbf{\Sigma}$.[5,6]

The GLS estimator for the slope parameters of this model is

$$\begin{aligned} \widehat{\boldsymbol{\beta}}_{\text{RE}} &= (\mathbf{X}^{*'}\mathbf{\Omega}^{-1}\mathbf{X}^*)^{-1}(\mathbf{X}^{*'}\mathbf{\Omega}^{-1}\mathbf{y}) \\ &= \left(\sum_i \mathbf{X}_i^{*'}\mathbf{\Sigma}^{-1}\mathbf{X}_i^*\right)^{-1}\left(\sum_i \mathbf{X}_i^{*'}\mathbf{\Sigma}^{-1}\mathbf{y}_i\right) \end{aligned}$$

To compute this estimator, we require $\mathbf{\Omega}^{-1/2} = (\mathbf{I}_n \otimes \mathbf{\Sigma})^{-1/2}$, which involves

$$\mathbf{\Sigma}^{-1/2} = \sigma_\epsilon^{-1}(\mathbf{I} - T^{-1}\theta \boldsymbol{\iota}_T \boldsymbol{\iota}_T')$$

where

$$\theta = 1 - \frac{\sigma_\epsilon}{\sqrt{\sigma_\epsilon^2 + T\sigma_u^2}}$$

and the *quasidemeaning* transformation defined by $\mathbf{\Sigma}^{-1/2}$ is then $\sigma_\epsilon^{-1}(y_{it} - \theta\overline{y}_i)$; that is, rather than subtracting the entire individual mean of y from each value, we should subtract some fraction of that mean, as defined by θ. The quasidemeaning transformation reduces to the within transformation when $\theta = 1$. Like pooled OLS, the GLS RE estimator is a matrix-weighted average of the within and between estimators, but we apply optimal weights, as based on

$$\lambda = \frac{\sigma_\epsilon^2}{\sigma_\epsilon^2 + T\sigma_u^2} = (1-\theta)^2$$

---

5. The operator $\otimes$ denotes the Kronecker product of the two matrices. For any matrices $\mathbf{A}_{K \times L}, \mathbf{B}_{M \times N}$, $\mathbf{A} \otimes \mathbf{B} = \mathbf{C}_{KM \times LN}$. To form the product matrix, each element of $\mathbf{A}$ scalar multiplies the entire matrix $\mathbf{B}$. See Greene (2003, 824–825).

6. I give the expressions for a balanced panel. Unbalanced panels merely complicate the algebra.

where λ is the weight attached to the covariance matrix of the between estimator. To the extent that λ differs from unity, pooled OLS will be inefficient, as it will attach too much weight on the between-units variation, attributing it all to the variation in $\mathbf{x}$ rather than apportioning some of the variation to the differences in ϵ_i across units.

The setting $\lambda = 1$ ($\theta = 0$) is appropriate if $\sigma_u^2 = 0$; that is, if there are no RE, then a pooled OLS model is optimal. If $\theta = 1$, $\lambda = 0$ and the FE estimator is appropriate. To the extent that λ differs from zero, the FE estimator will be inefficient, in that it applies zero weight to the between estimator. The GLS RE estimator applies the optimal λ in the unit interval to the between estimator, whereas the FE estimator arbitrarily imposes $\lambda = 0$. This imposition would be appropriate only if the variation in ϵ was trivial in comparison with the variation in u.

To implement the FGLS estimator of the model, all we need are consistent estimates of σ_ϵ^2 and σ_u^2. Because the FE model is consistent, its residuals can be used to estimate σ_ϵ^2. Likewise, the residuals from the pooled OLS model can be used to generate a consistent estimate of $(\sigma_\epsilon^2 + \sigma_u^2)$. These two estimators may be used to estimate θ and transform the data for the GLS model.[7] Because the GLS model uses quasidemeaning, it can include time-invariant variables (such as gender or race).

The FGLS estimator may be executed in Stata by using the command `xtreg` with the `re` (RE) option. The command will display estimates of σ_u^2, σ_ϵ^2, and what Stata calls `rho`: the fraction of the total variance due to ϵ_i. Breusch and Pagan (1980) have developed a Lagrange multiplier test for $\sigma_u^2 = 0$, which may be computed following an RE estimation via the command `xttest0` (see [XT] **xtreg** for details).

We can also estimate the parameters of the RE model with full maximum likelihood. Typing `xtreg, mle` requests that estimator. The application of maximum likelihood estimation continues to assume that the regressors and u are uncorrelated, adding the assumption that the distributions of u and ϵ are normal. This estimator will produce a likelihood-ratio test of $\sigma_u^2 = 0$ corresponding to the Breusch–Pagan test available for the GLS estimator.

To illustrate the one-way RE estimator and implement a test of the orthogonality assumption under which RE is appropriate and preferred, we estimate the parameters of the RE model that corresponds to the FE model above.

7. A possible complication: as generally defined, the two estimators above are not guaranteed to generate a positive estimate of σ_ϵ^2 in finite samples. Then the variance estimates without degrees-of-freedom corrections, which will still be consistent, may be used.

```
. xtreg fatal beertax spircons unrate perincK, re
Random-effects GLS regression                   Number of obs      =       336
Group variable (i): state                       Number of groups   =        48
R-sq:  within  = 0.2263                         Obs per group: min =         7
       between = 0.0123                                        avg =       7.0
       overall = 0.0042                                        max =         7
Random effects u_i ~ Gaussian                   Wald chi2(4)       =     49.90
corr(u_i, X)       = 0 (assumed)                Prob > chi2        =    0.0000

       fatal        Coef.   Std. Err.      z    P>|z|     [95% Conf. Interval]

     beertax     .0442768   .1204613     0.37   0.713     -.191823    .2803765
    spircons     .3024711   .0642954     4.70   0.000     .1764546    .4284877
      unrate    -.0491381   .0098197    -5.00   0.000    -.0683843   -.0298919
     perincK    -.0110727   .0194746    -0.57   0.570    -.0492423    .0270968
       _cons     2.001973   .3811247     5.25   0.000     1.254983    2.748964

     sigma_u    .41675665
     sigma_e    .15678965
         rho    .87601197   (fraction of variance due to u_i)
```

Compared with the FE model, where all four regressors were significant, we see that the `beertax` and `perincK` variables do not have significant effects on the fatality rate. The latter variable's coefficient switched sign.

9.1.5 Testing the appropriateness of RE

We can use a Hausman test (presented in section 8.11) to test the null hypothesis that the extra orthogonality conditions imposed by the RE estimator are valid. If the regressors are correlated with the u_i, the FE estimator is consistent but the RE estimator is not consistent. If the regressors are uncorrelated with the u_i, the FE estimator is still consistent, albeit inefficient, whereas the RE estimator is consistent and efficient. Therefore, we may consider these two alternatives in the Hausman test framework, fitting both models and comparing their common coefficient estimates in a probabilistic sense. If both FE and RE models generate consistent point estimates of the slope parameters, they will not differ meaningfully. If the orthogonality assumption is violated, the inconsistent RE estimates will significantly differ from their FE counterparts.

To implement the Hausman test, we fit each model and store its results by typing `estimates store` *set* after each estimation (*set* defines that set of estimates: for instance, *set* might be `fix` for the FE model). Then typing **hausman** *setconsist seteff* will invoke the Hausman test, where *setconsist* refers to the name of the FE estimates (which are consistent under the null and alternative) and *seteff* refers to the name of the RE estimates, which are consistent and efficient only under the null hypothesis. This test uses the difference of the two estimated covariance matrices (which is not guaranteed to be positive definite) to weight the difference between the FE and RE vectors of slope coefficients.

We illustrate the Hausman test with the two forms of the motor vehicle fatality equation:

```
. quietly xtreg fatal beertax spircons unrate perincK, fe
. estimates store fix
. quietly xtreg fatal beertax spircons unrate perincK, re
. estimates store ran
. hausman fix ran
                 ---- Coefficients ----
             |      (b)          (B)            (b-B)     sqrt(diag(V_b-V_B))
             |      fix          ran         Difference          S.E.
-------------+----------------------------------------------------------------
     beertax |   -.4840728     .0442768       -.5283495        .1090815
    spircons |    .8169652     .3024711        .514494         .0462668
      unrate |   -.0290499    -.0491381       .0200882              .
     perincK |    .1047103    -.0110727        .115783         .0067112
------------------------------------------------------------------------------
                          b = consistent under Ho and Ha; obtained from xtreg
           B = inconsistent under Ha, efficient under Ho; obtained from xtreg

    Test:  Ho:  difference in coefficients not systematic

                  chi2(4) = (b-B)'[(V_b-V_B)^(-1)](b-B)
                          =       130.93
                Prob>chi2 =       0.0000
                (V_b-V_B is not positive definite)
```

As we might expect from the different point estimates generated by the RE estimator, the Hausman test's null hypothesis—that the RE estimator is consistent—is soundly rejected. The state-level individual effects do appear to be correlated with the regressors.[8]

9.1.6 Prediction from one-way FE and RE

Following `xtreg`, the `predict` command may be used to generate a variety of series. The default result is `xb`, the linear prediction of the model. Stata normalizes the unit-specific effects (whether fixed or random) as deviations from the intercept term `_cons`; therefore, the `xb` prediction ignores the individual effect. We can generate predictions that include the RE or FE by specifying the `xbu` option; the individual effect itself may be predicted with option `u`;[9] and the ϵ_{it} error component (or "true" residual) may be predicted with option `e`. The three last predictions are available only in sample for either the FE or RE model, whereas the linear prediction `xb` and the "combined residual" (option `ue`) by default will be computed out of sample as well, just as with predictions from `regress`.

8. Here Stata signals that the difference of the estimated VCEs is not positive definite.

9. Estimates of u_i are not consistent with $N \rightarrow \infty$ and fixed T.

9.2 IV models for panel data

If the Hausman test indicates that the RE u_i cannot be considered orthogonal to the individual-level error, an IV estimator may be used to generate consistent estimates of the coefficients on the time-invariant variables. The Hausman–Taylor estimator (Hausman and Taylor 1981) assumes that some of the regressors in $\mathbf{x}_{it}$ and $\mathbf{z}_i$ are correlated with u but that none are correlated with ϵ. This estimator is available in Stata as `xthtaylor`. This approach begins by writing (9.2) as

$$y_{it} = \mathbf{x}_{1,it}\boldsymbol{\beta}_1 + \mathbf{x}_{2,it}\boldsymbol{\beta}_2 + \mathbf{z}_{1,i}\boldsymbol{\delta}_1 + \mathbf{z}_{2,i}\boldsymbol{\delta}_2 + u_i + \epsilon_{it}$$

where the $\mathbf{x}$ variables are time varying, the $\mathbf{z}$ variables are time invariant, the variables subscripted with a "1" are exogenous, and the variables subscripted with a "2" are correlated with the u_i. Identifying the parameters requires that k_1 (the number of $\mathbf{x}_{1,it}$ variables) be at least as large as ℓ_2 (the number of $\mathbf{z}_{2,i}$ variables). Applying the Hausman–Taylor estimator circumvents the problem that the $\mathbf{x}_{2,it}$ and $\mathbf{z}_{2,i}$ variables are correlated with u_i, but it requires that we find variables that are not correlated with the individual-level effect.

Stata also provides an IV estimator for the FE and RE models in which some of the $\mathbf{x}_{it}$ and $\mathbf{z}_i$ variables are correlated with the disturbance term ϵ_{it}. These are different assumptions about the nature of any suspected correlation between the regressor and the composite error term from those underlying the Hausman–Taylor estimator. The `xtivreg` command offers FE, RE, between-effects, and first-differenced IV estimators in a panel-data context.

9.3 Dynamic panel-data models

A serious difficulty arises with the one-way FE model in the context of a dynamic panel-data (DPD) model, one containing a lagged dependent variable (and possibly other regressors), particularly in the "small T, large N" context. As Nickell (1981) shows, this problem arises because the within-transform N, the lagged dependent variable, is correlated with the error term. As Nickell (1981) shows, the resulting correlation creates a large-sample bias in the estimate of the coefficient of the lagged dependent variable, which is not mitigated by increasing N, the number of individual units. In the simplest setup of a pure AR(1) model without additional regressors:

$$\begin{aligned} y_{it} &= \beta + \rho y_{i,t-1} + u_i + \epsilon_{it} \\ y_{it} - \overline{y}_{ix} &= \rho(y_{i,t-1} - \overline{L.y_i}) + (\epsilon_{it} - \epsilon_{i\cdot}) \end{aligned}$$

$\overline{L.y_i}$ is correlated with $(\epsilon_{it} - \epsilon_i)$ by definition. Nickell demonstrates that the inconsistency of $\widehat{\rho}$ as $N \to \infty$ is of order $1/T$, which may be sizable in a "small T" context. If $\rho > 0$, the bias is invariably negative, so the persistence of y will be underestimated. For reasonably large values of T, the limit of $(\widehat{\rho} - \rho)$ as $N \to \infty$ will be approximately $-(1+\rho)/(T-1)$, which is a sizable value. With $T = 10$ and $\rho = 0.5$, the bias will be -0.167, or about $1/3$ of the true value. Including more regressors does not remove this bias. If the regressors are correlated with the lagged dependent variable to some degree, their coefficients may be seriously biased as well. This bias is not caused by an autocorrelation in the error process ϵ and arises even if the error process is i.i.d. If the error process is autocorrelated, the problem is even more severe given the difficulty of deriving a consistent estimate of the AR parameters in that context. The same problem affects the one-way RE model. The u_i error component enters every value of y_{it} by assumption, so that the lagged dependent variable cannot be independent of the composite error process.

A solution to this problem involves taking first differences of the original model. Consider a model containing a lagged dependent variable and regressor $\mathbf{x}$:

$$y_{it} = \beta_1 + \rho y_{i,t-1} + \mathbf{x}_{it}\boldsymbol{\beta}_2 + u_i + \epsilon_{it}$$

The first difference transformation removes both the constant term and the individual effect:

$$\Delta y_{it} = \rho \Delta y_{i,t-1} + \Delta \mathbf{x}_{it}\boldsymbol{\beta}_2 + \Delta \epsilon_{it}$$

There is still correlation between the differenced lagged dependent variable and the disturbance process [which is now a first-order moving average process, or MA(1)]: the former contains $y_{i,t-1}$ and the latter contains $\epsilon_{i,t-1}$. But with the individual FE swept out, a straightforward IV estimator is available. We may construct instruments for the lagged dependent variable from the second and third lags of y, either in the form of differences or lagged levels. If ϵ is i.i.d., those lags of y will be highly correlated with the lagged dependent variable (and its difference) but uncorrelated with the composite-error process.[10] Even if we believed that ϵ might be following an AR(1) process, we could still follow this strategy, "backing off" one period and using the third and fourth lags of y (presuming that the time series for each unit is long enough to do so).

The DPD approach of Arellano and Bond (1991) is based on the notion that the IV approach noted above does not exploit all the information available in the sample. By doing so in a GMM context, we can construct more efficient estimates of the DPD model. The Arellano–Bond estimator can be thought of as an extension to the Anderson–Hsiao estimator implemented by `xtivreg, fd`. Arellano and Bond argue that the Anderson–Hsiao estimator, although consistent, fails to take all the potential orthogonality conditions into account. Consider the equations

$$\begin{aligned} y_{it} &= \mathbf{x}_{it}\boldsymbol{\beta}_1 + \mathbf{w}_{it}\boldsymbol{\beta}_2 + v_{it} \\ v_{it} &= u_i + \epsilon_{it} \end{aligned}$$

where $\mathbf{x}_{it}$ includes strictly exogenous regressors and $\mathbf{w}_{it}$ are predetermined regressors (which may include lags of y) and endogenous regressors, all of which may be correlated with u_i, the unobserved individual effect. First-differencing the equation removes the u_i and its associated omitted-variable bias. The Arellano–Bond estimator begins by specifying the model as a system of equations, one per period, and allows the instruments

10. The degree to which these instruments are not weak depends on the true value of ρ. See Arellano and Bover (1995) and Blundell and Bond (1998).

applicable to each equation to differ (for instance, in later periods, more lagged values of the instruments are available). The instruments include suitable lags of the levels of the endogenous variables, which enter the equation in differenced form, as well as the strictly exogenous regressors and any others that may be specified. This estimator can easily generate a great many instruments, since by period τ all lags prior to, say, $(\tau - 2)$ might be individually considered as instruments. If T is nontrivial, we may need to use the option that limits the maximum lag of an instrument to prevent the number of instruments from becoming too large. This estimator is available in Stata as `xtabond` (see [XT] **xtabond**).

A potential weakness in the Arellano–Bond DPD estimator was revealed in later work by Arellano and Bover (1995) and Blundell and Bond (1998). The lagged levels are often rather poor instruments for first-differenced variables, especially if the variables are close to a random walk. Their modification of the estimator includes lagged levels as well as lagged differences. The original estimator is often entitled difference GMM, whereas the expanded estimator is commonly termed system GMM. The cost of the system GMM estimator involves a set of additional restrictions on the initial conditions of the process generating y.

Both the difference GMM and system GMM estimators have one-step and two-step variants. The two-step estimates of the difference GMM standard errors have been shown to have a severe downward bias. To evaluate the precision of the two-step estimators for hypothesis tests, we should apply the "Windmeijer finite-sample correction" (see Windmeijer 2005) to these standard errors. Bond (2002) provides an excellent guide to the DPD estimators.

All the features described above are available in David Roodman's improved version of official Stata's estimator. His version, `xtabond2`, offers a much more flexible syntax than official Stata's `xtabond`, which does not allow the same specification of instrument sets, nor does it provide the system GMM approach or the Windmeijer correction to the standard errors of the two-step estimates. On the other hand, Stata's `xtabond` has a simpler syntax and is faster, so you may prefer to use it.

To illustrate the use of the DPD estimators, we first specify a model of `fatal` as depending on the prior year's value (`L.fatal`), the state's `spircons`, and a time trend (`year`). We provide a set of instruments for that model with the `gmm` option and list `year` as an `iv` instrument. We specify that the two-step Arellano–Bond estimator be used with the Windmeijer correction. The `noleveleq` option specifies the original Arellano–Bond estimator in differences:[11]

```
. use http://www.stata-press.com/data/imeus/traffic, clear
. tsset
       panel variable:  state, 1 to 56
        time variable:  year, 1982 to 1988
```

11. The estimated parameters of the difference GMM model do not include a constant term because it is differenced out.

```
. xtabond2 fatal L.fatal spircons year,
> gmmstyle(beertax spircons unrate perincK)
> ivstyle(year) twostep robust noleveleq
Favoring space over speed. To switch, type or click on mata: mata set matafavor
>  speed.
Warning: Number of instruments may be large relative to number of observations.
Suggested rule of thumb: keep number of instruments <= number of groups.

Arellano-Bond dynamic panel-data estimation, two-step difference GMM results
------------------------------------------------------------------------------
Group variable: state                           Number of obs      =       240
Time variable : year                            Number of groups   =        48
Number of instruments = 48                      Obs per group: min =         5
Wald chi2(3)    =      51.90                                   avg =      5.00
Prob > chi2     =      0.000                                   max =         5
------------------------------------------------------------------------------
             |              Corrected
             |      Coef.   Std. Err.      z    P>|z|     [95% Conf. Interval]
-------------+----------------------------------------------------------------
fatal        |
         L1. |   .3205569    .071963     4.45   0.000      .1795121    .4616018
    spircons |   .2924675   .1655214     1.77   0.077     -.0319485    .6168834
        year |   .0340283   .0118935     2.86   0.004      .0107175    .0573391
------------------------------------------------------------------------------
Hansen test of overid. restrictions: chi2(82) =  47.26    Prob > chi2 =  0.999

Arellano-Bond test for AR(1) in first differences: z =  -3.17  Pr > z =  0.002
Arellano-Bond test for AR(2) in first differences: z =   1.24  Pr > z =  0.216
------------------------------------------------------------------------------
```

This model is moderately successful in relating `spircons` to the dynamics of the fatality rate. The Hansen test of overidentifying restrictions is satisfactory, as is the test for AR(2) errors. We expect to reject the test for AR(1) errors in the Arellano–Bond model.

To contrast the difference GMM and system GMM approaches, we use the latter estimator by dropping the `noleveleq` option:

(Continued on next page)

```
. xtabond2 fatal L.fatal spircons year,
> gmmstyle(beertax spircons unrate perincK) ivstyle(year) twostep robust
Favoring space over speed. To switch, type or click on mata: mata set matafavor
> speed.
Warning: Number of instruments may be large relative to number of observations.
Suggested rule of thumb: keep number of instruments <= number of groups.
Arellano-Bond dynamic panel-data estimation, two-step system GMM results
------------------------------------------------------------------------------
Group variable: state                           Number of obs      =       288
Time variable : year                            Number of groups   =        48
Number of instruments = 48                      Obs per group: min =         6
Wald chi2(3)  =    1336.50                                     avg =      6.00
Prob > chi2   =      0.000                                     max =         6
------------------------------------------------------------------------------
             |              Corrected
             |      Coef.   Std. Err.      z    P>|z|     [95% Conf. Interval]
-------------+----------------------------------------------------------------
       fatal |
         L1. |   .8670531   .0272624    31.80   0.000     .8136198    .9204865
    spircons |  -.0333786   .0166285    -2.01   0.045    -.0659697   -.0007874
        year |   .0135718   .0051791     2.62   0.009     .0034209    .0237226
       _cons |  -26.62532   10.27954    -2.59   0.010    -46.77285   -6.477799
------------------------------------------------------------------------------
Hansen test of overid. restrictions: chi2(110) =  44.26   Prob > chi2 =  1.000
Arellano-Bond test for AR(1) in first differences: z =  -3.71  Pr > z =  0.000
Arellano-Bond test for AR(2) in first differences: z =   1.77  Pr > z =  0.077
------------------------------------------------------------------------------
```

Although the other summary measures from this estimator are acceptable, the marginally significant negative coefficient on `spircons` casts doubt on this specification.

9.4 Seemingly unrelated regression models

Often we want to estimate a similar specification for several different units, a production function or cost function for each industry. If the equation to be estimated for a given unit meets the zero-conditional-mean assumption of (4.2), we can estimate each equation independently. However, we may want to estimate the equations jointly: first, to allow cross-equation restrictions to be imposed or tested, and second, to gain efficiency, since we might expect the error terms across equations to be *contemporaneously correlated*. Such equations are often called *seemingly unrelated regressions* (SURs), and Zellner (1962) proposed an estimator for this problem: the SUR estimator. Unlike the FE and RE estimators, whose large-sample justification is based on "small T, large N" datasets in which $N \rightarrow \infty$, the SUR estimator is based on the large-sample properties of "large T, small N" datasets in which $T \rightarrow \infty$, so it may be considered a multiple time-series estimator.

Equation i of the SUR model is

$$y_i = \mathbf{x}_i \boldsymbol{\beta}_i + \epsilon_i, \;\; i = 1, \ldots, N$$

where y_i is the ith equation's dependent variable and $\mathbf{X}_i$ is the $T \times k_i$ matrix of observations on the regressors for the ith equation. The disturbance process $\boldsymbol{\epsilon} = (\epsilon_1', \epsilon_2', \ldots, \epsilon_N')'$ is assumed to have an expectation of zero and an $NT \times NT$ covariance matrix of $\boldsymbol{\Omega}$. We will consider only the case where we have T observations per equation, although we could fit the model with an unbalanced panel. Each equation may have a differing set of regressors, and apart from the constant term, there may be no variables in common across the $\mathbf{x}_i$. Applying SUR requires that the T observations per unit exceed N, the number of units, to render $\boldsymbol{\Omega}$ of full rank and invertible. If this constraint is not satisfied, we cannot use SUR. In practice, T should be much larger than N for the large-sample approximations to work well.

We assume that $E[\epsilon_{it}\epsilon_{js}] = \sigma_{ij},\ t = s$, and otherwise zero, which implies that we are allowing for the error terms in different equations to be *contemporaneously correlated*, but assuming that they are not correlated at other points (including within a unit: they are assumed independent). Thus for any two error vectors,

$$
\begin{aligned}
E[\epsilon_i \epsilon_j'] &= \sigma_{ij}\mathbf{I}_T \\
\boldsymbol{\Omega} &= \boldsymbol{\Sigma} \otimes \mathbf{I}_T
\end{aligned}
$$

where $\boldsymbol{\Sigma}$ is the $N \times N$ covariance matrix of the N error vectors and $\otimes$ is the Kronecker matrix product.

The efficient estimator for this problem is GLS, in which we can write $\mathbf{y}$ as the stacked set of $\mathbf{y}_i$ vectors and $\mathbf{X}$ as the block-diagonal matrix of $\mathbf{X}_i$. Since the GLS estimator is

$$\widehat{\boldsymbol{\beta}}_{\text{GLS}} = (\mathbf{X}'\boldsymbol{\Omega}^{-1}\mathbf{X})(\mathbf{X}'\boldsymbol{\Omega}^{-1}\mathbf{y})$$

and

$$\boldsymbol{\Omega}^{-1} = \boldsymbol{\Sigma}^{-1} \otimes \mathbf{I}$$

We can write the (infeasible) GLS estimator as

$$\widehat{\boldsymbol{\beta}}_{\text{GLS}} = \{\mathbf{X}'(\boldsymbol{\Sigma}^{-1} \otimes \mathbf{I})\mathbf{X}\}^{-1}\{\mathbf{X}'(\boldsymbol{\Sigma}^{-1} \otimes \mathbf{I})\mathbf{y}\}$$

which if expanded demonstrates that each block of the $\mathbf{X}_i'\mathbf{X}_j$ matrix is weighted by the scalar σ_{ij}^{-1}. The large-sample VCE of $\widehat{\boldsymbol{\beta}}_{\text{GLS}}$ is the first term of this expression.

When will this estimator provide a gain in efficiency over equation-by-equation OLS? First, if the $\sigma_{ij},\ i \neq j$ are actually zero, there is no gain. Second, if the $\mathbf{X}_i$ matrices are identical across equations—not merely having the same variable names, but containing the same numerical values—GLS is identical to equation-by-equation OLS, and there is no gain. Beyond these cases, the gain in efficiency depends on the magnitude of the cross-equation contemporaneous correlations of the residuals. The higher those correlations are, the greater the gain will be. Furthermore, if the $\mathbf{X}_i$ matrices' columns are highly correlated across equations, the gains will be smaller.

The feasible SUR estimator requires a consistent estimate of $\boldsymbol{\Sigma}$, the $N \times N$ contemporaneous covariance matrix of the equations' disturbance processes. We can estimate the representative element σ_{ij}, the contemporaneous correlation between ϵ_i, ϵ_j, from equation-by-equation OLS residuals as

$$s_{ij} = \frac{e_i' e_j}{T}$$

assuming that each unit's equation is estimated from T observations.[12] We use these estimates to perform the "Zellner step", where the algebra of partitioned matrices will show that the Kronecker products may be rewritten as products of the blocks in the expression for $\widehat{\boldsymbol{\beta}}_{\text{GLS}}$. The estimator may be iterated. The GLS estimates will produce a new set of residuals, which may be used in a second Zellner step, and so on. Iteration will make the GLS estimates equivalent to maximum likelihood estimates of the system.

The SUR estimator is available in Stata via the `sureg` command; see [R] **sureg**. SUR can be applied to panel-data models in the wide format.[13] SUR is a more attractive estimator than pooled OLS, or even FE, in that SUR allows each unit to have its own coefficient vector.[14] Not only does the constant term differ from unit to unit, but each of the slope parameters and σ_ϵ^2 differ across units. In contrast, the slope and variance parameters are constrained to be equal across units in pooled OLS, FE, or RE estimators. We can use standard F tests to compare the unrestricted SUR results with those that may be generated in the presence of linear constraints, such as cross-equation restrictions (see [R] **constraint**). Cross-equation constraints correspond to the restriction that a particular regressor's effect is the same for each panel unit. We can use the `isure` option to iterate the estimates, as described above.

We can test whether applying SUR has yielded a significant gain in efficiency by using a test for the diagonality of $\boldsymbol{\Sigma}$ proposed by Breusch and Pagan (1980).[15] Their LM statistic sums the squared correlations between residual vectors **i** and **j**, with a null hypothesis of diagonality (zero contemporaneous covariance between the errors of different equations). This test is produced by `sureg` when the `corr` option is specified.

We apply SUR to detrended annual output and factor input prices of five U.S. industries (SIC codes 32–35) for 1958–1996, stored in the wide format.[16] The descriptive statistics of the price series are given below.

```
. use http://www.stata-press.com/data/imeus/4klem_wide_defl, clear
(35KLEM: Jorgensen industry sector data)
. tsset
       time variable:  year, 1958 to 1996
```

12. A degrees-of-freedom correction could be used in the denominator, but relying on large-sample properties, it is not warranted.
13. If the data are set up in the long format more commonly used with panel data, the `reshape` command (see [D] **reshape**) may be used to place them in the wide format; see section 3.8.
14. See [XT] **xtgls** for a SUR estimator that imposes a common coefficient vector on a panel-data model.
15. This test should not be confused with these authors' test for heteroskedasticity described in section 6.2.1.
16. The price series have been detrended with a cubic polynomial time trend.

```
. summarize *d year, sep(5)
    Variable |       Obs        Mean    Std. Dev.       Min        Max
-------------+--------------------------------------------------------
       pi32d |        39     .611359      .02581    .566742   .6751782
       pk32d |        39    .7335128    .0587348   .5981754    .840534
       pl32d |        39    .5444872    .0198763   .4976022   .5784216
       pe32d |        39    .5592308    .0786871   .4531953   .7390293
       pm32d |        39    .5499744    .0166443   .5171617   .5823871
-------------+--------------------------------------------------------
       pi33d |        39    .4948205    .0149315   .4624915   .5163859
       pk33d |        39    .5190769     .035114   .4277323   .5760419
       pl33d |        39    .5200256    .0424153   .4325826   .6127931
       pe33d |        39    .5706154     .093766   .4387668   .8175654
       pm33d |        39    .5192564    .0151137   .4870717   .5421571
-------------+--------------------------------------------------------
       pi34d |        39    .5013333    .0178689   .4659021   .5258276
       pk34d |        39    .5157692    .0558735    .377311   .6376742
       pl34d |        39    .5073077    .0169301    .468933   .5492905
       pe34d |        39    .5774359    .0974223   .4349643   .8020797
       pm34d |        39    .5440256    .0180344   .5070866   .5773573
-------------+--------------------------------------------------------
       pi35d |        39    .5159487    .0168748   .4821945   .5484785
       pk35d |        39    .7182051    .1315394    .423117   1.061852
       pl35d |        39    .4984872    .0216141   .4493805   .5516838
       pe35d |        39    .5629231    .0865252   .4476493   .7584586
       pm35d |        39    .5684615    .0234541   .5317762   .6334837
-------------+--------------------------------------------------------
        year |        39        1977    11.40175       1958       1996
```

We regress each industry's output price on its lagged value and four factor input prices: those for capital (`k`), labor (`l`), energy (`e`), and materials (`m`). The `sureg` command requires the specification of each equation in parentheses. We build up the equations' specification by using a `forvalues` loop over the industry codes.

```
. forvalues i=32/35 {
  2.          local eqn "`eqn' (pi`i'd L.pi`i'd pk`i'd pl`i'd pe`i'd pm`i'd) "
  3. }
. sureg `eqn', corr
Seemingly unrelated regression
----------------------------------------------------------------------
Equation          Obs  Parms        RMSE    "R-sq"       chi2        P
----------------------------------------------------------------------
pi32d              38      5    .0098142    0.8492     219.14   0.0000
pi33d              38      5    .0027985    0.9615    1043.58   0.0000
pi34d              38      5    .0030355    0.9677    1182.37   0.0000
pi35d              38      5    .0092102    0.6751      78.10   0.0000
----------------------------------------------------------------------
```

	Coef.	Std. Err.	z	P>|z|	[95% Conf.	Interval]
pi32d						
pi32d						
L1.	-.0053176	.1623386	-0.03	0.974	-.3234953	.3128602
pk32d	-.0188711	.0344315	-0.55	0.584	-.0863556	.0486133
pl32d	-.5575705	.1166238	-4.78	0.000	-.786149	-.328992
pe32d	.0402698	.0592351	0.68	0.497	-.0758289	.1563684
pm32d	1.587711	.3252302	4.88	0.000	.9502717	2.225151
_cons	.0362004	.1104716	0.33	0.743	-.1803199	.2527208
pi33d						
pi33d						
L1.	.1627936	.0495681	3.28	0.001	.065642	.2599453
pk33d	-.0199381	.0250173	-0.80	0.425	-.0689712	.0290949
pl33d	-.0655277	.0225466	-2.91	0.004	-.1097181	-.0213372
pe33d	-.0657604	.008287	-7.94	0.000	-.0820027	-.0495181
pm33d	1.133285	.084572	13.40	0.000	.9675273	1.299043
_cons	-.0923547	.0185494	-4.98	0.000	-.1287109	-.0559985
pi34d						
pi34d						
L1.	.3146301	.0462574	6.80	0.000	.2239673	.405293
pk34d	.0137423	.009935	1.38	0.167	-.0057298	.0332145
pl34d	.0513415	.0373337	1.38	0.169	-.0218312	.1245142
pe34d	-.0483202	.0115829	-4.17	0.000	-.0710222	-.0256182
pm34d	.8680835	.0783476	11.08	0.000	.7145251	1.021642
_cons	-.1338766	.0241593	-5.54	0.000	-.1812279	-.0865252
pi35d						
pi35d						
L1.	.2084134	.1231019	1.69	0.090	-.0328619	.4496887
pk35d	-.0499452	.0125305	-3.99	0.000	-.0745046	-.0253858
pl35d	.0129142	.0847428	0.15	0.879	-.1531786	.179007
pe35d	.1071003	.0641549	1.67	0.095	-.018641	.2328415
pm35d	.0619171	.2051799	0.30	0.763	-.3402282	.4640624
_cons	.3427017	.1482904	2.31	0.021	.0520579	.6333454

Correlation matrix of residuals:

	pi32d	pi33d	pi34d	pi35d
pi32d	1.0000			
pi33d	-0.3909	1.0000		
pi34d	-0.2311	0.2225	1.0000	
pi35d	-0.1614	-0.1419	0.1238	1.0000

Breusch-Pagan test of independence: chi2(6) = 12.057, Pr = 0.0607

The summary output indicates that each equation explains almost all the variation in the industry's output price. The `corr` option displays the estimated VCE of residuals and tests for independence of the residual vectors. Sizable correlations—both positive and negative—appear in the correlation matrix, and the Breusch–Pagan test rejects its null of independence of these residual series at the 10% level.

We can test cross-equation constraints in the sureg framework with test, combining multiple hypotheses as expressions in parentheses. We consider the null hypothesis that each industry's coefficient on the energy price index is the same.

```
. test ([pi32d]pe32d = [pi33d]pe33d) ([pi32d]pe32d = [pi34d]pe34d)
>      ([pi32d]pe32d = [pi35d]pe35d)
 ( 1)  [pi32d]pe32d - [pi33d]pe33d = 0
 ( 2)  [pi32d]pe32d - [pi34d]pe34d = 0
 ( 3)  [pi32d]pe32d - [pi35d]pe35d = 0
           chi2(  3) =   11.38
         Prob > chi2 =    0.0098
```

The joint test decisively rejects these equality constraints. To illustrate using constrained estimation with sureg, we impose the restriction that the coefficient on the energy price index should be identical over industries. This test involves the definition of three constraints on the coefficient vector. Imposing constraints cannot improve the fit of each equation but may be warranted if the data accept the restriction.

```
. constraint define 1 [pi32d]pe32d = [pi33d]pe33d
. constraint define 2 [pi32d]pe32d = [pi34d]pe34d
. constraint define 3 [pi32d]pe32d = [pi35d]pe35d
. sureg 'eqn', notable c(1 2 3)
Seemingly unrelated regression
Constraints:
 ( 1)  [pi32d]pe32d - [pi33d]pe33d = 0
 ( 2)  [pi32d]pe32d - [pi34d]pe34d = 0
 ( 3)  [pi32d]pe32d - [pi35d]pe35d = 0
```

Equation	Obs	Parms	RMSE	"R-sq"	chi2	P
pi32d	38	5	.0098793	0.8472	236.78	0.0000
pi33d	38	5	.0029664	0.9567	719.32	0.0000
pi34d	38	5	.0030594	0.9672	1212.12	0.0000
pi35d	38	5	.0101484	0.6055	110.37	0.0000

These constraints considerably increase the RMSE (or Root MSE) values for each equation, as we would expect from the results of the test command.

9.4.1 SUR with identical regressors

The second case discussed above, in which SUR will generate the same point and interval estimates—the case of numerically identical regressors—arises often in economic theory and financial theory. For instance, the demand for each good should depend on the set of prices and income, or the portfolio share of assets held in a given class should depend on the returns to each asset and on total wealth. Here there is no reason to use anything other than OLS for efficiency. However, SUR estimation is often used in this case because it allows us to test cross-equation constraints or to estimate with those constraints in place.

If we try to apply SUR to a system with adding-up constraints, such as a *complete set* of cost share or portfolio share equations, the SUR estimator will fail because the error covariance matrix is singular. This assertion holds not only for the unobservable errors but also for the least-squares residuals. A bit of algebra will show that if there are adding-up constraints across equations—for instance, if the set of y_i variables is a complete set of portfolio shares or demand shares—the OLS residuals will sum to zero across equations, and their empirical covariance matrix will be singular *by construction*.

We may still want to use systems estimation to impose the cross-equation constraints arising from economic theory. Here we drop one of the equations and estimate the system of $N-1$ equations with SUR. The parameters of the Nth equation, in point and interval form, can be algebraically derived from those estimates. The FGLS estimates will be sensitive to which equation is dropped, but iterated SUR will restore the invariance property of the maximum likelihood estimator of the problem. For more details, see Greene (2003, 362–369). Poi (2002) shows how to fit singular systems of nonlinear equations.

9.5 Moving-window regression estimates

As with `mvsumm` and `mvcorr` (discussed in section 3.5.3), we may want to compute moving-window regression estimates in a panel context. As with `mvsumm`, we can compute regression estimates for nonoverlapping subsamples with Stata's `statsby` command. However, that command cannot deal with overlapping subsamples, as that would correspond to the same observation's being a member of several by-groups. The functionality to compute moving-window regression estimates is available from the author's `rollreg` routine, available from `ssc`.

With a moving-window regression routine, how should we design the window? One obvious scheme would mimic `mvsumm` and allow for a window of fixed width that is to be passed through the sample, one period at a time: the `move(#)` option.[17] In other applications, we may want an "expanding window": that is, starting with the first τ periods, we compute a set of estimates that consider observations $1 \ldots (\tau + 1)$, $1 \ldots (\tau + 2)$, and so on. This sort of window corresponds to the notion of the information set available to an economic agent at a point in time (and to the scheme used to generate instruments in a DPD model; see [XT] **xtabond**). Thus `rollreg` also offers that functionality via its `add(`τ`)` option. For completeness, the routine also offers the `drop(`τ`)` option, which implements a window that initially takes into account the last τ periods and then expands the window back toward the beginning of the sample. This sort of moving-window estimate can help us determine the usefulness of past information in generating an ex ante forecast, using more or less of that information in the computation. We must use one of these three options when executing `rollreg`.

17. One could imagine something like a 12-month window that is to be advanced to end-of-quarter months, but that could be achieved by merely discarding the intermediate window estimates from `rollreg`.

A moving-window regression will generate sequences of results corresponding to each estimation period. A Stata routine could store those sequences in the columns of a matrix (which perhaps makes them easier to present in tabular format) or as additional variables in the current dataset (which perhaps makes them easier to include in computations or in graphical presentations using `tsline`). The latter, on balance, seems handier and is implemented in `rollreg` via the mandatory `stub(`*string*`)` option, which specifies that new variables should be created with names beginning with *string*.

All the features of `rollreg` (including built-in graphics with the `graph()` option) are accessible with panel data when applied to one time series within the panel by using an `if` *exp* or `in` *range* qualifier. However, rolling regressions certainly have their uses with a full panel. For instance, a finance researcher may want to calculate a "CAPM beta" for each firm in a panel using a moving window of observations, simulating the information set used by the investor at each point in time. Therefore, `rollreg` has been designed to operate with panels where the same sequence of rolling regressions is computed for each time series within the panel.[18] In this context, the routine's graphical output is not available. Although `rollreg` does not produce graphics when multiple time series are included from a panel, it is easy to generate graphics using the results left behind by the routine. For example,

```
. use http://www.stata-press.com/data/imeus/invest2, clear
. keep if company<5
(20 observations deleted)
. tsset company time
       panel variable:  company, 1 to 4
        time variable:  time, 1 to 20
. rollreg market L(0/1).invest time, move(8) stub(mktM)
. local dv `r(depvar)'
. local rl `r(reglist)'
. local stub `r(stub)'
. local wantcoef invest
. local m "`r(rolloption)'(`r(rollobs)')"
. generate fullsample = .
(80 missing values generated)
. forvalues i = 1/4 {
  2.       qui regress `dv' `rl' if company==`i'
  3.       qui replace fullsample = _b[`wantcoef'] if company==`i' & time > 8
  4. }
. label var `stub'_`wantcoef' "moving beta"
. xtline `stub'_`wantcoef', saving("`wantcoef'.gph",replace)
> byopts(title(Moving coefficient of market on invest)
> subtitle("Full-sample coefficient displayed") yrescale legend(off))
> addplot(line fullsample time if fullsample < .)
(file invest.gph saved)
```

Here an 8-year moving window is used to generate the regression estimates of a model where the firm's market value is regressed on current and once-lagged investment ex-

18. I thank Todd Prono for suggesting that this feature be added to the routine.

penditures and a time trend. The trajectory of the resulting coefficient for current investment expenditures is graphed in figure 9.1 for each firm.

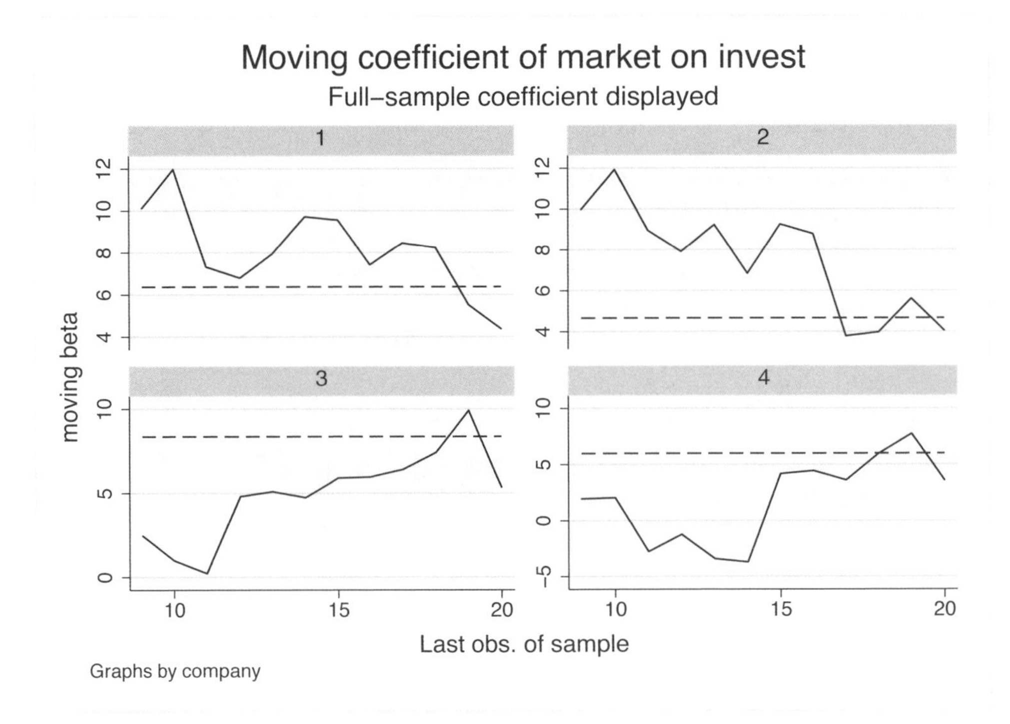

Figure 9.1: Moving-window regression estimates

Companies 1 and 2 display broadly similar trajectories, as do companies 3 and 4; the second pair is different from the first pair. A clear understanding of the temporal stability of the coefficient estimates is perhaps more readily obtained graphically. Although they are not displayed on this graph, `rollreg` also creates series of coefficients' standard errors, from which we can compute confidence intervals, as well as the `Root MSE` of the equation and its R^2.

Or we could use Stata's `rolling` prefix to specify that the moving-window regression be run over each firm.[19] Below we save the estimated coefficients (`_b`) in a new dataset, which we may then merge with the original dataset for further analysis or producing graphics.

```
. use http://www.stata-press.com/data/imeus/invest2, clear
. keep if company<5
(20 observations deleted)
```

19. The `add` and `drop` options of `rollreg` are available using the `rolling` prefix as options `recursive` and `rrecursive`, respectively.

```
. tsset company time
       panel variable:  company, 1 to 4
        time variable:  time, 1 to 20
. rolling _b, window(8) saving(roll_invest, replace) nodots:
> regress market L(0/1).invest time
file roll_invest.dta saved
. use http://www.stata-press.com/data/imeus/roll_invest, clear
(rolling: regress)
. tsset company start
       panel variable:  company, 1 to 4
        time variable:  start, 1 to 13
. describe
Contains data from roll_invest.dta
  obs:            52                          rolling: regress
 vars:             7                          9 Jun 2006 14:08
 size:         1,664 (99.8% of memory free)
-------------------------------------------------------------------------------
              storage  display     value
variable name   type   format      label      variable label
-------------------------------------------------------------------------------
company         float  %9.0g
start           float  %9.0g
end             float  %9.0g
_b_invest       float  %9.0g                  _b[invest]
_stat_2         float  %9.0g                  _b[L.invest]
_b_time         float  %9.0g                  _b[time]
_b_cons         float  %9.0g                  _b[_cons]
-------------------------------------------------------------------------------
Sorted by:  company  start
```

We could produce a graph of each firm's moving coefficient estimate for `invest` with the commands

```
. label var _b_invest "moving beta"
. xtline _b_invest,  byopts(title(Moving coefficient of market on invest))
```

using the `roll_invest` dataset produced by `rolling`.

Exercises

1. The `cigconsump` dataset contains 48 states' annual data for 1985–1995. Fit an FE model of demand for cigarettes, `packpc`, as a function of price (`avgprs`) and per capita income (`incpc`). What are the expected signs? Are they borne out by the estimates? If not, how might you explain the estimated coefficients? Can you reject the pooled OLS model of demand?

2. Store the estimates from the FE model, and refit the model with RE. How do these estimates compare? Does a Hausman test accept RE as the more appropriate estimator?

3. Refit the FE model in constant-elasticity form by using `lpackpc`, `lavgprs`, and `lincpc`. How do the results compare to those on the levels variables? Is this form of the model more in line with economic theory?

4. Refit the constant-elasticity form of the model as a dynamic model, including `L.packpc` as a regressor. Use the two-step robust DPD estimator of `xtabond2` with `lpackpc` as a GMM instrument and `year, L.avgprs` as IV instruments. Do the results support the dynamic formulation of the model? Is the model more in line with economic theory than the static form? Is it adequate for the test of overidentifying restrictions and second-order serial correlation?
5. The `cigconsumpNE` dataset contains the log demand, price, and per capita income variables for the six New England states in wide format. Use that dataset to fit the constant-elasticity form of the model for the six states as a seemingly unrelated regression model with `sureg`. Are there meaningful correlations across the equations' residuals? How do the results differ state by state?

10 Models of discrete and limited dependent variables

This chapter deals with models for discrete and limited dependent variables. Discrete dependent variables arise naturally from discrete-choice models in which individuals choose from a finite or countable number of distinct outcomes and from count processes that record how many times an event has occurred. Limited dependent variables have a restricted range, such as the wage or salary income of non–self-employed individuals, which runs from 0 to the highest level recorded.[1] Discrete and limited dependent variables cannot be modeled by linear regression. These models require more computational effort to fit and are harder to interpret.

This chapter discusses models of binary choice, which can be fitted by binomial logit or probit techniques. The following section takes up their generalization to ordered logit or ordered probit in which the response is one of a set of values from an ordered scale. I then present techniques appropriate for truncated and censored data and their extension to sample-selection models. The final section of the chapter considers bivariate probit and probit with selection.[2]

10.1 Binomial logit and probit models

In models of Boolean response variables, or *binary-choice* models, the response variable is coded as 1 or 0, corresponding to responses of true or false to a particular question:

- Did you watch the seventh game of the 2004 World Series?
- Were you pleased with the outcome of the 2004 presidential election?
- Did you purchase a new car in 2005?

---

1. Most surveys "top-code" certain responses like income, meaning that all responses greater than or equal to a value x are recorded as having the value x.

2. I will not discuss models of "count data" in which the response variable is the count of some item's occurrence for each observation. The methodology appropriate for these data is not a standard linear regression because it cannot take into account the constraint that the data (and the model's predictions) can take on only nonnegative integer values. Stata provides comprehensive facilities for modeling count data via *Poisson regression* and its generalization, the *negative binomial regression*; see [R] **poisson** and [R] **nbreg**, respectively. The "publisher's device" (incorrectly termed the colophon) of Stata Press refers to a Poisson model. See the title page of this book.

We could develop a behavioral model of each of these phenomena, including several explanatory factors (we should not call them regressors) that we expect to influence the respondent's answer to such a question. But we should readily spot the flaw in the *linear probability model*

$$r_i = \mathbf{x}_i \boldsymbol{\beta}_i + u_i \tag{10.1}$$

where we place the Boolean response variable in r and regress it upon a set of $\mathbf{x}$ variables. All the observations we have on r are either 0 or 1 and may be viewed as the ex post probabilities of responding "yes" to the question posed. But the predictions of a linear regression model are unbounded, and the model of (10.1), fitted with `regress`, can produce negative predictions and predictions exceeding unity, neither of which can be considered probabilities. Because the response variable is bounded, restricted to take on values of {0,1}, the model should generate a predicted *probability* that individual i will choose to answer "yes" rather than "no". In such a framework, if $\beta_j > 0$, individuals with high values of x_j will be more likely to respond "yes", but their probability of doing so must respect the upper bound. For instance, if higher disposable income makes a new car purchase more probable, we must be able to include a wealthy person in the sample and find that his or her predicted probability of new car purchase is no greater than 1. Likewise, a poor person's predicted probability must be bounded by 0.

Although we can fit (10.1) with OLS, the model is likely to produce point predictions outside the unit interval. We could arbitrarily constrain them to either 0 or 1, but this linear probability model has other problems: the error term cannot satisfy the assumption of homoskedasticity. For a given set of $\mathbf{x}$ values, there are only two possible values for the disturbance, $-\mathbf{x}\boldsymbol{\beta}$ and $(1 - \mathbf{x}\boldsymbol{\beta})$: the disturbance follows a binomial distribution. Given the properties of the binomial distribution, the variance of the disturbance process, conditioned on $\mathbf{x}$, is

$$\text{Var}[u|\mathbf{x}] = \mathbf{x}\boldsymbol{\beta}\ (1 - \mathbf{x}\boldsymbol{\beta})$$

No constraint can ensure that this quantity will be positive for arbitrary $\mathbf{x}$ values. Therefore, we cannot use regression with a binary-response variable but must follow a different strategy. Before developing that strategy, let us consider another formulation of the model from an economic standpoint.

10.1.1 The latent-variable approach

Using a latent variable is a useful approach to such an econometric model. Express the model of (10.1) as

$$y_i^* = \mathbf{x}_i \boldsymbol{\beta}_i + u_i \tag{10.2}$$

where y^* is an unobservable magnitude, which can be considered the net benefit to individual i of taking a particular course of action (e.g., purchasing a new car). We cannot observe that net benefit, but we can observe the outcome of the individual having followed the decision rule

$$
\begin{aligned}
y_i &= 0 \text{ if } y_i^* < 0 \\
y_i &= 1 \text{ if } y_i^* \geq 0
\end{aligned}
\tag{10.3}
$$

That is, we observe that the individual did ($y = 1$) or did not ($y = 0$) purchase a new car in 2005. We speak of y^* as a *latent variable*, linearly related to a set of factors $\mathbf{x}$ and a disturbance process u.

In the latent model, we model the probability of an individual making each choice. Using (10.2) and (10.3), we have

$$
\begin{aligned}
\Pr(y^* > 0|\mathbf{x}) &= \\
\Pr(u > -\mathbf{x}\boldsymbol{\beta}|\mathbf{x}) &= \\
\Pr(u < \mathbf{x}\boldsymbol{\beta}|\mathbf{x}) &= \\
\Pr(y = 1|\mathbf{x}) &= \Psi(y_i^*)
\end{aligned}
\tag{10.4}
$$

where $\Psi(\cdot)$ is a cumulative distribution function (CDF).

We can estimate the parameters of binary-choice models by using maximum likelihood techniques.[3] For each observation, the probability of observing y conditional on $\mathbf{x}$ may be written as

$$
\Pr(y|\mathbf{x}) = \{\Psi(\mathbf{x}_i\boldsymbol{\beta})\}^{y_i} \{1 - \Psi(\mathbf{x}_i\boldsymbol{\beta})\}^{1-y_i}, \quad y_i = 0, 1 \tag{10.5}
$$

The log likelihood for observation i may be written as

$$
\ell_i(\boldsymbol{\beta}) = y_i \log\{\Psi(\mathbf{x}_i\boldsymbol{\beta})\} + (1 - y_i)\log\{1 - \Psi(\mathbf{x}_i\boldsymbol{\beta})\}
$$

and the log likelihood of the sample is $L(\boldsymbol{\beta}) = \sum_{i=1}^{N} \ell_i(\boldsymbol{\beta})$, to be numerically maximized with respect to the k elements of $\boldsymbol{\beta}$.

The two common estimators of the binary-choice model are the *binomial probit* and *binomial logit* models. For the probit model, $\Psi(\cdot)$ is the CDF of the normal distribution function (Stata's `normal()` function).

For the logit model, $\Psi(\cdot)$ is the CDF of the logistic distribution:[4]

$$
\Pr(y = 1|\mathbf{x}) = \frac{\exp(\mathbf{x}\boldsymbol{\beta})}{1 + \exp(\mathbf{x}\boldsymbol{\beta})}
$$

The CDFs of the normal and logistic distributions are similar. In the latent-variable model, we must assume that the disturbance process has a known variance, σ_u^2. Unlike the linear regression problem, we do not have enough information in the data to estimate

3. For a discussion of maximum likelihood estimation, see Greene (2003, chap. 17) and Gould, Pitblado, and Sribney 2006.

4. The probability density function of the logistic distribution, which is needed to calculate marginal effects, is $\psi(z) = \exp(z)/\{1 + \exp(z)\}^2$.

its magnitude. Because we can divide (10.2) by any positive σ without altering the estimation problem, σ is not identified. σ is set to one for the probit model and $\pi/\sqrt{3}$ in the logit model.

The logistic distribution has fatter tails, resembling the Student t distribution with 7 degrees of freedom.[5] The two models will produce similar results if the distribution of sample values of y_i is not too extreme. However, a sample in which the proportion $y_i = 1$ (or the proportion $y_i = 0$) is very small will be sensitive to the choice of CDF. Neither of these cases is really amenable to the binary-choice model. If an unusual event is modeled by y_i, the "naïve model" that it will not happen in any event is hard to beat. The same is true for an event that is almost ubiquitous: the naïve model that predicts that all people have eaten a candy bar at some time in their lives is accurate.

We can fit these binary-choice models in Stata with the commands `probit` and `logit`. Both commands assume that the response variable is coded with zeros indicating a negative outcome and a positive, nonmissing value corresponding to a positive outcome (i.e., I purchased a new car in 2005). These commands do not require that the variable be coded {0,1}, although that is often the case.

10.1.2 Marginal effects and predictions

One major challenge in working with limited dependent variable models is the complexity of explanatory factors' marginal effects on the result of interest, which arises from the nonlinearity of the relationship. In (10.4), the latent measure is translated by $\Psi(y_i^*)$ to a probability that $y_i = 1$. Although (10.2) is a linear relationship in the $\boldsymbol{\beta}$ parameters, (10.4) is not. Therefore, although x_j has a linear effect on y_i^*, it will not have a linear effect on the resulting probability that $y = 1$:

$$\frac{\partial \Pr(y=1|\mathbf{x})}{\partial x_j} = \frac{\partial \Pr(y=1|\mathbf{x})}{\partial \mathbf{x}\boldsymbol{\beta}} \cdot \frac{\partial \mathbf{x}\boldsymbol{\beta}}{\partial x_j} = \Psi'(\mathbf{x}\boldsymbol{\beta}) \cdot \beta_j = \psi(\mathbf{x}\boldsymbol{\beta}) \cdot \beta_j \tag{10.6}$$

Via the chain rule, the effect of an increase in x_j on the probability is the product of two factors: the effect of x_j on the latent variable and the derivative of the CDF evaluated at y_i^*. The latter term, $\psi(\cdot)$, is the probability density function of the distribution.

In a linear regression model, the coefficient β_j measures the marginal effect $\partial y/\partial x_j$, and that effect is constant over the sample. In a binary-outcome model, a change in factor x_j does not induce a constant change in the $\Pr(y = 1|\mathbf{x})$ because $\Psi()$ is a nonlinear function of $\mathbf{x}$. As discussed above, one of the reasons that we use $\Psi()$ in the binary-outcome model is to keep the predicted probabilities inside the interval [0,1]. This boundedness property of $\Psi()$ implies that the marginal effects must go to zero as the absolute value of x_j gets large. Choosing smooth distribution functions, like the normal and logistic, implies that the marginal effects vary continuously with each x_j.

5. Other distributions, including nonsymmetric distributions, may be used in this context. For example, Stata's `cloglog` command (see [R] **cloglog**) fits the complementary log-log model $\Pr(y = 1|x) = 1 - \exp\{\exp(-\mathbf{x}\boldsymbol{\beta})\}$.

Binomial probit

Stata's `probit` command reports the maximum likelihood estimates of the coefficients. We can also use `dprobit` to display the marginal effect $\partial \Pr(y=1|\mathbf{x})/\partial x_j$, that is, the effect of an infinitesimal change in x_j.[6] We can use `probit` with no arguments following a `dprobit` command to "replay" the probit results in this format. Using `probit` this way does not affect the z statistics or p-values of the estimated coefficients. Because the model is nonlinear, the `dF/dx` reported by `dprobit` will vary through the sample space of the explanatory variables. By default, the marginal effects are calculated at the multivariate point of means but can be calculated at other points via the `at()` option.

After fitting the model with either `probit` or `logit`, we can use `mfx` to compute the marginal effects. A `probit` estimation followed by `mfx` calculates the `dF/dx` values (identical to those from `dprobit`). We can use `mfx`'s `at()` option to compute the effects at a particular point in the sample space. As discussed in section 4.7, `mfx` can also calculate elasticities and semielasticities.

By default, the `dF/dx` effects produced by `dprobit` or `mfx` are the marginal effects for an average individual. Some argue that it would be more preferable to compute the *average marginal effect*: that is, the average of each individual's marginal effect. The marginal effect computed at the average $\mathbf{x}$ is different from the average of the marginal effect computed at the individual $\mathbf{x}_i$. Increasingly, current practice is moving to looking at the distribution of the marginal effects computed for each individual in the sample. Stata does not have such a capability, but a useful `margeff` routine written by Bartus (2005) adds this capability for `probit`, `logit`, and several other Stata commands discussed in this chapter (although not `dprobit`). Its `dummies()` option signals the presence of categorical explanatory variables. If some explanatory variables are integer variables, the `count` option should be used.

After fitting a probit model, the `predict` command, with the default option `p`, computes the predicted probability of a positive outcome. Specifying the `xb` option calculates the predicted value of y_i^*.

The following example uses a modified version of the `womenwk` dataset, which contains information on 2,000 women, 657 of which are not recorded as wage earners. The indicator variable `work` is set to zero for the nonworking and to one for those reporting positive wages.

```
. use http://www.stata-press.com/data/imeus/womenwk, clear
. summarize work age married children education
    Variable |       Obs        Mean    Std. Dev.       Min        Max
-------------+--------------------------------------------------------
        work |      2000       .6715    .4697852          0          1
         age |      2000      36.208     8.28656         20         59
     married |      2000       .6705    .4701492          0          1
    children |      2000      1.6445    1.398963          0          5
   education |      2000      13.084    3.045912         10         20
```

6. Because an indicator variable cannot undergo an infinitesimal change, the default calculation for such a variable is the discrete change in the probability when the indicator is switched from 0 to 1.

We fit a probit model of the decision to work depending on the woman's age, marital status, number of children, and level of education.[7]

```
. probit work age married children education, nolog
Probit regression                                 Number of obs   =       2000
                                                  LR chi2(4)      =     478.32
                                                  Prob > chi2     =     0.0000
Log likelihood = -1027.0616                       Pseudo R2       =     0.1889

------------------------------------------------------------------------------
        work |      Coef.   Std. Err.      z    P>|z|     [95% Conf. Interval]
-------------+----------------------------------------------------------------
         age |   .0347211   .0042293     8.21   0.000     .0264318    .0430105
     married |   .4308575    .074208     5.81   0.000     .2854125    .5763025
    children |   .4473249   .0287417    15.56   0.000     .3909922    .5036576
   education |   .0583645   .0109742     5.32   0.000     .0368555    .0798735
       _cons |  -2.467365   .1925635   -12.81   0.000    -2.844782   -2.089948
------------------------------------------------------------------------------
```

Surprisingly, the effect of more children in the household increases the likelihood that the woman will work. `mfx` computes marginal effects at the multivariate point of means, or we could generate them by using `dprobit` for the estimation.

```
. mfx compute
Marginal effects after probit
      y  = Pr(work) (predict)
         =  .71835948
------------------------------------------------------------------------------
variable |      dy/dx    Std. Err.     z    P>|z|  [    95% C.I.   ]      X
---------+--------------------------------------------------------------------
     age |    .011721      .00142    8.25   0.000   .008935  .014507   36.208
married* |    .150478      .02641    5.70   0.000   .098716   .20224    .6705
children |   .1510059      .00922   16.38   0.000   .132939  .169073   1.6445
educat~n |   .0197024       .0037    5.32   0.000   .012442  .026963   13.084
------------------------------------------------------------------------------
(*) dy/dx is for discrete change of dummy variable from 0 to 1
```

The marginal effects imply that married women have a 15% higher probability of labor force participation, whereas a marginal change in age from the average of 36.2 years is associated with a 1% increase in participation. Bartus's `margeff` routine computes average marginal effects, each of which is slightly smaller than that computed at the point of sample means by `mfx`.

7. The `nolog` option is used to suppress the iteration log.

```
. margeff, dummies(married) count
Average marginal effects on Prob(work==1) after probit
Variables treated as counts:          age children education
```

work	Coef.	Std. Err.	z	P>|z|	[95% Conf.	Interval]
age	.0100178	.0011512	8.70	0.000	.0077615	.0122742
married	.1292759	.0225035	5.74	0.000	.0851698	.173382
children	.1181349	.0057959	20.38	0.000	.106775	.1294947
education	.0167698	.0030558	5.49	0.000	.0107806	.0227591

Binomial logit and grouped logit

When the logistic CDF is used in (10.5), the probability of $y = 1$, conditioned on x, is $\pi_i = \exp(\mathbf{x}_i\boldsymbol{\beta})/\{1 + \exp(\mathbf{x}_i\boldsymbol{\beta})\}$. Unlike the CDF of the normal distribution, which lacks a closed-form inverse, this function can be inverted to yield

$$\log\left(\frac{\pi_i}{1-\pi_i}\right) = \mathbf{x}_i\boldsymbol{\beta}$$

This expression is termed the *logit* of π_i, which is a contraction of the *log of the odds ratio*. The *odds ratio* reexpresses the probability in terms of the odds of $y = 1$. It does not apply to microdata in which y_i equals zero or one, but it is well defined for averages of such microdata. For instance, in the 2004 U.S. presidential election, the ex post probability of a Massachusetts resident voting for John Kerry according to cnn.com was 0.62, with a logit of $\log\{0.62/(1-0.62)\} = 0.4895$. The probability of that person voting for George W. Bush was 0.37, with a logit of -0.5322. Say that we had such data for all 50 states. It would be inappropriate to use linear regression on the probabilities `voteKerry` and `voteBush`, just as it would be inappropriate to run a regression on the `voteKerry` and `voteBush` indicator variables of individual voters. We can use `glogit` (grouped logit) to produce weighted least-squares estimates for the model on state-level data. As an alternative, we can use `blogit` to produce maximum likelihood estimates of that model on grouped (or "blocked") data, or we could use the equivalent commands `gprobit` and `bprobit` to fit a probit model to grouped data.

What if we have microdata in which voters' preferences are recorded as indicator variables, for example `voteKerry = 1` if that individual voted for John Kerry, and vice versa? Instead of fitting a probit model to that response variable, we can fit a logit model with the `logit` command. This command will produce coefficients that, like those of `probit`, express the effect on the latent variable y^* of a change in $\mathbf{x}_j$; see (10.6). As with `dprobit`, we can use `logistic` to compute coefficients that express the effects of the explanatory variables in terms of the odds ratio associated with that explanatory factor. Given the algebra of the model, the odds ratio is merely $\exp(\widehat{\boldsymbol{\beta}}_j)$ for the jth coefficient estimated by `logit` and may also be requested by specifying the `or` option on the `logit` command. *Logistic regression* is intimately related to the binomial logit model and is not an alternative econometric technique to `logit`. The documentation for `logistic` states that the computations are carried out by calling `logit`.

As with `probit`, by default `predict` after `logit` calculates the probability of a positive outcome. `mfx` produces marginal effects expressing the effect of an infinitesimal change in each x on the probability of a positive outcome, evaluated by default at the multivariate point of means. We can also calculate elasticities and semielasticities. We can use Bartus's `margeff` routine to calculate the average marginal effects over the sample observations after either `logit` or `logistic`.

10.1.3 Evaluating specification and goodness of fit

We can apply both the binomial logit and binomial probit estimators, so we might wonder which to use. The CDFs underlying these models differ most in the tails, producing similar predicted probabilities for nonextreme values of $\mathbf{x}\boldsymbol{\beta}$. Because the likelihood functions of the two estimators are not nested, there is no obvious way to test one against the other.[8] The coefficient estimates of `probit` and `logit` from the same model will differ because they are estimates of $(\boldsymbol{\beta}/\sigma_u)$. Whereas the variance of the standard normal distribution is unity, the variance of the logistic distribution is $\pi^2/3$, causing reported logit coefficients to be larger by a factor of about $\pi/\sqrt{3} = 1.814$. However, we often want the marginal effects generated by these models rather than their estimated coefficients. The magnitude of the marginal effects generated by `mfx` or Bartus's `margeff` routine are likely to be similar for both estimators.

We use `logit` to fit the same model of women's probability of working:

```
. logit work age married children education, nolog
Logistic regression                               Number of obs   =       2000
                                                  LR chi2(4)      =     476.62
                                                  Prob > chi2     =     0.0000
Log likelihood = -1027.9144                       Pseudo R2       =     0.1882

------------------------------------------------------------------------------
        work |      Coef.   Std. Err.      z    P>|z|     [95% Conf. Interval]
-------------+----------------------------------------------------------------
         age |   .0579303    .007221     8.02   0.000     .0437774    .0720833
     married |   .7417775   .1264704     5.87   0.000     .4939001    .9896549
    children |   .7644882   .0515287    14.84   0.000     .6634938    .8654827
   education |   .0982513   .0186522     5.27   0.000     .0616936    .1348089
       _cons |  -4.159247   .3320397   -12.53   0.000    -4.810033   -3.508462
------------------------------------------------------------------------------
```

Although the logit coefficients' magnitudes differ considerably from their probit counterparts, the marginal effects at the multivariate point of means are similar to those computed after `probit`.

8. An approach similar to the Davidson–MacKinnon J test described in section 4.5.5 has been proposed but has been shown to have low power.

```
. mfx compute
Marginal effects after logit
      y  = Pr(work) (predict)
         =  .72678588
```

variable	dy/dx	Std. Err.	z	P>|z|	[95% C.I.	]	X
age	.0115031	.00142	8.08	0.000	.008713	.014293	36.208
married*	.1545671	.02703	5.72	0.000	.101592	.207542	.6705
children	.151803	.00938	16.19	0.000	.133425	.170181	1.6445
educat~n	.0195096	.0037	5.27	0.000	.01226	.02676	13.084

(*) dy/dx is for discrete change of dummy variable from 0 to 1

We illustrate the `at()` option, evaluating the estimated logit function at `children = 0`. The magnitudes of each of the marginal effects are increased at this point in the x space, with the effect of an additional year of education being almost 5% higher (0.0241 versus 0.0195) for the childless woman.

```
. mfx compute, at(children=0)
warning: no value assigned in at() for variables age married education;
   means used for age married education
Marginal effects after logit
      y  = Pr(work) (predict)
         =  .43074191
```

variable	dy/dx	Std. Err.	z	P>|z|	[95% C.I.	]	X
age	.0142047	.00178	7.97	0.000	.01071	.0177	36.208
married*	.1762562	.02825	6.24	0.000	.120897	.231615	.6705
children	.1874551	.01115	16.82	0.000	.165609	.209301	0
educat~n	.0240915	.00458	5.26	0.000	.015115	.033068	13.084

(*) dy/dx is for discrete change of dummy variable from 0 to 1

We can test for appropriate specification of a subset model, as in the regression context, with the `test` command. The test statistics for exclusion of one or more explanatory variables are reported as χ^2 rather than F statistics because Wald tests from ML estimators have large-sample χ^2 distributions. We can apply the other postestimation commands—tests of linear expressions with `test` or `lincom` and tests of nonlinear expressions with `testnl` or `nlcom`—the same way as with `regress`.

How can we judge the adequacy of a binary-choice model fitted with `probit` or `logit`? Just as the "ANOVA F" tests a regression specification against the *null model* in which all regressors are omitted, we may consider a null model for the binary-choice specification to be $\Pr(y = 1) = \overline{y}$. Because the mean of an indicator variable is the sample proportion of 1s, it may be viewed as the unconditional probability that $y = 1$.[9] We can contrast that with the conditional probabilities generated by the model that takes into account the explanatory factors $\mathbf{x}$. Because the likelihood function for the null model can readily be evaluated in either the probit or logit context, both

9. For instance, the estimate of the constant in a constant-only probit model is `invnormal(`$\overline{y}$`)`.

commands produce a likelihood-ratio test[10] [`LR chi2(`$k-1$`)`] where $(k-1)$ is the number of explanatory factors in the model (presuming the existence of a constant term). As mentioned above, the null model is hard to beat if $\overline{y}$ is very close to 0 or 1.

Although this likelihood-ratio test provides a statistical basis to reject the null model versus the fitted model, there is no measure of goodness of fit analogous to R^2 for linear regression. Stata produces a measure called `Pseudo R2` for both commands and for all commands estimated by maximum likelihood; see [R] **maximize**. Let L_1 be the log-likelihood value for the fitted model, as presented on the estimation output after convergence. Let L_0 be the log-likelihood value for the null model excluding all explanatory variables. This quantity is not displayed but is available after estimation as `e(ll_0)`. The `LR chi2(`$k-1$`)` likelihood-ratio test is merely $2(L_1 - L_0)$, and it has a large-sample $\chi^2(k-1)$ distribution under the null hypothesis that the explanatory factors are jointly uninformative.

If we rearrange the log-likelihood values, we may define the `pseudo R2` as $(1-L_1/L_0)$, which like regression R^2 is on a $[0,1]$ scale, with 0 indicating that the explanatory variables failed to increase likelihood and 1 indicating that the model perfectly predicts each observation. We cannot interpret this pseudo-R^2, as we can for linear regression, as the proportion of variation in y explained by x, but in other aspects it does resemble an R^2 measure.[11] Adding more explanatory factors to the model does not always result in perfect prediction, as it does in linear regression. In fact, perfect prediction may inadvertently occur because one or more explanatory factors are perfectly correlated with the response variable. Stata's documentation in `probit` and `logit` discusses this issue, which Stata will detect and report.

Several other measures based on the predictions of the binary-choice model have been proposed, but all have their weaknesses, particularly if there is a high proportion of 0s or 1s in the sample. `estat gof` and `estat clas` compute many of these measures. With a constant term included, the binomial logit model will produce $\overline{\widehat{y}} = \overline{y}$, as does regression: the average of predicted probabilities from the model equals the sample proportion $\overline{y}$, but that outcome is not guaranteed in the binomial probit model.

10.2 Ordered logit and probit models

Chapter 7 discussed the issues related to using *ordinal independent variables*, which indicate a ranking of responses, rather than a cardinal measure, such as the codes of a Likert scale of agreement with a statement. Since the values of such an ordered response are arbitrary, we should not treat an ordinal variable as if it can be measured in a cardinal sense and entered into a regression, either as a regressor or as a response variable. If we want to model an ordinal variable as a function of a set of explanatory factors, we can use a generalization of the binary-choice framework known as *ordered probit* or *ordered logit* estimation techniques.

10. I introduce the concept of likelihood-ratio tests in section 4.5. For more information, see Greene (2003, chap. 17).

11. Stata's documentation attributes this measure to Judge et al. (1985), but other sources describe it as the likelihood-ratio index of McFadden (1974).

In the latent-variable approach to the binary-choice model, we observe $y_i = 1$ if $y_i^* > 0$. The ordered-choice model generalizes this concept to the notion of multiple thresholds. For instance, a variable recorded on a five-point Likert scale will have four thresholds over the latent variable. If $y^* \leq \kappa_1$, we observe $y = 1$; if $\kappa_1 < y^* \leq \kappa_2$, we observe $y = 2$; if $\kappa_2 < y^* \leq \kappa_3$, we observe $y = 3$, and so on, where the $\boldsymbol{\kappa}$ values are the thresholds. In a sense, this is imprecise measurement: we cannot observe y^* directly, but only the range in which it falls. Imprecise measurement is appropriate for many forms of microeconomic data that are "bracketed" for privacy or summary reporting purposes. Alternatively, the observed choice might reveal only an individual's relative preference.

The parameters to be estimated are a set of coefficients $\boldsymbol{\beta}$ corresponding to the explanatory factors in x, as well as a set of $(I-1)$ threshold values $\boldsymbol{\kappa}$ corresponding to the I alternatives. In Stata's implementation of these estimators in `oprobit` and `ologit`, the actual values of the response variable are not relevant. Larger values are taken to correspond to higher outcomes. If there are I possible outcomes (e.g., 5 for the Likert scale), a set of threshold coefficients or *cutpoints* $\{\kappa_1, \kappa_2, \ldots, \kappa_{I-1}\}$ is defined, where $\kappa_0 = -\infty$ and $\kappa_I = \infty$. The model for the jth observation defines

$$\Pr(y_j = i) = \Pr(\kappa_{i-1} < \mathbf{x}_j\boldsymbol{\beta} + u_j < \kappa_i)$$

where the probability that individual j will choose outcome i depends on the product $\mathbf{x}_j\boldsymbol{\beta}$ falling between cutpoints $(i-1)$ and i. This is a direct generalization of the two-outcome binary-choice model, which has one threshold at zero. As in the binomial probit model, we assume that the error is normally distributed with variance unity (or distributed logistic with variance $\pi^2/3$ for ordered logit).

Prediction is more complex in ordered probit (logit) because there are I possible predicted probabilities corresponding to the I possible values of the response variable. The default option for `predict` is to compute predicted probabilities. If I new variable names are given in the command, they will contain the probability that $i = 1$, the probability that $i = 2$, and so on.

The marginal effects of an ordered probit (logit) model are also more complex than their binomial counterparts because an infinitesimal change in $\mathbf{x}_j$ will not only change the probability within the current cell (for instance, if $\kappa_2 < \widehat{y}^* \leq \kappa_3$) but will also make it more likely that the individual crosses the threshold into the adjacent category. Thus if we predict the probabilities of being in each category at a different point in the sample space (for instance, for a family with three rather than two children), we will find that those probabilities have changed, and the larger family may be more likely to choose the jth response and less likely to choose the $(j-1)$st response. We can calculate the average marginal effects with `margeff`.

We illustrate the ordered probit and logit techniques with a model of corporate bond ratings. The dataset contains information on 98 U.S. corporations' bond ratings and financial characteristics where the bond ratings are AAA (excellent) to C (poor). The integer codes underlying the ratings increase in the quality of the firm's rating, such that an increase in the response variable indicates that the firm's bonds are a more attractive investment opportunity. The bond rating variable (`rating83c`) is coded as integers 2–5, with 5 corresponding to the highest quality (AAA) bonds and 2 to the lowest. The tabulation of `rating83c` shows that the four ratings categories contain a similar number of firms. We model the 1983 bond rating as a function of the firm's income-to-asset ratio in 1983 (`ia83`: roughly, return on assets) and the change in that ratio from 1982 to 1983 (`dia`). The income-to-asset ratio, expressed as a percentage, varies widely around a mean of 10%.

```
. use http://www.stata-press.com/data/imeus/panel84extract, clear
. summarize rating83c ia83 dia
    Variable |       Obs        Mean    Std. Dev.       Min        Max
-------------+--------------------------------------------------------
   rating83c |        98    3.479592     1.17736          2          5
        ia83 |        98    10.11473    7.441946  -13.08016   30.74564
         dia |        98    .7075242    4.711211  -10.79014   20.05367
. tabulate rating83c
      Bond |
   rating, |
      1983 |      Freq.     Percent        Cum.
-----------+-----------------------------------
    BA_B_C |         26       26.53       26.53
       BAA |         28       28.57       55.10
      AA_A |         15       15.31       70.41
       AAA |         29       29.59      100.00
-----------+-----------------------------------
     Total |         98      100.00
```

We fit the model with `ologit`; the model's predictions are quantitatively similar if we use `oprobit`.

```
. ologit rating83c ia83 dia, nolog
Ordered logistic regression                       Number of obs   =         98
                                                  LR chi2(2)      =      11.54
                                                  Prob > chi2     =     0.0031
Log likelihood = -127.27146                       Pseudo R2       =     0.0434

------------------------------------------------------------------------------
   rating83c |      Coef.   Std. Err.      z    P>|z|     [95% Conf. Interval]
-------------+----------------------------------------------------------------
        ia83 |   .0939166   .0296196     3.17   0.002     .0358633    .1519699
         dia |  -.0866925   .0449789    -1.93   0.054    -.1748496    .0014646
-------------+----------------------------------------------------------------
       /cut1 |  -.1853053   .3571432                     -.8852931    .5146825
       /cut2 |   1.185726   .3882098                      .4248489    1.946603
       /cut3 |   1.908412   .4164895                      1.092108    2.724717
------------------------------------------------------------------------------
```

`ia83` has a significant positive effect on the bond rating, but somewhat surprisingly the change in that ratio (`dia`) has a negative effect. The model's ancillary parameters `_cut1` to `_cut3` indicate the thresholds for the ratings categories.

Following the `ologit` estimation, we use `predict` to compute the predicted probabilities of achieving each rating. We then examine the firms who were classified as most likely to have an "AAA" (excellent) rating and "BA_B_C" (poor quality) rating, respectively. Firm 31 has a 75% predicted probability of being rated "AAA", whereas firm 67 has a 72% predicted probability of being rated "BA" or below. The former probability is in accordance with the firm's rating, whereas the latter is a substantial misclassification. However, many factors enter into a bond rating, and that firm's level and change of net income combined to produce a very low prediction.

```
. predict spBA_B_C spBAA spAA_A spAAA
(option pr assumed; predicted probabilities)
. summarize spAAA, mean
. list sp* rating83c if spAAA==r(max)
```

	spBA_B_C	spBAA	spAA_A	spAAA	rati~83c
31.	.0388714	.0985567	.1096733	.7528986	AAA

```
. summarize spBA_B_C, mean
. list sp* rating83c if spBA_B_C==r(max)
```

	spBA_B_C	spBAA	spAA_A	spAAA	rati~83c
67.	.7158453	.1926148	.0449056	.0466343	AAA

Economic research also uses response variables, which represent unordered discrete alternatives, or *multinomial* models. For a discussion of how to fit and interpret unordered discrete-choice models in Stata, see Long and Freese (2006).

10.3 Truncated regression and tobit models

I now discuss a situation where the response variable is not binary or necessarily integer but has limited range. This situation is a bit trickier, because the restrictions on the range of a limited dependent variable (LDV) may not be obvious. We must fully understand the context in which the data were generated, and we must identify the restrictions. Modeling LDVs by OLS will be misleading.

10.3.1 Truncation

Some LDVs are generated by truncated processes. For *truncation*, the sample is drawn from a subset of the population so that only certain values are included in the sample. We lack observations on both the response variable and explanatory variables. For instance, we might have a sample of individuals who have a high school diploma, some college experience, or one or more college degrees. The sample has been generated by interviewing those who completed high school. This is a truncated sample, relative to the population, in that it excludes all individuals who have not completed high school. The excluded individuals are not likely to have the same characteristics as those in our sample. For instance, we might expect average or median income of dropouts to be lower than that of graduates.

The effect of truncating the distribution of a random variable is clear. The expected value or mean of the truncated random variable moves away from the truncation point, and the variance is reduced. Descriptive statistics on the level of education in our sample should make that clear: with the minimum years of education set to 12, the mean education level is higher than it would be if high school dropouts were included, and the variance will be smaller. In the subpopulation defined by a truncated sample, we have no information about the characteristics of those who were excluded. For instance, we do not know whether the proportion of minority high school dropouts exceeds the proportion of minorities in the population.

We cannot use a sample from this truncated population to make inferences about the entire population without correcting for those excluded individuals' not being randomly selected from the population at large. Although it might appear that we could use these truncated data to make inferences about the subpopulation, we cannot even do that. A regression estimated from the subpopulation will yield coefficients that are biased toward zero—or *attenuated*—as well as an estimate of σ_u^2 that is biased downward. If we are dealing with a truncated normal distribution, where $y = \mathbf{x}_i\boldsymbol{\beta} + u_i$ is observed only if it exceeds τ, we can define

$$\begin{aligned} \alpha_i &= \frac{\tau - \mathbf{x}_i\boldsymbol{\beta}}{\sigma_u} \\ \lambda(\alpha_i) &= \frac{\phi(\alpha_i)}{\{1 - \Phi(\alpha_i)\}} \end{aligned}$$

where σ_u is the standard error of the untruncated disturbance u, $\phi(\cdot)$ is the normal density function, and $\Phi(\cdot)$ is the normal CDF. The expression $\lambda(\alpha_i)$ is termed the *inverse Mills ratio* (IMR).

Standard manipulation of normally distributed random variables shows that

$$E[y_i|y_i > \tau, \mathbf{x}_i] = \mathbf{x}_i\boldsymbol{\beta} + \sigma_u\lambda(\alpha_i) + u_i \tag{10.7}$$

The above equation implies that a simple OLS regression of y on $\mathbf{x}$ suffers from the exclusion of the term $\lambda(\alpha_i)$. This regression is misspecified, and the effect of that misspecification will differ across observations, with a heteroskedastic error term whose variance depends on $\mathbf{x}_i$. To deal with these problems, we include the IMR as an additional regressor, so we can use a truncated sample to make consistent inferences about the subpopulation.

If we can justify the assumption that the regression errors in the *population* are normally distributed, we can estimate an equation for a truncated sample with the Stata command `truncreg`.[12] Under the assumption of normality, we can make inferences for the population from the truncated regression model. The `truncreg` option `ll(#)` indicates that values of the response variable less than or equal to # are truncated. We might have a sample of college students with `yearsEduc` truncated from below at 12

12. More details on the truncated regression model with normal errors are available in Greene (2003, 756–761).

years. Upper truncation can be handled with the `ul(#)` option; for instance, we may have a sample of individuals whose income is recorded up to $200,000. We can specify both lower and upper truncation by combining the options. In the example below, we consider a sample of married women from the `laborsub` dataset whose hours of work (`whrs`) are truncated from below at zero. Other variables of interest are the number of preschool children (`kl6`), number of school-aged children (`k618`), age (`wa`), and years of education (`we`).

```
. use http://www.stata-press.com/data/imeus/laborsub, clear
. summarize whrs kl6 k618 wa we
    Variable |       Obs        Mean    Std. Dev.       Min        Max
-------------+--------------------------------------------------------
        whrs |       250      799.84    915.6035          0       4950
         kl6 |       250        .236    .5112234          0          3
        k618 |       250       1.364    1.370774          0          8
          wa |       250       42.92    8.426483         30         60
          we |       250      12.352    2.164912          5         17
```

To illustrate the consequences of ignoring truncation, we fit a model of hours worked with OLS, including only working women.

```
. regress whrs kl6 k618 wa we if whrs>0
      Source |       SS       df       MS              Number of obs =     150
-------------+------------------------------           F(  4,   145) =    2.80
       Model |  7326995.15     4  1831748.79           Prob > F      =  0.0281
    Residual |  94793104.2   145  653745.546           R-squared     =  0.0717
-------------+------------------------------           Adj R-squared =  0.0461
       Total |   102120099   149  685369.794           Root MSE      =  808.55

------------------------------------------------------------------------------
        whrs |      Coef.   Std. Err.      t    P>|t|     [95% Conf. Interval]
-------------+----------------------------------------------------------------
         kl6 |  -421.4822   167.9734    -2.51   0.013    -753.4748   -89.48953
        k618 |  -104.4571   54.18616    -1.93   0.056    -211.5538    2.639668
          wa |  -4.784917   9.690502    -0.49   0.622     -23.9378    14.36797
          we |   9.353195   31.23793     0.30   0.765    -52.38731     71.0937
       _cons |   1629.817   615.1301     2.65   0.009     414.0371    2845.597
------------------------------------------------------------------------------
```

We now refit the model with `truncreg`, taking into account that 100 of the 250 observations have zero recorded `whrs`:

(*Continued on next page*)

```
. truncreg whrs kl6 k618 wa we, ll(0) nolog
(note: 100 obs. truncated)

Truncated regression
Limit:   lower =          0                         Number of obs =     150
         upper =       +inf                         Wald chi2(4)  =   10.05
Log likelihood = -1200.9157                         Prob > chi2   = 0.0395

------------------------------------------------------------------------------
        whrs |      Coef.   Std. Err.      z    P>|z|     [95% Conf. Interval]
-------------+----------------------------------------------------------------
eq1          |
         kl6 |  -803.0042   321.3614    -2.50   0.012    -1432.861   -173.1474
        k618 |   -172.875   88.72898    -1.95   0.051    -346.7806    1.030578
          wa |  -8.821123   14.36848    -0.61   0.539    -36.98283    19.34059
          we |   16.52873   46.50375     0.36   0.722    -74.61695    107.6744
       _cons |    1586.26    912.355     1.74   0.082    -201.9233    3374.442
-------------+----------------------------------------------------------------
sigma        |
       _cons |   983.7262   94.44303    10.42   0.000     798.6213    1168.831
------------------------------------------------------------------------------
```

Some of the attenuated coefficient estimates from **regress** are no more than half as large as their counterparts from **truncreg**. The parameter **sigma _cons**, comparable to **Root MSE** in the OLS regression, is considerably larger in the truncated regression, reflecting its downward bias in a truncated sample. We can use the coefficient estimates and marginal effects from **truncreg** to make inferences about the entire population, whereas we should not use the results from the misspecified regression model for any purpose.

10.3.2 Censoring

Censoring is another common mechanism that restricts the range of dependent variables. Censoring occurs when a response variable is set to an arbitrary value when the variable is beyond the *censoring point*. In the truncated case, we observe neither the dependent nor the explanatory variables for individuals whose y_i lies in the truncation region. In contrast, when the data are censored we do not observe the value of the dependent variable for individuals whose y_i is beyond the censoring point, but we do observe the values of the explanatory variables. A common example of censoring is "top coding", which occurs when a variable that takes on values of x or more is recorded as x. For instance, many household surveys top code reported income at $150,000 or $200,000.

There is some discussion in the literature about how to interpret some LDVs that appear to be censored. As Wooldridge (2002) points out, censoring is a problem with how the data were recorded, not how they were generated. For instance, in the above top-coding example, if the survey administrators chose not to top code the data, the data would not be censored. In contrast, some LDVs result from corner solutions to choice problems. For example, the amount an individual spends on a new car in a given year may be zero or positive. Wooldridge (2002) argues that this LDV is a corner solution, not a censored variable. He also shows that the object of interest for a corner solution model can be different from that for a censored model. Fortunately, both the censoring and corner-solution motivations give rise to the same ML estimator. Furthermore, the same Stata postestimation tools can be used to interpret the results from censored and corner-solution models.

A solution to the problem with censoring at 0 was first proposed by Tobin (1958) as the *censored regression* model; it became known as "Tobin's probit" or the *tobit* model.[13] The model can be expressed in terms of a latent variable:

$$\begin{aligned} y_i^* &= \mathbf{x}_i\boldsymbol{\beta} + u_i \\ y_i &= \begin{cases} 0 & \text{if } y_i^* \leq 0 \\ y_i^* & \text{if } y_i^* > 0 \end{cases} \end{aligned} \tag{10.8}$$

y_i contains either zeros for nonpurchasers or a positive dollar amount for those who chose to buy a car last year. The model combines aspects of the binomial probit for the distinction of $y_i = 0$ versus $y_i > 0$ and the regression model for $E[y_i|y_i > 1, \mathbf{x}_i]$. Of course, we could collapse all positive observations on y_i and treat this as a binomial probit (or logit) estimation problem, but doing so would discard the information on the dollar amounts spent by purchasers. Likewise, we could throw away the $y_i = 0$ observations, but we would then be left with a truncated distribution, with the various problems that creates.[14] To take account of all the information in y_i properly, we must fit the model with the `tobit` estimation method, which uses maximum likelihood to combine the probit and regression components of the log-likelihood function. We can express the log likelihood of a given observation as

$$\begin{aligned} \ell_i(\boldsymbol{\beta}, \sigma_u) &= I(y_i = 0) \log \left\{ 1 - \Phi\left(\frac{\mathbf{x}_i\boldsymbol{\beta}}{\sigma_u}\right) \right\} + \\ & I(y_i > 0) \left\{ \log \phi \left(\frac{y_i - \mathbf{x}_i\boldsymbol{\beta}}{\sigma_u} \right) - \frac{1}{2} \log \left(\frac{\sigma_u^2}{u} \right) \right\} \end{aligned}$$

where $I(\cdot) = 1$ if its argument is true and is zero otherwise. We can write the likelihood function, summing ℓ_i over the sample, as the sum of the probit likelihood for those observations with $y_i = 0$ and the regression likelihood for those observations with $y_i > 0$.

We can define tobit models with a threshold other than zero. We can specify censoring from below at any point on the y scale with the `ll(#)` option for *left censoring*. Similarly, the standard tobit formulation may use an upper threshold (censoring from above, or *right censoring*) using the `ul(#)` option to specify the upper limit. Stata's `tobit` command also supports the *two-limit tobit* model where observations on y are censored from both left and right by specifying both the `ll(#)` and `ul(#)` options.

Even with one censoring point, predictions from the tobit model are complex, since we may want to calculate the regression-like `xb` with `predict`, but we could also compute

13. The term "censored regression" is now more commonly used for a generalization of the tobit model in which the censoring values may vary from observation to observation. See [R] **cnreg**.

14. The regression coefficients estimated from the positive y observations will be attenuated relative to the tobit coefficients, with the degree of bias toward zero increasing in the proportion of "limit observations" in the sample.

the predicted probability that y (conditional on x) falls within a particular interval (which may be open ended on the left or right).[15] We can do so with the `pr(`a,b`)` option, where arguments a,b specify the limits of the interval; the missing-value code (`.`) is taken to mean infinity (of either sign). Another `predict` option, `e(`a,b`)`, calculates the $E[\mathbf{x}_i\widehat{\boldsymbol{\beta}}+u_i|a<\mathbf{x}_i\widehat{\boldsymbol{\beta}}+u_i<b]$. Last, the `ystar(`$a,b$`)` option computes the prediction from (10.8): a censored prediction, where the threshold is taken into account.

The marginal effects of the tobit model are also complex. The estimated coefficients are the marginal effects of a change in x_j on y^*, the unobservable latent variable

$$\frac{\partial E[y^*|\mathbf{x}]}{\partial x_j}=\beta_j$$

but that information is rarely useful. The effect on the observable y is

$$\frac{\partial E[y|\mathbf{x}]}{\partial x_j}=\beta_j\times\Pr(a<y_i^*<b)$$

where a,b are defined as above for `predict`. For instance, for left censoring at zero, $a=0,b=+\infty$. Since that probability is at most unity (and will be reduced by a larger proportion of censored observations), the marginal effect of x_j is attenuated from the reported coefficient toward zero. An increase in an explanatory variable with a positive coefficient implies that a left-censored individual is less likely to be censored. The predicted probability of a nonzero value will increase. For an uncensored individual, an increase in x_j will imply that $E[y|y>0]$ will increase. So, for instance, a decrease in the mortgage interest rate will allow more people to be homebuyers (since many borrowers' incomes will qualify them for a mortgage at lower interest rates) and allow prequalified homebuyers to purchase a more expensive home. The marginal effect captures the combination of those effects. Since newly qualified homebuyers will be purchasing the cheapest homes, the effect of the lower interest rate on the average price at which homes are sold will incorporate both effects. We expect that it will increase the average transactions price, but because of attenuation, by a smaller amount than the regression function component of the model would indicate. We can calculate the marginal effects with `mfx` or, for average marginal effects, with Bartus's `margeff`.

For an empirical example, we return to the `womenwk` dataset used to illustrate binomial probit and logit. We generate the log of the wage (`lw`) for working women and set `lwf` equal to `lw` for working women and zero for nonworking women.[16] We first fit the model with OLS, ignoring the censored nature of the response variable:

15. For more information, see Greene (2003, 764–773).

16. This variable creation could be problematic if recorded wages less than $1.00 were present in the data, but in these data the minimum wage recorded is $5.88.

```
. use http://www.stata-press.com/data/imeus/womenwk, clear
. regress lwf age married children education
      Source |       SS       df       MS              Number of obs =    2000
-------------+------------------------------           F(  4,  1995) =  134.21
       Model |  937.873188     4  234.468297           Prob > F      =  0.0000
    Residual |  3485.34135  1995  1.74703827           R-squared     =  0.2120
-------------+------------------------------           Adj R-squared =  0.2105
       Total |  4423.21454  1999  2.21271363           Root MSE      =  1.3218

------------------------------------------------------------------------------
         lwf |      Coef.   Std. Err.      t    P>|t|     [95% Conf. Interval]
-------------+----------------------------------------------------------------
         age |   .0363624    .003862     9.42   0.000     .0287885    .0439362
     married |   .3188214   .0690834     4.62   0.000     .1833381    .4543046
    children |   .3305009   .0213143    15.51   0.000     .2887004    .3723015
   education |   .0843345   .0102295     8.24   0.000     .0642729    .1043961
       _cons |  -1.077738   .1703218    -6.33   0.000    -1.411765   -.7437105
------------------------------------------------------------------------------
```

Refitting the model as a tobit and indicating that `lwf` is left censored at zero with the `ll()` option yields

```
. tobit lwf age married children education, ll(0)
Tobit regression                                  Number of obs   =       2000
                                                  LR chi2(4)      =     461.85
                                                  Prob > chi2     =     0.0000
Log likelihood = -3349.9685                       Pseudo R2       =     0.0645

------------------------------------------------------------------------------
         lwf |      Coef.   Std. Err.      t    P>|t|     [95% Conf. Interval]
-------------+----------------------------------------------------------------
         age |    .052157   .0057457     9.08   0.000     .0408888    .0634252
     married |   .4841801   .1035188     4.68   0.000     .2811639    .6871964
    children |   .4860021   .0317054    15.33   0.000     .4238229    .5481812
   education |   .1149492   .0150913     7.62   0.000     .0853529    .1445454
       _cons |  -2.807696   .2632565   -10.67   0.000    -3.323982   -2.291409
-------------+----------------------------------------------------------------
      /sigma |   1.872811    .040014                      1.794337    1.951285
------------------------------------------------------------------------------
  Obs. summary:        657  left-censored observations at lwf<=0
                      1343     uncensored observations
                         0 right-censored observations
```

The tobit estimates of `lwf` show positive, significant effects for age, marital status, the number of children, and the number of years of education. We expect each of these factors to increase the probability that a woman will work as well as increase her wage conditional on employment status. Following tobit estimation, we first generate the marginal effects of each explanatory variable on the probability that an individual will have a positive log(wage) by using the `pr(`a, b`)` option of `predict`.

```
. mfx compute, predict(pr(0,.))
Marginal effects after tobit
      y  = Pr(lwf>0) (predict, pr(0,.))
         =  .81920975

variable |     dy/dx    Std. Err.     z    P>|z|  [    95% C.I.   ]      X
---------+-------------------------------------------------------------------
     age |  .0073278      .00083    8.84   0.000   .005703  .008952   36.208
married* |  .0706994      .01576    4.48   0.000   .039803  .101596    .6705
children |  .0682813      .00479   14.26   0.000   .058899  .077663   1.6445
educat~n |  .0161499      .00216    7.48   0.000   .011918  .020382   13.084

(*) dy/dx is for discrete change of dummy variable from 0 to 1
```

We then calculate the marginal effect of each explanatory variable on the expected log wage, given that the individual has not been censored (i.e., was working). These effects, unlike the estimated coefficients from `regress`, properly take into account the censored nature of the response variable.

```
. mfx compute, predict(e(0,.))
Marginal effects after tobit
      y  = E(lwf|lwf>0) (predict, e(0,.))
         =  2.3102021

variable |     dy/dx    Std. Err.     z    P>|z|  [    95% C.I.   ]      X
---------+-------------------------------------------------------------------
     age |  .0314922      .00347    9.08   0.000   .024695   .03829   36.208
married* |  .2861047      .05982    4.78   0.000   .168855  .403354    .6705
children |  .2934463      .01908   15.38   0.000   .256041  .330852   1.6445
educat~n |  .0694059      .00912    7.61   0.000   .051531  .087281   13.084

(*) dy/dx is for discrete change of dummy variable from 0 to 1
```

Since the tobit model has a probit component, its results are sensitive to the assumption of homoskedasticity. Robust standard errors are not available for Stata's `tobit` command, although bootstrap or jackknife standard errors may be computed with the `vce` option. The tobit model imposes the constraint that the same set of factors x determine both whether an observation is censored (e.g., whether an individual purchased a car) and the value of a noncensored observation (how much a purchaser spent on the car). Furthermore, the marginal effect is constrained to have the same sign in both parts of the model. A generalization of the tobit model, often termed the *Heckit* model (after James Heckman), can relax this constraint and allow different factors to enter the two parts of the model. We can fit this generalized tobit model with Stata's `heckman` command, as described in the next section of this chapter.

10.4 Incidental truncation and sample-selection models

For truncation, the sample is drawn from a subset of the population and does not contain observations on the dependent or independent variables for any other subset of the population. For example, a truncated sample might include only individuals with a permanent mailing address and exclude the homeless. For *incidental truncation*, the sample is representative of the entire population, but the observations on the dependent variable are truncated according to a rule whose errors are correlated with the errors from the equation of interest. We do not observe y because of the outcome of some other variable, which generates the *selection indicator*, s.

To understand the issue of sample selection, consider a population model in which the relationship between y and a set of explanatory factors $\mathbf{x}$ can be written as a linear model with additive error u. That error is assumed to satisfy the zero-conditional-mean assumption of (4.2). Now consider that we observe only some of the observations on y_i—for whatever reason—and that indicator variable s_i equals 1 when we observe both y_i and $\mathbf{x}_i$ and is zero otherwise. If we merely run a regression on the observations

$$y_i = \mathbf{x}_i\boldsymbol{\beta} + u_i \tag{10.9}$$

on the full sample, those observations with missing values of y_i (or any elements of $\mathbf{x}_i$) will be dropped from the analysis. We can rewrite this regression as

$$s_i y_i = s_i \mathbf{x}_i \boldsymbol{\beta} + s_i u_i \tag{10.10}$$

The OLS estimator $\widehat{\boldsymbol{\beta}}$ of (10.10) will yield the same estimates as that of (10.9). They will be unbiased and consistent if the error term $s_i u_i$ has zero mean and is uncorrelated with each element of x_i. For the population, these conditions can be written as

$$\begin{aligned} E[su] &= 0 \\ E[(s\mathbf{x})(su)] &= E[s\mathbf{x}u] = 0 \end{aligned}$$

because $s^2 = s$. This condition differs from that of a standard regression equation (without selection), where the corresponding zero-conditional-mean assumption requires only that $E[\mathbf{x}u] = 0$. In the presence of selection, the error process u must be uncorrelated with $s\mathbf{x}$.

Consider the source of the sample-selection indicator s_i. If that indicator is purely a function of the explanatory variables in $\mathbf{x}$, we have *exogenous sample selection*. If the explanatory variables in $\mathbf{x}$ are uncorrelated with u, and s is a function of $\mathbf{x}$s, then it too will be uncorrelated with u, as will the product $s\mathbf{x}$. OLS regression estimated on a subset will yield unbiased and consistent estimates. For instance, if gender is one of the explanatory variables, we can estimate separate regressions for men and women with no difficulty. We have selected a subsample based on observable characteristics; e.g., s_i identifies the set of observations for females.

We can also consider selection of a random subsample. If our full sample is a random sample from the population and we use Stata's `sample` command to draw a 10%, 20%, or 50% subsample, estimates from that subsample will be consistent as long as estimates from the full sample are consistent. In this case, s_i is set randomly.

If s_i is set by a rule, such as $s_i = 1$ if $y_i \leq c$, then as in section 10.3.1, OLS estimates will be biased and inconsistent. We can rewrite the rule as $s_i = 1$ if $u_i \leq (c - x_i\beta)$, which makes it clear that s_i must be correlated with u_i. As shown above, we must use the truncated regression model to derive consistent estimates.

Incidental truncation means that we observe y_i based not on its value but rather on the observed outcome of another variable. For instance, we observe hourly wage when an individual participates in the labor force. We can imagine fitting a binomial probit or logit model that predicts the individual's probability of participation. In this circumstance, s_i is set to zero or one based on the factors underlying that decision

$$y_i = \mathbf{x}_i\boldsymbol{\beta} + u \tag{10.11}$$
$$s_i = I(\mathbf{z}_i\gamma + v \geq 0) \tag{10.12}$$

where we assume that the explanatory factors in $\mathbf{x}$ satisfy the zero-conditional-mean assumption $E[\mathbf{x}u] = 0$. The $I(\cdot)$ function equals 1 if its argument is true and is zero otherwise. We observe y_i if $s_i = 1$. The *selection function* contains a set of explanatory factors $\mathbf{z}$, which must be a superset of $\mathbf{x}$. For us to identify the model, $\mathbf{z}$ contains all $\mathbf{x}$ but must also contain more factors that do not appear in $\mathbf{x}$.[17] The error term in the selection equation, v, is assumed to have a zero-conditional mean: $E[\mathbf{z}v] = 0$, which implies that $E[\mathbf{x}v] = 0$. We assume that v follows a standard normal distribution.

Incidental truncation arises when there is a nonzero correlation between u and v. If both these processes are normally distributed with zero means, the conditional expectation $E[u|v] = \rho v$, where ρ is the correlation of u and v. From (10.11),

$$E[y|\mathbf{z}, v] = \mathbf{x}\beta + \rho v \tag{10.13}$$

We cannot observe v, but s is related to v by (10.12). Equation (10.13) then becomes

$$E[y|\mathbf{z}, s] = \mathbf{x}\boldsymbol{\beta} + \rho E[v|\mathbf{z}, s]$$

The conditional expectation $E[v|\mathbf{z}, s]$ for $s_i = 1$, the case of observability, is merely λ, the IMR defined in section 10.3.1. Therefore, we must augment (10.11) with that term:

$$E[y|\mathbf{z}, s = 1] = \mathbf{x}\boldsymbol{\beta} + \rho\lambda(\mathbf{z}\boldsymbol{\gamma}) \tag{10.14}$$

If $\rho \neq 0$, OLS estimates from the incidentally truncated sample will not consistently estimate β unless the IMR term is included. Conversely, if $\rho = 0$, that OLS regression will yield consistent estimates.

The IMR term includes the unknown population parameters γ, which may be fitted by a binomial probit model

$$\Pr(s = 1|\mathbf{z}) = \Phi(\mathbf{z}\boldsymbol{\gamma})$$

from the entire sample. With estimates of $\boldsymbol{\gamma}$, we can compute the IMR term for each observation for which y_i is observed ($s_i = 1$) and fit the model of (10.14). This two-step procedure, based on the work of Heckman (1976), is often termed the *Heckit model*. Instead, we can use a full maximum-likelihood procedure to jointly estimate $\boldsymbol{\beta}$, $\boldsymbol{\gamma}$, and ρ.

17. As Wooldridge (2006) discusses, when $\mathbf{z}$ contains the same variables as $\mathbf{x}$ the parameters are theoretically identified, but this identification is usually too weak to be practically applied.

The Heckman selection model in this context is driven by the notion that some of the **z** factors for an individual are different from the factors in **x**. For instance, in a wage equation, the number of preschool children in the family is likely to influence whether a woman participates in the labor force but might be omitted from the wage determination equation: it appears in **z** but not **x**. We can use such factors to identify the model. Other factors are likely to appear in both equations. A woman's level of education and years of experience in the labor force will likely influence her decision to participate as well as the equilibrium wage that she will earn in the labor market. Stata's `heckman` command fits the full maximum-likelihood version of the Heckit model with the following syntax:

`heckman` *depvar* [*indepvars*] [*if*] [*in*], `select(`*varlist2*`)`

where *indepvars* specifies the regressors in x and *varlist2* specifies the list of Z factors expected to determine the selection of an observation as observable. Unlike with `tobit`, where the `depvar` is recorded at a threshold value for the censored observations, we should code the `depvar` as missing (`.`) for those observations that are not selected.[18] The model is fitted over the entire sample and gives an estimate of the crucial correlation ρ, along with a test of the hypothesis that $\rho = 0$. If we reject that hypothesis, a regression of the observed *depvar* on *indepvars* will produce inconsistent estimates of β.[19]

The `heckman` command can also generate the *two-step* estimator of the selection model (Heckman 1979) if we specify the `twostep` option. This model is essentially the regression of (10.7) in which the IMR has been estimated as the prediction of a binomial probit (10.12) in the first step and used as a regressor in the second step. A significant coefficient of the IMR, denoted `lambda`, indicates that the selection model must be used to avoid inconsistency. The `twostep` approach, computationally less burdensome than the full maximum-likelihood approach used by default in `heckman`, may be preferable in complex selection models.[20]

The example below revisits the `womenwk` dataset used to illustrate `tobit`. To use these data in `heckman`, we define `lw` as the log of the wage for working women and as missing for nonworking women. We assume that marital status affects selection (whether a woman is observed in the labor force) but does not enter the log(wage) equation. All factors in both the log(wage) and selection equations are significant. By using the selection model, we have relaxed the assumption that the factors determining participation and the wage are identical and of the same sign. The effect of more children increases the probability of selection (participation) but decreases the predicted wage, conditional on participation. The likelihood-ratio test for $\rho = 0$ rejects its null, so that

18. An alternative syntax of `heckman` allows for a second dependent variable: an indicator that signals which observations of *depvar* are observed.

19. The output produces an estimate of `/athrho`, the hyperbolic arctangent of ρ. That parameter is entered in the log-likelihood function to enforce the constraint that $-1 < \rho < 1$. The point and interval estimates of ρ are derived from the inverse transformation.

20. For more details on the two-step versus maximum likelihood approaches, see Wooldridge (2002, 560–566).

estimation of the log(wage) equation without taking selection into account would yield inconsistent results.

```
. heckman lw education age children,
> select(age married children education) nolog
Heckman selection model                         Number of obs     =      2000
(regression model with sample selection)        Censored obs      =       657
                                                Uncensored obs    =      1343

                                                Wald chi2(3)      =    454.78
Log likelihood = -1052.857                      Prob > chi2       =    0.0000

------------------------------------------------------------------------------
             |      Coef.   Std. Err.      z    P>|z|     [95% Conf. Interval]
-------------+----------------------------------------------------------------
lw           |
   education |   .0397189   .0024525    16.20   0.000     .0349121    .0445256
         age |   .0075872   .0009748     7.78   0.000     .0056767    .0094977
    children |  -.0180477   .0064544    -2.80   0.005    -.0306981   -.0053973
       _cons |   2.305499   .0653024    35.30   0.000     2.177509     2.43349
-------------+----------------------------------------------------------------
select       |
         age |   .0350233   .0042344     8.27   0.000     .0267241    .0433225
     married |   .4547724   .0735876     6.18   0.000     .3105434    .5990014
    children |   .4538372   .0288398    15.74   0.000     .3973122    .5103621
   education |   .0565136   .0110025     5.14   0.000     .0349492    .0780781
       _cons |  -2.478055   .1927823   -12.85   0.000    -2.855901   -2.100208
-------------+----------------------------------------------------------------
     /athrho |   .3377674   .1152251     2.93   0.003     .1119304    .5636045
    /lnsigma |  -1.375543   .0246873   -55.72   0.000    -1.423929   -1.327156
-------------+----------------------------------------------------------------
         rho |   .3254828   .1030183                      .1114653    .5106469
       sigma |   .2527024   .0062385                      .2407662    .2652304
      lambda |   .0822503   .0273475                      .0286501    .1358505
------------------------------------------------------------------------------
LR test of indep. eqns. (rho = 0):   chi2(1) =     5.53   Prob > chi2 = 0.0187
------------------------------------------------------------------------------
```

We also use the `heckman` two-step procedure, which makes use of the IMR from a probit equation for selection.

```
. heckman lw education age children,
> select(age married children education) twostep
Heckman selection model -- two-step estimates   Number of obs      =      2000
(regression model with sample selection)        Censored obs       =       657
                                                Uncensored obs     =      1343
                                                Wald chi2(6)       =    737.21
                                                Prob > chi2        =    0.0000

------------------------------------------------------------------------------
             |      Coef.   Std. Err.      z    P>|z|     [95% Conf. Interval]
-------------+----------------------------------------------------------------
lw           |
   education |   .0427067    .003106    13.75   0.000     .0366191    .0487944
         age |    .009322   .0014343     6.50   0.000     .0065108    .0121333
    children |  -.0019549   .0115202    -0.17   0.865    -.0245341    .0206242
       _cons |   2.124787   .1249789    17.00   0.000     1.879833    2.369741
-------------+----------------------------------------------------------------
select       |
         age |   .0347211   .0042293     8.21   0.000     .0264318    .0430105
     married |   .4308575    .074208     5.81   0.000     .2854125    .5763025
    children |   .4473249   .0287417    15.56   0.000     .3909922    .5036576
   education |   .0583645   .0109742     5.32   0.000     .0368555    .0798735
       _cons |  -2.467365   .1925635   -12.81   0.000    -2.844782   -2.089948
-------------+----------------------------------------------------------------
mills        |
      lambda |   .1822815   .0638285     2.86   0.004       .05718     .307383
-------------+----------------------------------------------------------------
         rho |    0.66698
       sigma |  .27329216
      lambda |  .18228151   .0638285
------------------------------------------------------------------------------
```

Although it also provides consistent estimates of the selection model's parameters, we see a qualitative difference in the log(wage) equation: the number of children is not significant in this formulation of the model. The maximum likelihood formulation, when computationally feasible, is attractive—not least because it can generate interval estimates of the selection model's ρ and σ parameters.

10.5 Bivariate probit and probit with selection

Another example of a limited-dependent-variable framework in which a correlation of equations' disturbances plays an important role is the *bivariate probit* model. In its simplest form, the model may be written as

$$\begin{aligned} y_1^* &= \mathbf{x}_1\boldsymbol{\beta}_1 + u_1 \\ y_2^* &= \mathbf{x}_2\boldsymbol{\beta}_2 + u_2 \\ \begin{pmatrix} u_1 \\ u_2 \end{pmatrix} &\sim N\left\{\begin{pmatrix} 0 \\ 0 \end{pmatrix}, \begin{bmatrix} 1 & \rho \\ \rho & 1 \end{bmatrix}\right\} \end{aligned} \tag{10.15}$$

The observable counterparts to the two latent variables y_1^*, y_2^* are y_1, y_2. These variables are observed as 1 if their respective latent variables are positive and zero otherwise.

One formulation of this model, termed the *seemingly unrelated bivariate probit* model in `biprobit`, is similar to the SUR model that I presented in section 9.4. As in the regression context, we can view the two probit equations as a system and estimate them jointly if $\rho \neq 0$, but it will not affect the consistency of individual probit equations' estimates.

However, consider one common formulation of the bivariate probit model because it is similar to the selection model described above. Consider a two-stage process in which the second equation is observed conditional on the outcome of the first. For example, some fraction of patients diagnosed with circulatory problems undergoes multiple-bypass surgery ($y_1 = 1$). For each patient, we record whether he or she died within 1 year of the surgery ($y_2 = 1$). The y_2 variable is available only for those patients who are postoperative. We do not have records of mortality among those who chose other forms of treatment. In this context, the reliance of the second equation on the first is an issue of *partial observability*, and if $\rho \neq 0$ it will be necessary to take both equations' factors into account to generate consistent estimates. That correlation of errors may be likely in that unexpected health problems that caused the physician to recommend bypass surgery may recur and kill the patient.

As another example, consider a bank deciding to extend credit to a small business. The decision to offer a loan can be viewed as $y_1 = 1$. Conditional on that outcome, the borrower will or will not default on the loan within the following year, where a default is recorded as $y_2 = 1$. Those potential borrowers who were denied cannot be observed defaulting because they did not receive a loan in the first stage. Again the disturbances impinging upon the loan offer decision may well be correlated (here negatively) with the disturbances that affect the likelihood of default.

Stata can fit these two bivariate probit models with the `biprobit` command. The *seemingly unrelated bivariate probit* model allows $\mathbf{x}_1 \neq \mathbf{x}_2$, but the alternative form that we consider here allows only one *varlist* of factors that enter both equations. In the medical example, this *varlist* might include the patient's body mass index (a measure of obesity), indicators of alcohol and tobacco use, and age—all of which might affect both the recommended treatment and the 1-year survival rate. With the `partial` option, we specify that the partial observability model of Poirier (1981) be fitted.

10.5.1 Binomial probit with selection

Closely related to the bivariate probit with partial observability is the *binomial probit with selection* model. This formulation, first presented by Van de Ven and Van Pragg (1981), has the same basic setup as (10.15) above: the latent variable y_1^* depends on factors $\mathbf{x}$, and the binary outcome $y_1 = 1$ arises when $y_1^* > 0$. However, y_{1j} is observed only when

$$y_{2j} = (\mathbf{x}_2\gamma + u_{2j} > 0)$$

that is, when the selection equation generates a value of 1. This result could be viewed, in the earlier example, as y_2 indicating whether the patient underwent bypass surgery. We observe the following year's health outcome only for those patients who had the surgical procedure. As in (10.15), there is a potential correlation (ρ) between the errors of the two equations. If that correlation is nonzero, estimates of the y_1 equation will be biased unless we account for the selection. Here that suggests that focusing only on the patients who underwent surgery (for whom $y_2 = 1$) and studying the factors that contributed to survival is not appropriate if the selection process is nonrandom. In the medical example, selection is likely nonrandom in that those patients with less serious circulatory problems are not as likely to undergo heart surgery.

In the second example, we consider small business borrowers' likelihood of getting a loan and for successful borrowers, whether they defaulted on the loan. We can observe only a default if they were selected by the bank to receive a loan ($y_2 = 1$). Conditional on receiving a loan, they did or did not fulfill their obligations, as recorded in y_1. If we focus only on loan recipients and whether they defaulted, we are ignoring the selection issue. Presumably, a well-managed bank is not choosing among loan applicants at random. Both deterministic and random factors influencing the extension of credit and borrowers' subsequent performance are likely to be correlated. Unlike the bivariate probit with partial observability, the probit with sample selection explicitly considers $\mathbf{x}_1 \neq \mathbf{x}_2$. The factors influencing the granting of credit and the borrowers' performance must differ to identify the model. Stata's `heckprob` command has a syntax similar to that of `heckman`, with an *indepvars* of the factors in x_1 and a `select(`*varlist2*`)` option specifying the explanatory factors driving the selection outcome.

I illustrate one form of this model with the Federal Reserve Bank of Boston HMDA dataset[21] (Munnell et al. 1996), a celebrated study of racial discrimination in banks' home mortgage lending. Of the 2,380 loan applications in this subset of the dataset, 88% were granted, as `approve` indicates. For those 2,095 loans that were approved and originated, we may observe whether they were purchased in the secondary market by Fannie Mae (FNMA) or Freddie Mac (FHLMC), the quasigovernment mortgage finance agencies. The variable `fanfred` indicates that 33% (698) of those loans were sold to Fannie or Freddie. We seek to explain whether certain loans were attractive enough to the secondary market to be resold as a function of the loan amount (`loanamt`), an indicator of above-average vacant properties in that census tract (`vacancy`), an indicator of above-average median income in that tract (`med_income`), and the appraised value of the dwelling (`appr_value`). The secondary market activity is observable only if the loan was originated. The selection equation contains an indicator for black applicants, applicants' income, and their debt-to-income ratio (`debt_inc_r`) as predictors of loan approval.

```
. use http://www.stata-press.com/data/imeus/hmda, clear
. replace fanfred=. if deny
(285 real changes made, 285 to missing)
. rename s6 loanamt
. rename vr vacancy
```

21. Under the Home Mortgage Disclosure Act of 1975, as amended, institutions regulated by HMDA must report information on the disposition of every mortgage application and purchase as well as provide data on the race, income, and gender of the applicant or mortgagor.

```
. rename mi med_income
. rename s50 appr_value
. rename s17 appl_income
. replace appl_income = appl_income/1000
(2379 real changes made)
. rename s46 debt_inc_r
. summarize approve fanfred loanamt vacancy med_income appr_value
> black appl_income debt_inc_r, sep(0)
    Variable |       Obs        Mean    Std. Dev.       Min        Max
-------------+--------------------------------------------------------
     approve |      2380    .8802521    .3247347          0          1
     fanfred |      2095    .3331742    .4714608          0          1
     loanamt |      2380    139.1353    83.42097          2        980
     vacancy |      2380    .4365546    .4960626          0          1
  med_income |      2380    .8294118    .3762278          0          1
  appr_value |      2380    198.5426    152.9863         25       4316
       black |      2380     .142437    .3495712          0          1
 appl_income |      2380     13.9406    116.9485          0   999.9994
  debt_inc_r |      2380    33.08136    10.72573          0        300
```

We fit the model with heckprob:

```
. heckprob fanfred loanamt vacancy med_income appr_value,
> select(approve= black appl_income debt_inc_r) nolog
Probit model with sample selection              Number of obs      =      2380
                                                Censored obs       =       285
                                                Uncensored obs     =      2095

                                                Wald chi2(4)       =     80.69
Log likelihood = -2063.066                      Prob > chi2        =    0.0000

------------------------------------------------------------------------------
             |      Coef.   Std. Err.      z    P>|z|     [95% Conf. Interval]
-------------+----------------------------------------------------------------
fanfred      |
     loanamt |  -.0026434   .0008029    -3.29   0.001    -.0042169   -.0010698
     vacancy |  -.2163306   .0609798    -3.55   0.000    -.3358488   -.0968124
  med_income |   .2671338   .0893349     2.99   0.003     .0920407    .4422269
  appr_value |  -.0014358   .0005099    -2.82   0.005    -.0024351   -.0004364
       _cons |   .1684829   .1182054     1.43   0.154    -.0631954    .4001612
-------------+----------------------------------------------------------------
approve      |
       black |  -.7343534    .081858    -8.97   0.000    -.8947921   -.5739147
 appl_income |  -.0006596    .000236    -2.80   0.005    -.0011221   -.0001971
  debt_inc_r |  -.0262367   .0036441    -7.20   0.000     -.033379   -.0190944
       _cons |   2.236424   .1319309    16.95   0.000     1.977844    2.495004
-------------+----------------------------------------------------------------
     /athrho |  -.6006626    .271254    -2.21   0.027    -1.132311   -.0690146
-------------+----------------------------------------------------------------
         rho |  -.5375209   .1928809                     -.8118086   -.0689052
------------------------------------------------------------------------------
LR test of indep. eqns. (rho = 0):   chi2(1) =     4.99   Prob > chi2 = 0.0255
------------------------------------------------------------------------------
```

The model is successful, indicating that the secondary market sale is more likely to take place for smaller-value loans (or properties). The probability is affected negatively by nearby vacant properties and positively by higher income in the neighborhood. In the selection equation, the original researchers' findings of a strong racial effect on loan approvals is borne out by the sign and significance of the `black` coefficient. Applicants' income has an (unexpected) negative effect on the probability of approval, although the debt-to-income ratio has the expected negative sign. The likelihood-ratio test of independent equations conclusively rejects that null hypothesis with an estimated `rho` of -0.54 between the two equations' errors, indicating that ignoring the selection into approved status would render the estimates of a univariate probit equation for `fanfred` equation biased and inconsistent.

Exercises

1. In section 10.3.1, we estimated an OLS regression and a truncated regression from the `laborsub` sample of 250 married women, 150 of whom work. This dataset can be treated as censored in that we have full information on nonworking women's characteristics. Refit the model with `tobit` and compare the results to those of OLS.
2. In section 10.3.2, we fitted a tobit model for the log of the wage from `womenwk`, taking into account a zero wage recorded by 1/3 of the sample. Create a wage variable in which wages above $25.00 per hour are set to that value and missing `wage` is set to zero. Generate the log of the transformed wage, and fit the model as a two-limit tobit. How do the `tobit` coefficients and their marginal effects differ from those presented in section 10.3.2?
3. Using the dataset http://www.stata-press.com/data/r9/school.dta, fit a bivariate probit model of `private` (whether a student is enrolled in private school) and `vote` (whether the parent voted in favor of public school funding). Model the first response variable as depending on `years` and `logptax`, the tax burden; and estimate the second response variable as depending on those factors plus `loginc`. Are these equations successful? What do the estimate of ρ and the associated Wald test tell you?
4. Using the HMDA dataset from section 10.5.1, experiment with alternative specifications of the model for loan approval (`approve` = 1). Should factors such as the loan amount or the ratio of the loan amount to the appraised value of the property be entered in the loan approval equation? Test an alternative `heckprob` model with your revised loan approval equation.

A Getting the data into Stata

This appendix discusses problems you may have in inputting and managing economic and financial data. You can download source data from a web site, acquire it in spreadsheet format, or import it from some other statistical package. The two sections deal with those variations.

A.1 Inputting data from ASCII text files and spreadsheets

Before carrying out econometric analysis with Stata, many researchers must face several thorny issues in converting their foreign data into Stata-usable form. These issues range from the mundane (e.g., a text-file dataset may have coded missing values as `99`) to the challenging (e.g., a text-file dataset may be in a *hierarchical* format, with master records and detail records). Although I cannot possibly cover all the ways in which external data may be organized and transformed for use in Stata, several rules apply:

- Familiarize yourself with the various Stata commands for data input. Each has its use, and in the spirit of "don't pound nails with a screwdriver", data handling is much simpler if you use the correct tool. Reading [U] **21 Inputting data** is well worth your time.

- When you need to manipulate a text file, use a text editor, not a word processor or spreadsheet.

- Get the data into Stata as early as you can, and perform all manipulations via well-documented do-files that you can edit and reuse. I will not discuss `input` or the Data Editor, which allow you to interactively enter data, or various copy-and-paste strategies involving simultaneous use of a spreadsheet and Stata. Such a strategy is not reproducible and should be avoided.

- Keep track of multiple steps of a data input and manipulation process through good documentation. If you ever need to replicate or audit the data manipulation process, you will regret not properly documenting your research.

- If you are working with anything but a simple rectangular data array, you will need to use `append`, `merge`, or `reshape`. Review chapter 3 to understand their capabilities.

A.1.1 Handling text files

Text files—often described as ASCII files—are the most common source of raw data in economic research. Text files may have any file extension: they may be labeled `.raw` (as Stata would prefer), `.txt`, `.csv`, or `.asc`. A text file is just that: text. Word-processing programs like Microsoft Word are inappropriate tools for working with text files because they have their own native, binary format and generally use features such as proportional spacing that will cause columns to be misaligned.

Every operating system supports a variety of text editors, many of which are freely available. A useful summary of text editors of interest to Stata users is edited by Nicholas J. Cox and is available as a web page from `ssc` as the package `texteditors`. A good text editor—one without the memory limitations present in Stata's ado-file editor or the built-in routines in some operating systems—is much faster than a word processor when scrolling through a large data file. Many text editors colorize Stata commands, making them useful for developing Stata programs. Text editors are also useful for working with large microeconomic survey datasets that come with machine-readable codebooks, which are often many megabytes. Searching those codebooks for particular keywords with a robust text editor is efficient.

Free format versus fixed format

Text files may be *free format* or *fixed format*. A free-format file contains several fields per record, separated by *delimiters*: characters that are not to be found within the fields. A purely numeric file (or one with simple string variables such as U.S. state codes) may be *space delimited*; that is, successive fields in the record are separated by one or more space characters:

```
AK 12.34  0.09  262000
AL 9.02 0.075 378000
AZ 102.4  0.1  545250
```

The columns in the file need not be aligned. These data may be read from a text file (by default with extension `.raw`) with Stata's `infile` command, which assigns names (and if necessary data types) to the variables:

```
. clear
. infile str2 state members prop potential using appA_1
(3 observations read)
. list
```

	state	members	prop	potent~l
1.	AK	12.34	.09	262000
2.	AL	9.02	.075	378000
3.	AZ	102.4	.1	545250

We must indicate that the first variable is a string variable of maximum length two characters (`str2`), or every record will generate an error that `state` cannot be read as a number. We may even have a string variable with contents of various length in the record:

```
. clear
. infile str2 state members prop potential str20 state_name key using appA_2
(3 observations read)
. list

     +-----------------------------------------------------+
     | state   members   prop   potent~l   state_~e   key |
     |-----------------------------------------------------|
  1. |    AK     12.34    .09     262000     Alaska     1 |
  2. |    AL      9.02   .075     378000    Alabama     2 |
  3. |    AZ     102.4     .1     545250    Arizona     3 |
     +-----------------------------------------------------+
```

However, this scheme will break down as soon as we hit New Hampshire. Stata will read the space within the state name as a delimiter. If you use string variables with embedded spaces in a space-delimited file, you must delimit the variable names (usually with quotation marks in the text file):

```
. clear
. type appA_3.raw
AK 12.34  0.09  262000 Alaska 1
AL 9.02 0.075 378000 Alabama 2
AZ 102.4  0.1  545250 Arizona 3
NH 14.9  0.02  212000 "New Hampshire" 4

. infile str2 state members prop potential str20 state_name key using appA_3
(4 observations read)
. list

     +----------------------------------------------------------+
     | state   members   prop   potent~l      state_name   key |
     |----------------------------------------------------------|
  1. |    AK     12.34    .09     262000          Alaska     1 |
  2. |    AL      9.02   .075     378000         Alabama     2 |
  3. |    AZ     102.4     .1     545250         Arizona     3 |
  4. |    NH      14.9    .02     212000   New Hampshire     4 |
     +----------------------------------------------------------+
```

So what should you do if your text file is space delimited and contains string variables with embedded spaces? No mechanical transformation will generally solve this problem. For instance, using a text editor to change multiple spaces to one space and then each single space to a tab character will not help because it will then place a tab between "New" and "Hampshire".

If you download the data from a web page that offers formatting choices, you should choose *tab-delimited* rather than *space-delimited* format. The other option, comma-delimited text, or *comma-separated values* (`.csv`), has its own difficulties. Consider field contents (without quotation marks), such as "College Station, TX", "J. Arthur Jones, Jr.", "F. Lee Bailey, Esq.", or "Ronald Anderson, S.J." If every city name is followed by a comma, there is no problem, since the city and state can then be read as separate variables: but if some are written without commas ("Brighton MA"), the problem returns. In any case, parsing proper names with embedded commas is problematic, but using tab-delimited text avoids most of these problems.

The insheet command

To read tab-delimited text files, we should use `insheet` rather than `infile`. Despite its name, `insheet` does not read binary spreadsheet files (e.g., `.xls`), and it reads a tab-delimited (or comma-delimited) text file, whether or not a spreadsheet program was used to create it. For instance, most database programs have an option for generating a tab-delimited or comma-delimited export file, and many datasets available for web download are in one of these formats.

The `insheet` command is handy, as long as one observation in your target Stata dataset is contained on one record with tab or comma delimiters. Stata will automatically try to determine the delimiter (but options `tab` and `comma` are available), or you can specify any ASCII character as a delimiter with the `delimiter(`*char*`)` option. For instance, some European database exports use semicolon (;) delimiters because standard European numeric formats use the comma as the decimal separator. If the first line of the `.raw` file contains valid Stata variable names, these names will be used. If you are extracting the data from a spreadsheet, they will often have that format. To use the sample dataset above, now tab delimited with a header record of variable names, you could type

```
. clear

. insheet using appA_4
(6 vars, 4 obs)

. list

     +----------------------------------------------------------------+
     | state   members    prop   potent~l       state_name   key |
     |----------------------------------------------------------------|
  1. |    AK     12.34     .09     262000           Alaska     1 |
  2. |    AL      9.02    .075     378000          Alabama     2 |
  3. |    AZ     102.4      .1     545250          Arizona     3 |
  4. |    NH      14.9     .02     212000    New Hampshire     4 |
     +----------------------------------------------------------------+
```

The issue of embedded spaces or commas no longer arises in tab-delimited data. The first line of the file defines the variable names.

Pay particular attention to informational or error messages produced by the data input commands. If you know how many observations are in the text file, check to see that the number Stata reports is correct. Likewise, you can use `summarize` to discern whether the number of observations, minimum, and maximum for each numeric variable are sensible. You can usually spot data-entry errors if a particular variable takes on nonsensical values, which usually means that one or more fields on that record have been omitted and should trigger an error message. For instance, leaving out a numeric field on a particular record will move an adjacent string field into that variable. Stata will then complain that it cannot read the string as a number. A distinct advantage of the tab- or comma-delimited formats is that missing values may be coded with two successive delimiters. As discussed in chapter 2, we can use `assert` to good advantage to ensure that reasonable values appear in the data.

You can use `infile` with `if` *exp* and `in` *range* qualifiers to selectively input data, but not `insheet`. For instance, with a large text-file dataset, you could use `in 1/1000` to read only the first 1,000 observations and verify that the input process is working properly. Using `if gender=="M"`, we could read only the male observations; by using `if uniform() <= 0.15` we could draw a 15% sample from the input data. You cannot use these qualifiers with `insheet`: but unless the text-file dataset is huge and the computer slow, you could always read the entire dataset and apply `keep` or `drop` conditions to mimic the action of `infile`.

A.1.2 Accessing data stored in spreadsheets

Above, I said that you should not copy and paste to transfer data from another application directly to Stata because you cannot replicate the process. For instance, you cannot guarantee that the first and last rows or columns of a spreadsheet were selected and copied to the clipboard without affecting the data. If the data are in a spreadsheet, copy the appropriate portion of that spreadsheet and paste it into a new blank sheet (in Excel, use Paste Special to ensure that only values are stored). If you are going to add Stata variable names, leave the first row blank so that you can fill them in later. Save that sheet, and that sheet alone, as Text Only—Tab delimited to a new filename. Using the file extension `.raw` will simplify reading the file into Stata.

Both Excel and Stata read calendar dates as successive integers from an arbitrary starting point. For Stata to read the dates into a Stata date variable, they must be formatted with a four-digit year, preferably in a format with delimiters (e.g., 12/6/2004 or 6-Dec-2004). It is much easier to make these changes in the spreadsheet program before reading the data into Stata. Macintosh OS X users of Excel should note that Excel's default is the 1904 Date System. If the spreadsheet was produced in Excel for Windows, and you used the steps above to create a new sheet with the desired data, the dates will be off by 4 years (the difference between Excel for Macintosh and Excel for Windows defaults). Uncheck the preference Use 1904 Date System before saving the file as text.

A.1.3 Fixed-format data files

Many text-file datasets are composed of *fixed-format* records, which obey a strict columnar format in which a variable appears in a specific location in each record of the dataset. Such datasets are accompanied by *codebooks*, which define each variable's name, data type, location in the record, and possibly other information, such as missing values, value labels, or frequencies for integer variables.[1] Here is a fragment of the codebook

1. Stata itself can produce a codebook from a Stata dataset via the `codebook` command.

for the study "National Survey of Hispanic Elderly People, 1988", available from the Inter-University Consortium for Political and Social Research.[2]

```
VAR 0001       ICPSR STUDY NUMBER-9289      NO MISSING DATA CODES
               REF 0001         LOC    1 WIDTH  4          DK    1 COL  3- 6
VAR 0002       ICPSR EDITION NUMBER-2       NO MISSING DATA CODES
               REF 0002         LOC    5 WIDTH  1          DK    1 COL  7
VAR 0003       ICPSR PART NUMBER-001        NO MISSING DATA CODES
               REF 0003         LOC    6 WIDTH  3          DK    1 COL  8-10
VAR 0004       ICPSR ID                     NO MISSING DATA CODES
               REF 0004         LOC    9 WIDTH  4          DK    1 COL 11-14
VAR 0005       ORIGINAL ID                  NO MISSING DATA CODES
               REF 0005         LOC   13 WIDTH  4          DK    1 COL 15-18
VAR 0006       PROXY                        NO MISSING DATA CODES
               REF 0006         LOC   17 WIDTH  1          DK    1 COL 19
VAR 0007       TIME BEGUN-HOUR                          MD=99
               REF 0007         LOC   18 WIDTH  2          DK    1 COL 20-21
VAR 0008       TIME BEGUN-MINUTE                        MD=99
               REF 0008         LOC   20 WIDTH  2          DK    1 COL 22-23
VAR 0009       TIME BEGUN-AM/PM                          MD=9
               REF 0009         LOC   22 WIDTH  1          DK    1 COL 24
VAR 0010       AGE                          NO MISSING DATA CODES
               REF 0010         LOC   23 WIDTH  3          DK    1 COL 25-27
VAR 0011       HISPANIC GROUP               NO MISSING DATA CODES
               REF 0011         LOC   26 WIDTH  1          DK    1 COL 28
VAR 0012       HISPANIC GROUP-OTHER                     MD=99
               REF 0012         LOC   27 WIDTH  2          DK    1 COL 29-30
VAR 0013       MARITAL STATUS               NO MISSING DATA CODES
               REF 0013         LOC   29 WIDTH  1          DK    1 COL 31

          Q.A3.  ARE YOU NOW MARRIED, WIDOWED, DIVORCED, SEPARATED, OR
                  HAVE YOU NEVER MARRIED?
                  ------------------------------------------------------------
               1083  1.  MARRIED
                815  2.  WIDOWED
                160  3.  DIVORCED
                 99  4.  SEPARATED
                 14  5.  NOT MARRIED, LIVING WITH PARTNER
                128  6.  NEVER MARRIED

VAR 0014       MARITAL STATUS-YEARS               MD=97 OR GE  98
               REF 0014         LOC   30 WIDTH  2          DK    1 COL 32-33
VAR 0015       RESIDENCE TYPE                            MD=7
               REF 0015         LOC   32 WIDTH  1          DK    1 COL 34
VAR 0016       RESIDENCE TYPE-OTHER                  MD=GE  99
               REF 0016         LOC   33 WIDTH  2          DK    1 COL 35-36
VAR 0017       OWN/RENT                                  MD=7
               REF 0017         LOC   35 WIDTH  1          DK    1 COL 37
VAR 0018       OWN/RENT-OTHER                           MD=99
               REF 0018         LOC   36 WIDTH  2          DK    1 COL 38-39
VAR 0019       LIVE ALONE                   NO MISSING DATA CODES
               REF 0019         LOC   38 WIDTH  1          DK    1 COL 40
VAR 0020       HOW LONG LIVE ALONE                  MD=7 OR GE  8
               REF 0020         LOC   39 WIDTH  1          DK    1 COL 41
VAR 0021       PREFER LIVE ALONE                    MD=7 OR GE  8
               REF 0021         LOC   40 WIDTH  1          DK    1 COL 42
```

2. Study no. 9289, http://webapp.icpsr.umich.edu/cocoon/ICPSR-STUDY/09289.xml.

The codebook specifies the column in which each variable starts (`LOC`) and the number of columns it spans (`WIDTH`).[3] In this fragment of the codebook, only `integer` numeric variables appear. The missing-data (`MD`) codes for each variable are also specified. The listing above provides the full codebook detail for variable 13, marital status, quoting the question posed by the interviewer, coding of the six possible responses, and the frequency counts of each response.

In fixed-format data files, fields need not be separated: above, for example, the single-column fields of variables 0019, 0020, and 0021 are stored as three successive integers. We must tell Stata to interpret each of those digits as a separate variable, which we can do with a *data dictionary*: a separate Stata file, with file extension `.dct`, specifying the necessary information to read a fixed-format data file. The information in the codebook may be translated, line for line, into the Stata data dictionary. The Stata data dictionary need not be comprehensive. You might not want to read certain variables from the raw data file, so you would merely ignore those columns. This ability to select data might be particularly important when you are working with Intercooled Stata and its limit of 2,047 variables. Many survey datasets contain many more than 2,000 variables. By judiciously specifying only the subset of variables that are of interest in your research, you may read such a text file by using Intercooled Stata.

Stata supports two different formats of data dictionaries. The simpler format, used by `infix`, requires only that the starting and ending columns of each variable be given along with any needed data type information. To illustrate, I specify the information needed to read a subset of fields in this codebook into Stata variables, using the description of the data dictionary in [D] **infix (fixed format)**:

(*Continued on next page*)

3. The `COL` field should not be considered.

```
. clear
. infix using 09289-infix
infix dictionary using 09289-0001-Data.raw {
* dictionary to read extract of ICPSR study 9289
  int v1      1-4
  int v2      5
  int v3      6-8
  int v4      9-12
  int v5      13-16
  int v6      17
  int v7      18-19
  int v8      20-21
  int v9      22
  int v10     23-25
  int v11     26
  int v12     27-28
  int v13     29
  int v14     30-31
  int v15     32
  int v16     33-34
  int v17     35
  int v18     36-37
  int v19     38
  int v20     39
  int v21     40
}
(2299 observations read)
```

We could instead set up a dictionary file for the fixed-format version of `infile`. This is the more powerful command, as it allows us to attach variable labels and specify value labels. However, rather than specifying the column range of each field, we must indicate where it starts and its field width, given as the *%infmt* for that variable. With a codebook like that displayed above, we have the field widths available. We could also calculate the field widths from the starting and ending column numbers. We must not only specify which are string variables but also give their data storage type. The storage type could differ from the *%infmt* for that variable. You might read a six-character code into a 10-character field, knowing that other data use the latter width for that variable.

```
. clear
. infile using 09289-0001-Data
infile dictionary using 09289-0001-Data.raw {
_lines(1)
_line(1)
_column(1)     int   V1               %4f    "ICPSR STUDY NUMBER-9289"
_column(5)     int   V2      :V2      %1f    "ICPSR EDITION NUMBER-2"
_column(6)     int   V3               %3f    "ICPSR PART NUMBER-001"
_column(9)     int   V4               %4f    "ICPSR ID"
_column(13)    int   V5               %4f    "ORIGINAL ID"
_column(17)    int   V6      :V6      %1f    "PROXY"
_column(18)    int   V7      :V7      %2f    "TIME BEGUN-HOUR"
_column(20)    int   V8      :V8      %2f    "TIME BEGUN-MINUTE"
_column(22)    int   V9      :V9      %1f    "TIME BEGUN-AM-PM"
_column(23)    int   V10     :V10     %3f    "AGE"
_column(26)    int   V11     :V11     %1f    "HISPANIC GROUP"
_column(27)    int   V12     :V12     %2f    "HISPANIC GROUP-OTHER"
_column(29)    int   V13     :V13     %1f    "MARITAL STATUS"
```

```
_column(30)    int   V14      :V14      %2f     "MARITAL STATUS-YEARS"
_column(32)    int   V15      :V15      %1f     "RESIDENCE TYPE"
_column(33)    int   V16      :V16      %2f     "RESIDENCE TYPE-OTHER"
_column(35)    int   V17      :V17      %1f     "OWN-RENT"
_column(36)    int   V18      :V18      %2f     "OWN-RENT-OTHER"
_column(38)    int   V19      :V19      %1f     "LIVE ALONE"
_column(39)    int   V20      :V20      %1f     "HOW LONG LIVE ALONE"
_column(40)    int   V21      :V21      %1f     "PREFER LIVE ALONE"
}
(2299 observations read)
```

The _column() directives in this dictionary are used where dictionary fields are not adjacent. Indeed, you could skip back and forth along the input record since the columns read need not be in ascending order. But then we could achieve the same result with the order command after data input. We can define variable labels and value labels for each variable by using infile. In both examples above, the dictionary file specifies the name of the data file, which need not be the same as that of the dictionary file. For instance, highway.dct could read highway.raw, and if so, we need not specify the latter filename. But we might want to use the same dictionary to read more than one .raw file, and we can do that by changing the name specified in the .dct file. After loading the data, we can describe its contents:

```
. describe
Contains data
  obs:         2,299
 vars:            21
 size:       105,754 (98.9% of memory free)
------------------------------------------------------------------------------
              storage  display     value
variable name   type   format      label      variable label
------------------------------------------------------------------------------
V1              int    %8.0g                  ICPSR STUDY NUMBER-9289
V2              int    %8.0g       V2         ICPSR EDITION NUMBER-2
V3              int    %8.0g                  ICPSR PART NUMBER-001
V4              int    %8.0g                  ICPSR ID
V5              int    %8.0g                  ORIGINAL ID
V6              int    %8.0g       V6         PROXY
V7              int    %8.0g       V7         TIME BEGUN-HOUR
V8              int    %8.0g       V8         TIME BEGUN-MINUTE
V9              int    %8.0g       V9         TIME BEGUN-AM-PM
V10             int    %8.0g       V10        AGE
V11             int    %8.0g       V11        HISPANIC GROUP
V12             int    %8.0g       V12        HISPANIC GROUP-OTHER
V13             int    %8.0g       V13        MARITAL STATUS
V14             int    %8.0g       V14        MARITAL STATUS-YEARS
V15             int    %8.0g       V15        RESIDENCE TYPE
V16             int    %8.0g       V16        RESIDENCE TYPE-OTHER
V17             int    %8.0g       V17        OWN-RENT
V18             int    %8.0g       V18        OWN-RENT-OTHER
V19             int    %8.0g       V19        LIVE ALONE
V20             int    %8.0g       V20        HOW LONG LIVE ALONE
V21             int    %8.0g       V21        PREFER LIVE ALONE
------------------------------------------------------------------------------
Sorted by:
     Note:  dataset has changed since last saved
```

The dictionary indicates that value labels are associated with the variables but does not define those labels. We use a command such as

```
. label define V13 1 "MARRIED" 2 "WIDOWED" 3 "DIVORCED" 4 "SEPARATED"
> 5 "NOT MAR COHABITG" 6 "NEVER MARRIED"
```

to create those labels.

Another advantage of the more elaborate `infile` data dictionary format comes when you are working with a large survey dataset with variables that are real or floating-point values, such as a wage rate in dollars and cents or a percent interest rate such as 6.125%. To save space, the decimal points are excluded from the text file, and the codebook indicates how many decimal digits are included in the field. You could read these data as integer values and perform the appropriate division in Stata, but a simpler solution would be to build this information into the data dictionary. By specifying that a variable has an *%infmt* of, for example, `%6.2f`, a value such as `1234` may be read properly as an hourly wage of $12.34.

Stata's data dictionary syntax can handle many more complicated text datasets, including those with multiple records per observation, or those with header records that are to be ignored. See [D] **infile (fixed format)** for full details.

A.2 Importing data from other package formats

The previous section discussed how foreign data files could be brought into Stata. Often the foreign data are already in the format of some other statistical package or application. For instance, several economic and financial data providers make SAS-formatted datasets readily available, whereas socioeconomic datasets are often provided in SPSS format. The easiest and cheapest way to deal with these package formats is to use Stat/Transfer, a product of Circle Systems, which you can purchase from StataCorp.

If you do not have Stat/Transfer, you will need a working copy of the other statistical package and know how to export a dataset from that format to ASCII format.[4] But this is a rather cumbersome solution, because (like Stata) packages such as SAS and SPSS have their own missing-data formats, value labels, data types, and the like. Although you can export the raw data to ASCII format, these attributes of the data will have to be recreated in Stata. For a large survey dataset with many hundred (or several thousand!) variables, that prospect is unpalatable. A transformation utility like Stat/Transfer performs all those housekeeping chores, placing any attributes attached to the data (extended missing-value codes, value labels, etc.) in the Stata-format file. Of course, the mapping between packages is not always one to one. In Stata, a value label stands alone and can be attached to any variable or set of variables, whereas in other packages it is generally an attribute of a variable and must be duplicated for similar variables.

4. If SAS datasets are available in the SAS Transport (`.xpt`) format, they may be read by Stata's `fdause` command.

One difference between Stata and SAS and SPSS is Stata's flexible set of data types. Stata, like the C language in which its core code is written, offers five numeric data types (see [D] **data types**): integer types `byte`, `int`, and `long` and floating-point types `float` and `double`, in addition to string types `str1`-`str244`. Most other packages do not support this broad array of data types but store all numeric data in one data type: for example, "Raw data come in many different forms, but SAS simplifies this issue. In SAS there are just two data types: numeric and character" (Delwiche and Slaughter 1998, 4). This simplicity is costly, because an indicator variable requires only 1 byte of storage, whereas a double-precision floating-point variable requires 8 bytes to hold up to 15 decimal digits of accuracy. Stata allows you to specify the data type based on the contents of each variable, which can result in considerable savings in disk space and execution time when reading or writing those variables to disk. You can instruct Stat/Transfer to optimize a target Stata-format file in the transfer process, or you can use Stata's `compress` command to automatically perform that optimization. In any case, you should always take advantage of this optimization, since it will reduce the size of files and require less of your computer's memory to work with them.

Stat/Transfer lets you generate a subset of a large file while transferring it from SAS or SPSS format. Above I mentioned the possibility of reading only certain variables from a text file to avoid Intercooled Stata's limitation of 2,047 variables. You can always use Stat/Transfer to transfer a large survey data file from SAS to Stata format, but if there are more than 2,047 variables in the file, the target file must be specified as a Stata/SE file. If you do not have Stata/SE, you will have to use Stat/Transfer to read a list of variables that you would like to keep (or a list of variables to drop), which will generate a subset file on the fly. Because Stat/Transfer can generate a machine-readable list of variable names, you can edit that list to produce the keep list or drop list.

Although I have spoken of SAS and SPSS, Stat/Transfer can exchange datasets with many other packages, including GAUSS, Excel, MATLAB, and others; see http://stattransfer.com for details. Stat/Transfer is available for Windows, Mac OS X, and Unix.

To transfer data between databases that support Structured Query Language (SQL), Stata can perform Open Data Base Connectivity (ODBC) operations with databases that support ODBC (see [D] **odbc** for details). Most SQL databases and non-SQL data structures, such as Excel and Microsoft Access, support ODBC, so you can use ODBC to deal with foreign data. The computer system on which you are running Stata must be equipped with ODBC drivers. Excel and Microsoft Access are installed by default on Windows systems with Microsoft Office, but for Mac OS X or Linux systems, you may have to buy a third-party driver to connect to your particular data source.[5] If you have database connectivity, Stata's `odbc` is a full-featured solution that allows you to query external databases and insert or update records in those databases.

5. `odbc` is not currently available in Stata for Unix, other than Linux.

B The basics of Stata programming

This appendix discusses some key aspects of programming in Stata. As I discussed in section 3.9.1, you can place any sequence of Stata commands into a text file or do-file and execute them by using Stata's `do` command (or the Do-file Editor). The last two sections discuss Stata *programs*, which are stand-alone routines that create new commands in the Stata language. A program can also be contained in a text file, called an automatic do-file or ado-file, and can be invoked by the name of the ado-file that defines it. Most of this appendix describes how you can apply Stata's programming tools in using do-files. As discussed in section 3.9.1, you can often use a Stata do-file to work more efficiently by simplifying repetitive tasks and reducing the need to retype computed quantities.

To enhance your knowledge of Stata programming, you should have a copy of the *Stata Programming Reference Manual*, as many key commands related to do-file construction appear in that volume. The manual also documents more advanced commands used in writing ado-files. To learn from the masters, you can take one or more of the Stata NetCourses about programming to learn good programming techniques. If you participate in Statalist—even as a passive reader—you can take advantage of the many good problem-solving tips that are exchanged in the list's questions and answers.[1] You can also dissect Stata's own code. More than 80% of official Stata's commands are written in its own programming language, as are virtually all user-written routines available from the SSC archive.[2] Although you may not be interested in writing your own ado-files, it is invaluable to read through some of Stata's code and borrow from the techniques used there in your own do-files. Official Stata's ado-files are professionally written and tested and reflect best practices in Stata programming at the time of their construction. Since Stata's programming language is continually evolving, even parts of official Stata may be legacy code.

You can use the `findfile` command to locate any Stata command's ado-file on your machine.[3] To examine an ado-file, you can use the `viewsource` command.[4] You may use the `ssc type` command to inspect any ado-file hosted on the SSC archive.

1. See http://www.stata.com/statalist/.
2. But not all: support for plugins in Stata 8.1 has made it possible for users to write C-language code to increase computational efficiency. Likewise, some Stata 9 routines are now written in Mata code, which need not be accessible in source form. Future development will likely use Mata rather than C-language plugins.
3. Before Stata 9, you could use the `which` command for this task. An extended version of `which` is available from `ssc`: Thomas Steichen's `witch`.
4. Before Stata 9, the author's routine `adotype`, available from `ssc`, could be used for this purpose.

You can edit any ado-file up to 32 KB in the Do-file Editor by using the user-written routine `adoedit` provided by Dan Blanchette. You can use any external text editor on an ado-file since it is merely a text file.[5] Do not modify official Stata's ado-files or place your own files in the official ado-directories! Stata users often clone an official or user-written command by adding 2 to its name. If I want to create my own version of `summarize`, I would make a copy of `summarize.ado` as `summarize2.ado` in one of my own directories on the ado-path. I would then edit the copy to change the `program define` line to `program define summarize2, ...`. In that way, I can compare the answers produced by `summarize` and my modified `summarize2` routine and not tamper with any official Stata elements.

B.1 Local and global macros

If you are familiar with lower-level programming languages, such as FORTRAN, C, or Pascal, you may find Stata's terminology for various objects rather confusing. In those languages, you refer to a *variable* with statements such as `x = 2`. Although you might have to declare `x` before using it—for instance, as `integer` or `float`—the notion of a variable in those languages refers to an entity that can be assigned one value, either numeric or string. In contrast, the Stata variable refers to one column of the data matrix that contains `maxobs` values, one per observation.

So what corresponds to a FORTRAN or C variable in Stata's command language? Either a Stata *macro* or a *scalar*, to be discussed below.[6] But that correspondence is not one to one, since a Stata macro may contain multiple elements. In fact, a macro may contain any combination of alphanumeric characters and can hold more than 8,000 characters in all versions of Stata. The Stata macro is really an alias that has both a name and a value. When its name is *dereferenced*, it returns its value. That operation may be carried out at any time. Alternatively, the macro's value may be modified by another command. The following is an example of the first concept:

```
. local country US UK DE FR
. local ctycode 111 112 136 134
. display "`country'"
US UK DE FR
. display "`ctycode'"
111 112 136 134
```

The Stata command for defining the macro is `local` (see [P] **macro**). A macro may be either local or global in its scope, defining where its name will be recognized. A *local macro* is created in a do-file or in an ado-file and ceases to exist when that do-file terminates, either normally or abnormally. A *global macro* exists for the duration of the Stata program or interactive session. There are good reasons to use global macros, but like any global definition, they may have unintended consequences, so we will discuss local macros in most of the examples below.

5. See `ssc describe texteditors`.

6. Stata's scalars were purely numeric through version 8.0, as described in `scalar`. The ability to store strings in scalars was added in the executable update of 1 July 2004.

The first `local` command names the macro—as `country`—and then defines its value to be the list of four two-letter country codes. The following `local` statement does the same for macro `ctycode`. To work with the value of the macro, we must dereference it. `‘macroname’` refers to the value of the macro named `macroname`. The macro's name is preceded by the left tick character (‘) and followed by the apostrophe (’). Most errors in using macros are caused by not following this rule. To dereference the macro, correct punctuation is vital. In the example's `display` statements, we must wrap the dereferenced macro in double quotes since `display` expects a double-quoted string argument or the value of a scalar expression such as `display log(14)`.

In both cases, the `local` statement is written without an equals sign (=). You can use an equals sign following the macro's name, but do not make a habit of doing so unless it is required. The equals sign causes the rest of the expression to be evaluated, rather than merely aliased to the macro's name. This behavior is a common cause of head-scratching, when a user will complain, "My do-file worked when I had eight regressors, but not when I had nine." Defining a macro with an equals sign will cause Stata to evaluate the rest of the command as a numeric expression or as a character string. A character string, representing the contents of a string variable, cannot contain more than 244 characters.[7] In evaluating `local mybadstring = "This is an example of a string that will not all end up in the macro that it was intended to populate, which clearly, definitively, and unambiguously indicates that writing short, concise strings is a definite advantage to people who use macros in Stata"`, where the quotation marks are now required, Stata will truncate the string `mybadstring` at 244 characters without error or warning.

We should use an equals sign in a `local` statement when we must evaluate the macro's value. In this example, we show a macro used as a counter, which fails to do what we had in mind:

```
. local count 0
. local country US UK DE FR
. foreach c of local country {
  2.     local count ‘count’+1
  3.     display "Country ‘count’ : ‘c’"
  4. }
Country 0+1 : US
Country 0+1+1 : UK
Country 0+1+1+1 : DE
Country 0+1+1+1+1 : FR
```

7. Prior to version 9.1, the limit was 80 characters for Small Stata and Intercooled Stata.

We must use the equals sign to request *evaluation* rather than *concatenation*:

```
. local count 0

. local country US UK DE FR

. foreach c of local country {
  2.     local count = `count'+1
  3.     display "Country `count' : `c'"
  4. }
Country 1 : US
Country 2 : UK
Country 3 : DE
Country 4 : FR
```

The corrected example's `local` statement contains the name of the macro twice: first without punctuation to define its name, and on the right-hand side of the equals sign with its current value dereferenced by `` `count' ``. It is crucial to understand why the statement is written this way. Here we are redefining the macro in the first instance and referencing its current value in the second.

At other times, we want to construct a macro within a loop, repeatedly redefining its value, so we should always avoid the equals sign:

```
. local count 0

. local country US UK DE FR

. foreach c of local country {
  2.     local count = `count'+1
  3.     local newlist "`newlist' `count' `c'"
  4. }

. display "`newlist'"
 1 US 2 UK 3 DE 4 FR
```

The `local` *newlist* statement is unusual in that it defines the local macro *newlist* as a string containing its own current contents, space, value of count, space, value of c. The `foreach` statement defines the local macro `c` with the value of each biliteral country code in turn. The first time through the loop, *newlist* does not exist, so how can we refer to its current value? Easily: every Stata macro has a null value unless it has explicitly been given a nonnull value. Thus the value takes on the string `" 1 US"` the first time, and then the second time through concatenates that string with the new string `" 2 UK"`, and so on. In this example, using the equals sign in the `local` *newlist* statement truncates *newlist* at 244 characters. This truncation would not cause trouble in this example, but it would be a serious problem if we had a longer list of countries or if country names were spelled out.

From these examples, you can see that Stata's macros are useful in constructing lists, or as counters and loop indices, but they play a much larger role in Stata do-files and ado-files and in the return values of virtually all Stata commands as I discussed in section 4.3.6. Macros are one of the key elements of Stata's language that allow you to avoid repetitive commands and the retyping of computed results. For instance, the macro defined by `local country US UK DE FR` may be used to generate a set of graphs with country-specific content and labels,

```
. local country US UK DE FR
. foreach c of local country {
  2.        tsline gdp if cty=="`c'", title("GDP for `c'")
  3. }
```

or to produce one graph with panels for each country,

```
. local country US UK DE FR
. foreach c of local country {
  2.        tsline gdp if cty=="`c'", title("GDP for `c'")
>           nodraw name(`c',replace)
  3. }
. graph combine `country', ti("Gross Domestic Product, 1971Q1-1995Q4")
```

Using macros makes the do-file easier to maintain because changes require only altering the contents of the local macro. To produce these graphs for a different set of countries, you alter just one command: the list of codes. You can thus make your do-file general, and you can easily reuse or adapt that set of Stata commands for use in similar tasks.

B.1.1 Global macros

Global macros are distinguished from local macros by the way they are created (with the `global` statement) and their means of reference. We obtain the value of the global macro `george` as `$george`, with the dollar sign taking the place of the punctuation surrounding the local macro's name when it is dereferenced. Global macros are often used to store items parametric to a program, such as a character string containing today's date to be embedded in all filenames created by the program or the name of a default directory in which your datasets and do-files are to be accessed.

Unless you really need a global macro—a symbol with *global scope*—you should use a local macro. It is easy to forget that a global symbol was defined in do-file A. By the time you run do-file G or H in that session of Stata, you may find that they do not behave as expected, since they now pick up the value of the global symbol. Such problems are difficult to debug. Authors of Fortran or C programs have always been encouraged to "keep definitions local unless they must be visible outside the module." That is good advice for Stata programmers as well.

B.1.2 Extended macro functions and list functions

Stata has a versatile library of functions that you can apply to macros: the *extended functions* (see `help extended_fcn`, or [P] **macro**). These functions allow you to easily retrieve and manipulate the contents of macros. For instance,

```
. local country US UK DE FR
. local wds: word count `country'
. display "There are `wds' countries:"
There are 4 countries:
```

```
. forvalues i = 1/`wds' {
  2.         local wd: word `i' of `country'
  3.         display "Country `i' is `wd'"
  4. }
Country 1 is US
Country 2 is UK
Country 3 is DE
Country 4 is FR
```

Here we use the `word count` and `word # of` extended functions, both of which operate on strings. We do not enclose the macro's value (`` `country' ``) in double quotes, for it then would be considered one word.[8] This do-file will work for any definition of the country list in `local country` without the need to define a separate `count` variable.

Many extended macro functions (`help extended_fcn`) perform useful tasks, such as extracting the variable label or value label from a variable or determining its data type or display format; extracting the row or column names from a Stata matrix; or generating a list of the files in a particular directory that match a particular pattern (e.g., `*.dta`). The handy `subinstr` function allows you to substitute a particular pattern in a macro, either the first time the pattern is encountered or always.

Other functions let you manipulate lists held in local macros; see `help macrolists` or [P] **macro lists**. You can use them to identify the unique elements of a list or the duplicate entries; to sort a list; and to combine lists with Boolean operators such as "and" (`&`) or "or" (`|`). List functions allow one list's contents to be subtracted from another, identifying the elements of list A that are not duplicated in list B. You can test lists for equality, defined for lists as containing the identical elements in the same order, or for weak equality, which does not consider ordering. A `list` *macrolist_directive* (`posof`) may be used to determine whether a particular entry exists in a list, and if so, in which position in the list. An excellent discussion of many of these issues may be found in Cox (2003).

B.2 Scalars

The distinction between macros and Stata's *scalars* is no longer numeric content since both macros and scalars may contain string values. However, the length of a string scalar is limited to the length of a string variable (244 bytes: see `help limits`), whereas a macro's length is for most purposes unlimited.[9] Stata's scalars are normally used in a numeric context. When a numeric quantity is stored in a macro it must be converted from its internal (binary) representation into a printable form. That conversion is done with maximum accuracy but incurs some overhead if the numeric quantity is not an integer. By storing the result of a computation—for instance, a variable's mean or standard deviation—in a scalar, you need not convert its value and the result is held in Stata's full numeric precision. A scalar is also much more useful for storing one numeric

8. In this context a word is a space-delimited token in the string.

9. Actually a macro is limited to 67,784 characters in Intercooled Stata but can handle more than 1 million characters in Stata/SE.

result rather than storing that value in a Stata variable containing `maxobs` copies of the same number. Most of Stata's statistical and estimation commands return various numeric results as scalars. A scalar may be referred to in any Stata command by its name:

```
. scalar root2 = sqrt(2.0)
. generate double rootGDP = gdp*root2
```

The difference between a macro and a scalar appears when it is referenced. The macro must be dereferenced to refer to its value, whereas the scalar is merely named.[10] However, a scalar can appear only in an expression where a Stata variable or a numeric expression could be used. For instance, you cannot specify a scalar as part of an `in` *range* qualifier since its value will not be extracted. It may be used in an `if` *exp* qualifier since that contains a numeric expression.

Stata's scalars may play a useful role in a complicated do-file. By defining scalars at the beginning of the program and referring to them throughout the code, you make the program parametric. Doing so avoids the difficulties of changing various constants in the program's statements everywhere they appear. You may often need to repeat a complex data transformation task for a different category, such as when you want to work with 18- to 24-year-old subjects rather than 25- to 39-year-old subjects. Your do-files contain the qualifiers for minimum and maximum age throughout the program. If you define those age limits as scalars at the program's outset, the do-file becomes much simpler to modify and maintain.

B.3 Loop constructs

One of Stata's most powerful features is that it lets you write a versatile Stata program without many repetitive statements. Many Stata commands contribute to this flexibility. As discussed in section 2.2.8, using `egen` with a `by` prefix makes it possible to avoid many explicit statements such as `compute mean of age for race==1` or `compute mean of age for race==2`. Two of Stata's most useful commands are found in the *Stata Programming Reference Manual*: `forvalues` and `foreach`. These versatile tools have essentially supplanted other mechanisms in Stata for looping. You could use `while` to construct a loop, but you must furnish the counter as a local macro. The `for` command is now obsolete and is no longer described in the manuals. The `for` command allowed only one command to be included in a loop structure (or multiple commands with a tortuous syntax) and rendered nested loops almost impossible.

In contrast, the `forvalues` and `foreach` commands use a syntax familiar to users of C or other modern programming languages. The command is followed by a left brace (`{`), one or more following command lines, and a terminating line containing only a right brace (`}`). In Stata 8 and 9, you must separate the braces from the body of the loop.

10. Stata can work with scalars of the same name as Stata variables. Stata will not become confused, but you well may, so you should avoid using the same names for both entities.

You may place as many commands in the loop body as you wish. A simple numeric loop may thus be constructed as

```
. forvalues i = 1/4 {
  2.         generate double lngdp`i' = log(gdp`i')
  3.         summarize lngdp`i'
  4. }
    Variable |       Obs        Mean    Std. Dev.       Min        Max
-------------+--------------------------------------------------------
     lngdp1 |       400    7.931661      .59451   5.794211   8.768936
    Variable |       Obs        Mean    Std. Dev.       Min        Max
-------------+--------------------------------------------------------
     lngdp2 |       400    7.942132    .5828793   4.892062   8.760156
    Variable |       Obs        Mean    Std. Dev.       Min        Max
-------------+--------------------------------------------------------
     lngdp3 |       400    7.987095     .537941   6.327221   8.736859
    Variable |       Obs        Mean    Std. Dev.       Min        Max
-------------+--------------------------------------------------------
     lngdp4 |       400    7.886774    .5983831   5.665983   8.729272
```

Here the local macro **i** is defined as the loop index. Following an equals sign, we give the **range** of values that **i** is to take on as a Stata *numlist*. A range may be as simple as 1/4; or 10(5)50, indicating 10 to 50 in steps of 5; or 100(-10)20, from 100 to 20 counting down by 10s. Other syntaxes for the range are available. See [P] **forvalues** for details.

This example provides one of the most important uses of forvalues: looping over variables where the variables have been given names with an integer component so that you do not need separate statements to transform each of the variables. The integer component need not be a suffix. We could loop over variables named cty*N*gdp just as readily. Or, say that we have variable names with more than one integer component:

```
. forvalues y = 1995(2)1999 {
  2.         forvalues i = 1/4 {
  3.                 summarize gdp`i'_`y'
  4.         }
  5. }
    Variable |       Obs        Mean    Std. Dev.       Min        Max
-------------+--------------------------------------------------------
   gdp1_1995 |       400    3226.703    1532.497    328.393   6431.328
    Variable |       Obs        Mean    Std. Dev.       Min        Max
-------------+--------------------------------------------------------
   gdp2_1995 |       400    3242.162    1525.788   133.2281   6375.105
    Variable |       Obs        Mean    Std. Dev.       Min        Max
-------------+--------------------------------------------------------
   gdp3_1995 |       400    3328.577    1457.716   559.5993   6228.302
    Variable |       Obs        Mean    Std. Dev.       Min        Max
-------------+--------------------------------------------------------
   gdp4_1995 |       400    3093.778    1490.646   288.8719   6181.229
```

Variable	Obs	Mean	Std. Dev.	Min	Max
gdp1_1997	400	3597.038	1686.571	438.5756	7083.191
Variable	Obs	Mean	Std. Dev.	Min	Max
gdp2_1997	400	3616.478	1677.353	153.0657	7053.826
Variable	Obs	Mean	Std. Dev.	Min	Max
gdp3_1997	400	3710.242	1603.25	667.2679	6948.194
Variable	Obs	Mean	Std. Dev.	Min	Max
gdp4_1997	400	3454.322	1639.356	348.2078	6825.981
Variable	Obs	Mean	Std. Dev.	Min	Max
gdp1_1999	400	3388.038	1609.122	344.8127	6752.894
Variable	Obs	Mean	Std. Dev.	Min	Max
gdp2_1999	400	3404.27	1602.077	139.8895	6693.86
Variable	Obs	Mean	Std. Dev.	Min	Max
gdp3_1999	400	3495.006	1530.602	587.5793	6539.717
Variable	Obs	Mean	Std. Dev.	Min	Max
gdp4_1999	400	3248.467	1565.178	303.3155	6490.291

As we see here, a nested loop is readily constructed with two `forvalues` statements.

B.3.1 foreach

As useful as `forvalues` may be, the `foreach` command is even more useful in constructing efficient do-files. This command interacts perfectly with some of Stata's most common constructs: the macro, the varlist, and the numlist. Like `forvalues`, a local macro is defined as the loop index. Rather than cycling through a set of numeric values, `foreach` specifies that the loop index iterate through the elements of a local (or global) macro, the variable names of a varlist, or the elements of a numlist. The list can also be an arbitrary *list of elements* on the command line or a newvarlist of valid names for variables not present in the dataset.

This syntax allows `foreach` to be used flexibly with any set of items, regardless of pattern. Several of the examples above used `foreach` with the elements of a local macro defining the list. I illustrate its use here with a varlist from the `lifeexp` *Graphics Reference Manual* dataset. We compute summary statistics, compute correlations with `popgrowth`, and generate scatterplots for each element of a varlist versus `popgrowth`:

```
. foreach v of varlist lexp-safewater {
  2.         summarize `v'
  3.         correlate popgrowth `v'
  4.         scatter popgrowth `v'
  5. }

    Variable |       Obs        Mean    Std. Dev.       Min        Max
-------------+--------------------------------------------------------
        lexp |        68    72.27941    4.715315         54         79
(obs=68)

             | popgro~h     lexp
-------------+------------------
   popgrowth |   1.0000
        lexp |  -0.4360   1.0000

    Variable |       Obs        Mean    Std. Dev.       Min        Max
-------------+--------------------------------------------------------
       gnppc |        63    8674.857    10634.68        370      39980
(obs=63)

             | popgro~h    gnppc
-------------+------------------
   popgrowth |   1.0000
       gnppc |  -0.3580   1.0000

    Variable |       Obs        Mean    Std. Dev.       Min        Max
-------------+--------------------------------------------------------
   safewater |        40        76.1    17.89112         28        100
(obs=40)

             | popgro~h safewa~r
-------------+------------------
   popgrowth |   1.0000
   safewater |  -0.4280   1.0000
```

The following example automates the construction of a `recode` statement. The resulting statement could just be typed out for four elements, but imagine its construction if we had 180 country codes! `local ++i` is a shorthand way of incrementing the counter variable within the loop.[11]

```
. local ctycode 111 112 136 134

. local i 0

. foreach c of local ctycode {
  2.         local ++i
  3.         local rc "`rc' (`i'=`c')"
  4. }

. display "`rc'"
 (1=111) (2=112) (3=136) (4=134)

. recode cc `rc', gen(newcc)
(400 differences between cc and newcc)
```

11. Serious Stata programmers would avoid that line and write the following line as `local rc "`rc' (`=`++i''=`c')"`.

```
. tabulate newcc
   RECODE of |
          cc |      Freq.     Percent        Cum.
-------------+-----------------------------------
         111 |        100       25.00       25.00
         112 |        100       25.00       50.00
         134 |        100       25.00       75.00
         136 |        100       25.00      100.00
-------------+-----------------------------------
       Total |        400      100.00
```

You can also use the **foreach** statement with nested loops. You can combine **foreach** and **forvalues** in a nested loop structure, as illustrated here:

```
. local country US UK DE FR
. local yrlist 1995 1999
. forvalues i = 1/4 {
  2.            local cname: word `i' of `country'
  3.            foreach y of local yrlist {
  4.                    rename gdp`i'_`y' gdp`cname'_`y'
  5.            }
  6. }
. summ gdpUS*
    Variable |       Obs        Mean    Std. Dev.       Min        Max
-------------+--------------------------------------------------------
  gdpUS_1995 |       400    3226.703    1532.497    328.393   6431.328
  gdpUS_1999 |       400    3388.038    1609.122   344.8127   6752.894
```

It is a good idea to use indentation—either spaces or tabs—to align the loop body statements as shown here. Stata does not care, as long as the braces appear as required, but it makes the do-file much more readable and easier to revise later.

In summary, the **foreach** and **forvalues** statements are essential components of any do-file writer's toolkit. Whenever you see a set of repetitive statements in a Stata do-file, it probably means that its author did not understand how one of these loop constructs could have made the program, its upkeep, and her life simpler. For an excellent discussion of the loop commands, see Cox (2002a).

B.4 Matrices

Stata has always provided a full-featured matrix language that supports a broad range of matrix operations on real matrices, as described in [P] **matrix**. Stata 9 also provides a dedicated matrix language, Mata, which operates in a separate environment within Stata. I discuss Mata later in this appendix. First, I discuss the traditional matrix language as implemented within Stata with the **matrix** commands.

With Stata's traditional **matrix** commands, matrix size is limited. In Intercooled Stata, you cannot have more than 800 rows or 800 columns in a matrix.[12] Thus many

12. Limits in Stata/SE are considerably larger, but large matrices use much computer memory. Mata provides a more efficient solution.

matrix tasks cannot be handled easily by using traditional `matrix` commands. For instance, `mkmat` (see [P] **matrix mkmat**) can create a Stata matrix from a *varlist* of variables, but the number of observations that may be used is limited to 800 in Intercooled Stata. If you do not plan to use Mata, two points should be made. First, Stata contains specialized operators, such as `matrix accum`, that can compute cross-product matrices from any number of observations. A regression of 10,000 observations on five variables (including constant) involves a 5×5 cross-products matrix, regardless of N. Variations on this command such as `matrix glsaccum`, `matrix vecaccum`, and `matrix opaccum` generate other useful summaries. In that sense, the limitation on matrix dimension is not binding. The `matrix accum` command and `corr()` matrix function are also useful for generating correlation matrices from the data. For instance, `mat accum C =` *varlist*, `dev nocons` will compute a covariance matrix, and `mat Corr = corr(C)` will transform it into a correlation matrix. The `correlate` command can display a correlation matrix but cannot be used to save its elements.

Second, the brute-force approach is rarely appropriate when working with complex matrix expressions. For example, the SUR estimator (discussed in section 9.4) is presented in textbooks as a GLS estimator involving large block-diagonal X and large Ω matrices of enormous dimension. Given the algebra of partitioned matrices, every statistical package that performs SUR writes this expression as the product of several terms, one per equation in the system. In that expression, each term is no more than one equation's regression. A huge matrix computation can be simplified to a loop over the individual equations. Although you might be tempted to copy the matrix expression straight from the textbook or journal article into code, that method will usually not work—not only in Stata's traditional matrix commands or in Mata but in any matrix language, as limited by the computer's available memory. It may take some effort when implementing complicated matrix expressions to reduce the problem to a workable size.

If you are not developing your own programs (ado-files) or learning to use Mata, Stata matrices are likely to be useful with saved results and as a way of organizing information for presentation. Many of Stata's statistical commands and all estimation commands generate one or more matrices behind the scenes. As discussed in section 4.3.6, `regress`—like all Stata estimation commands—produces matrices `e(b)` and `e(V)` as the row vector of estimated coefficients (a $1 \times k$ matrix) and the estimated variance–covariance matrix of the coefficients (a $k \times k$ symmetric matrix), respectively. You can examine those matrices with the `matrix list` command or copy them for use in your do-file with the `matrix` statement. The command `matrix beta = e(b)` will create a matrix `beta` in your program as a copy of the last estimation command's coefficient vector.

References to matrix elements appear in square brackets. Since there is no vector data type, all Stata matrices have two subscripts, and both subscripts must be given in any reference. You can specify a range of rows or a range of columns in an expression: see [P] **matrix** for details. Stata's traditional matrices are unique in that their elements may be addressed both conventionally by their row and column numbers (counting from 1, not 0) and by their row and column names. The command `mat vv`

= v["gdp2","gdp3"] will extract the estimated covariance of the coefficients on gdp2 and gdp3 as a 1×1 matrix.

Stata's matrices are often useful for housekeeping purposes, such as accumulating results that are to be presented in tabular form. The tabstat command may generate descriptive statistics for a set of by-groups. Likewise, you can use statsmat (Cox and Baum, available from ssc) to generate a matrix of descriptive statistics for a set of variables or for one variable over by-groups. You can then use Baum and de Azevedo's outtable to generate a LaTeX table. You can use Michael Blasnik's mat2txt to generate tab-delimited output. You can change Stata matrices' row and column labels with matrix rownames, matrix colnames, and several macro extended functions (described in section B.1.2), which allow you to control the row and column headings on tabular output. Stata's traditional matrix operators make it possible to assemble a matrix from several submatrices. For instance, you may have one matrix for each country in a multicountry dataset. In summary, judicious use of Stata's traditional matrix commands ease the burden of many housekeeping tasks and make it easy to update material in tabular form without retyping.

B.5 return and ereturn

Each of Stata's commands reports its results, sometimes *noisily*, as when a nonzero return code is accompanied by an error message (help _rc), but usually silently. Stored results from Stata commands can be useful. Using stored results can greatly simplify your work with Stata, since you can create a do-file to use the results of a previous statement in a computation, title, graph label, or even a conditional statement.

Each Stata command belongs to a class—r-class, e-class, or less commonly s-class. These classes apply both to those commands that are *built in* (such as summarize) and to the 80% of official Stata commands that are implemented in the ado-file language.[13] The e-class commands are estimation commands, which return e(b) and e(V)—the estimated parameter vector and its variance–covariance matrix, respectively—to the calling program, as well as other information (see help ereturn). Almost all other official Stata commands are r-class commands, which return *results* to the calling program (help return). Let us deal first with the simpler r-class commands.

Virtually every Stata command—including those you might not think of as generating results—places items in the return list that may be displayed by the command of the same name.[14] For instance, consider describe:

13. If this distinction interests you, findfile will report that a command is either built in (i.e., compiled C or Mata code) or located in a particular ado-file on your hard disk.
14. Significant exceptions are generate and egen.

```
. use http://www.stata-press.com/data/imeus/abdata, clear
. describe
Contains data from http://www.stata-press.com/data/r9/abdata.dta
  obs:         1,031
 vars:            30                          3 Mar 2005 01:13
 size:       105,162 (98.9% of memory free)
-------------------------------------------------------------------------------
              storage  display     value
variable name   type   format      label      variable label
-------------------------------------------------------------------------------
c1              str9   %9s
ind             float  %9.0g
year            float  %9.0g
emp             float  %9.0g
wage            float  %9.0g
cap             float  %9.0g
indoutpt        float  %9.0g
n               float  %9.0g
w               float  %9.0g
k               float  %9.0g
ys              float  %9.0g
rec             float  %9.0g
yearm1          float  %9.0g
id              float  %9.0g
nL1             float  %9.0g
nL2             float  %9.0g
wL1             float  %9.0g
kL1             float  %9.0g
kL2             float  %9.0g
ysL1            float  %9.0g
ysL2            float  %9.0g
yr1976          byte   %8.0g                  year==  1976.0000
yr1977          byte   %8.0g                  year==  1977.0000
yr1978          byte   %8.0g                  year==  1978.0000
yr1979          byte   %8.0g                  year==  1979.0000
yr1980          byte   %8.0g                  year==  1980.0000
yr1981          byte   %8.0g                  year==  1981.0000
yr1982          byte   %8.0g                  year==  1982.0000
yr1983          byte   %8.0g                  year==  1983.0000
yr1984          byte   %8.0g                  year==  1984.0000
-------------------------------------------------------------------------------
Sorted by:  id  year
. return list
scalars:
           r(changed) =  0
          r(widthmax) =  9148
             r(k_max) =  2048
             r(N_max) =  89028
             r(width) =  98
                 r(k) =  30
                 r(N) =  1031
. local sb: sortedby
. display "dataset sorted by : `sb'"
dataset sorted by : id year
```

The return list for the `describe` command contains scalars, as described in appendix B.2. `r(N)` and `r(k)` provide the number of observations and variables present in the dataset in memory. `r(changed)` is an indicator variable that will be set to 1 as soon as a change is made in the dataset. I also demonstrate here how to retrieve information about the dataset's *sort order* by using one of the extended macro functions discussed in section B.1.2. Any scalars defined in the return list may be used in a following statement without displaying the return list. A subsequent r-class command will replace the contents of the return list with its return values so that if you want to use any of these items, you should save them to local macros or named scalars. For a more practical example, consider `summarize`:

```
. summarize emp, detail
                             emp
-------------------------------------------------------------
      Percentiles      Smallest
 1%         .142           .104
 5%         .431           .122
10%         .665           .123       Obs                1031
25%         1.18           .125       Sum of Wgt.        1031

50%        2.287                      Mean           7.891677
                        Largest       Std. Dev.      15.93492
75%        7.036         101.04
90%       17.919        103.129       Variance       253.9217
95%         32.4        106.565       Skewness       3.922732
99%         89.2        108.562       Kurtosis       19.46982
. return list
scalars:
                r(N) =  1031
            r(sum_w) =  1031
             r(mean) =  7.891677013539667
              r(Var) =  253.9217371514514
               r(sd) =  15.93492193741317
         r(skewness) =  3.922731923543387
         r(kurtosis) =  19.46982480250623
              r(sum) =  8136.319000959396
              r(min) =  .1040000021457672
              r(max) =  108.5619964599609
               r(p1) =  .1420000046491623
               r(p5) =  .4309999942779541
              r(p10) =  .6650000214576721
              r(p25) =  1.179999947547913
              r(p50) =  2.286999940872192
              r(p75) =  7.035999774932861
              r(p90) =  17.91900062561035
              r(p95) =  32.40000152587891
              r(p99) =  89.19999694824219
. scalar iqr = r(p75) - r(p25)
. display "IQR = " iqr
IQR = 5.8559998
. scalar semean = r(sd)/sqrt(r(N))
. display "Mean = " r(mean) " S.E. = " semean
Mean = 7.891677 S.E. = .49627295
```

The `detail` option displays the full range of results available—here, all in the form of scalars—after the `summarize` command. We compute the interquartile range (IQR) of the summarized variable and its standard error of mean as scalars and display those quantities. We often need the mean of a variable for more computations but do not wish to display the results of `summarize`. Here the `meanonly` option of `summarize` suppresses both the output and the calculation of the variance or standard deviation of the series. The scalars `r(N)`, `r(mean)`, `r(min)`, and `r(max)` are still available.

When working with time-series or panel data, you will often need to know whether the data have been `tsset`, and if so, what variable is serving as the calendar variable and which is the panel variable (if defined). For instance,

```
. use http://www.stata-press.com/data/imeus/abdata, clear
. tsset
       panel variable:  id, 1 to 140
        time variable:  year, 1976 to 1984
. return list
scalars:
                r(tmax) =  1984
                r(tmin) =  1976
                r(imax) =  140
                r(imin) =  1

macros:
            r(panelvar) : "id"
             r(timevar) : "year"
               r(unit1) : "."
               r(tsfmt) : "%9.0g"
               r(tmaxs) : "1984"
               r(tmins) : "1976"
```

Here the returned scalars include the first and last periods in this panel dataset (1976 and 1984) and the range of the `id` variable, which is designated as `r(panelvar)`. The macros also include the time-series calendar variable `r(timevar)` and the range of that variable in a form that can be readily manipulated, for instance, for graph titles.

Many statistical commands are r-class since they do not fit a model. `correlate` will return one estimated correlation coefficient, irrespective of the number of variables in the command's *varlist*: the correlation of the last and next-to-last variables.[15] The `ttest` command is also r-class, so we can access its return list to retrieve all the quantities it computes:

15. If a whole set of correlations is required for further use, use `mat accum C =` *varlist*, `dev nocons` followed by `mat Corr = corr(C)`.

```
. generate lowind = (ind<6)
. ttest emp, by(lowind)
Two-sample t test with equal variances
------------------------------------------------------------------------------
   Group |     Obs        Mean    Std. Err.   Std. Dev.   [95% Conf. Interval]
---------+--------------------------------------------------------------------
       0 |     434    8.955942    .9540405    19.87521    7.080816    10.83107
       1 |     597     7.11799    .5019414    12.26423    6.132201    8.103779
---------+--------------------------------------------------------------------
combined |    1031    7.891677     .496273    15.93492    6.917856    8.865498
---------+--------------------------------------------------------------------
    diff |            1.837952    1.004043               -.1322525    3.808157
------------------------------------------------------------------------------
    diff = mean(0) - mean(1)                                      t =   1.8306
Ho: diff = 0                                     degrees of freedom =     1029

    Ha: diff < 0                 Ha: diff != 0                 Ha: diff > 0
 Pr(T < t) = 0.9663         Pr(|T| > |t|) = 0.0675          Pr(T > t) = 0.0337
. return list
scalars:
                 r(sd) =  15.93492193741317
               r(sd_2) =  12.26422618476487
               r(sd_1) =  19.87520847697869
                 r(se) =  1.004042693732077
                r(p_u) =  .0337282628926395
                r(p_l) =  .9662717371073605
                  r(p) =  .0674565257852791
                  r(t) =  1.83055206312211
               r(df_t) =  1029
               r(mu_2) =  7.117989959978378
                r(N_2) =  597
               r(mu_1) =  8.955942384452314
                r(N_1) =  434
```

The return list contains scalars representing each of the displayed values from `ttest` except the total number of observations, which can be computed as `r(N_1)+r(N_2)`, the standard errors of the group means, and the confidence interval limits.

B.5.1 ereturn list

Even more information is provided after any e-class (estimation) command as displayed by `ereturn list`. Most e-class commands return four types of Stata objects: scalars, such as `e(N)`, summarizing the estimation process; macros, providing such information as the name of the response variable (`e(depvar)`) and the estimator used (`e(model)`); matrices `e(b)` and `e(V)` as described above; and a Stata *function*, `e(sample)`, which will return 1 for each observation included in the estimation sample and zero otherwise. For example, consider a simple regression:

```
. regress emp wage cap
      Source |       SS       df       MS              Number of obs =    1031
-------------+------------------------------           F(  2,  1028) = 1160.71
       Model |   181268.08     2   90634.04            Prob > F      =  0.0000
    Residual |  80271.3092  1028  78.0849311           R-squared     =  0.6931
-------------+------------------------------           Adj R-squared =  0.6925
       Total |  261539.389  1030  253.921737           Root MSE      =  8.8366

------------------------------------------------------------------------------
         emp |      Coef.   Std. Err.      t    P>|t|     [95% Conf. Interval]
-------------+----------------------------------------------------------------
        wage |  -.3238453   .0487472    -6.64   0.000    -.4195008   -.2281899
         cap |   2.104883   .0440642    47.77   0.000     2.018417    2.191349
       _cons |   10.35982   1.202309     8.62   0.000     8.000557    12.71908
------------------------------------------------------------------------------

. ereturn list
scalars:
                  e(N) =  1031
               e(df_m) =  2
               e(df_r) =  1028
                  e(F) =  1160.711019312048
                 e(r2) =  .6930813769821942
               e(rmse) =  8.83656783747737
                e(mss) =  181268.0800475577
                e(rss) =  80271.30921843699
               e(r2_a) =  .6924842590385798
                 e(ll) =  -3707.867843699609
               e(ll_0) =  -4316.762338658647
macros:
              e(title) : "Linear regression"
             e(depvar) : "emp"
                e(cmd) : "regress"
         e(properties) : "b V"
            e(predict) : "regres_p"
              e(model) : "ols"
          e(estat_cmd) : "regress_estat"
matrices:
                  e(b) :  1 x 3
                  e(V) :  3 x 3
functions:
             e(sample)

. local regressors: colnames e(b)

. display "Regressors: `regressors'"
Regressors: wage cap _cons
```

Two particularly useful scalars on this list are `e(df_m)` and `e(df_r)`: the *model* and *residual* degrees of freedom, respectively—the numerator and denominator d.f. for `e(F)`. The `e(rmse)` allows you to retrieve the `Root MSE` of the equation. Two of the scalars do not appear in the printed output: `e(ll) and e(ll_0)`, the likelihood function evaluated for the fitted model and for the null model, respectively.[16] Although the name of the response variable is available in macro `e(depvar)`, the names of the regressors are not shown here. They may be retrieved from the matrix `e(b)`, as shown in the example.

16. For OLS regression with a constant term, the null model is that considered by the ANOVA F: the intercept-only model with all slope coefficients set equal to zero.

Since the estimated parameters are returned in a $1 \times k$ row vector, the variable names are column names of that matrix.

Many official Stata commands, as well as many user-written routines, use the information available from `ereturn list`. How can a command like `estat ovtest` (see [R] **regress postestimation**), as described in section 5.2.7, compute the necessary quantities after `regress`? It can retrieve all relevant information—the names of the regressors, dependent variable, and the net effect of all `if` *exp* or `in` *range* conditions (from `e(sample)`)—from the results left behind as e-class scalars, macros, matrices, or functions by the e-class command. Any do-file you write can perform the same magic if you use `ereturn list` to find the names of each quantity left behind for your use and store the results you need in local macros or scalars immediately after the e-class command. As noted above, retaining scalars as scalars helps to maintain full precision. You should not store scalar quantities in Stata variables unless there is good reason to do so.

The e-class commands may be followed by any of the `estimates` suite of commands described in section 4.3.6. Estimates may be saved in sets, manipulated, and combined in tabular form, as described in section 4.4. Most estimation commands can be followed by any of the `estat` commands, which generate postestimation statistics. For instance, `estat vce` will display the variance–covariance matrix of the estimated parameters (`e(V)`) with flexibility over its formatting. `estat ic` computes Akaike's information criterion (AIC) and Schwarz's Bayesian information criteria (BIC). Each estimation command documents the commands that may be used following estimation. For example, [R] **regress postestimation** describes commands that may be given after `regress`. Some of these commands are types of `estat`, whereas others are standard postestimation commands, such as `predict`, `test`, and `mfx`.

B.6 The program and syntax statements

This section discusses the basics of a more ambitious task: writing your own ado-file, or Stata *command*. The distinction between our prior examples of do-file construction and an ado-file is that if you have written `myprog.do`, you run it with the Stata command `do myprog`. But if you have written `myrealprog.ado`, you may execute it as the Stata command `myrealprog` as long as your new command is defined on the ado-path.

There are more profound differences. Ado-file programs may accept *arguments* in the form of a *varlist*, `if` *exp* or `in` *range* conditions, or options. Nevertheless, we do not have to go far beyond the do-file examples above to define a new Stata command. We learned above that the `summarize` command does not compute the standard error of the mean. We might need that quantity for several variables, and despite other ways of computing it with existing commands, let's write a program to do so. Here we define the program in a do-file. In practice, we would place the program in its own file, `semean.ado`:

```
. capture program drop semean
. *! semean  v1.0.1  CFBaum  04aug2005
. program define semean, rclass
  1.          version 9.0
  2.          syntax varlist(max=1 numeric)
  3.          quietly summarize `varlist'
  4.          scalar semean = r(sd)/sqrt(r(N))
  5.          display _n "Mean of `varlist' = " r(mean) " S.E. = " semean
  6.          return scalar semean = semean
  7.          return scalar mean = r(mean)
  8.          return local var `varlist'
  9. end
. use http://www.stata-press.com/data/imeus/abdata, clear
. semean emp
Mean of emp = 7.891677 S.E. = .49627295
. return list
scalars:
              r(mean) =  7.891677013539667
            r(semean) =  .4962729540865196
macros:
               r(var) : "emp"
```

We start with a `capture program drop` *progname* command. Once a program has been loaded into Stata's memory, it is usually retained for the duration of the session. Since we will be repeatedly defining our program during its development, we want to make sure that we're working with the latest version. The following comment line starting with `*!` (termed *star-bang* in geekish) is a special comment that will show up in the `file` command. It is always a good idea to document an ado-file with a sequence number, author name, and date.

The `program` statement identifies the program name as `semean`. We have checked to see that `semean` is not the name of an existing Stata command. Since `findit semean` locates no program by that name, the name is not used by an official Stata command, a routine in the *Stata Technical Bulletin* or *Stata Journal*, or by any `ssc` routine. We define the program as `rclass`. Unless a program is defined as `rclass` or `eclass`, it cannot return values. The following `version` line states that the ado-file requires Stata 9 and ensures that the program will obey Stata 9 syntax when executed by Stata 10 or Stata 11.

The following line, `syntax`, allows a Stata program to parse its command line and extract the program's arguments for use within the program. In this simple example, we use only one element of `syntax`: specifying that the program has a mandatory *varlist* with at most one numeric element. Stata will enforce the constraint that one name appear on the command line referring to an existing numeric variable. The following lines echo those of the do-file example above by computing the scalar `semean` as the standard error of the mean. The following lines use `return` to place two scalars (`semean` and `mean`) and one macro (the variable name) in the return array.

This is all well and good, but to be useful, a statistical command should accept `if` *exp* and `in` *range* qualifiers. We might also want to use this program as a calculator, without printed output. We could always invoke it `quietly`, but an option to suppress output would be useful. Not much work is needed to add these useful features to our program. The definition of `if` *exp* and `in` *range* qualifiers and program options is all handled by the `syntax` statement. In the improved program, `[if]` and `[in]` denote that each of these qualifiers may be used. Square brackets `[]` in `syntax` signify an optional component of the command. The `[, noPRInt]` indicates that the command has a "noprint" option and that it is truly optional (you can define nonoptional or required options on a Stata command). Here is the revised program:

```
. capture program drop semean
. *! semean  v1.0.2  CFBaum  04aug2005
. program define semean, rclass
  1.         version 9.0
  2.         syntax varlist(max=1 numeric) [if] [in] [, noPRInt]
  3.         marksample touse
  4.         quietly summarize `varlist' if `touse'
  5.         scalar semean = r(sd)/sqrt(r(N))
  6.         if ("`print'" != "noprint") {
  7.                 display _n "Mean of `varlist' = " r(mean)
>               " S.E. = " semean
  8.         }
  9.         return scalar semean = semean
 10.         return scalar mean = r(mean)
 11.         return scalar N = r(N)
 12.         return local var `varlist'
 13. end
```

Since with an `if` *exp* or `in` *range* qualifier, something less than the full sample will be analyzed, we have returned `r(N)` to indicate the sample size used in the computations. The `marksample touse` command makes the `if` *exp* or `in` *range* qualifier operative if one was given on the command line. The command marks those observations that should enter the computations in an indicator variable `touse`, equal to 1 for the desired observations. The `touse` variable is a *tempvar*, or temporary variable, which like a local macro will disappear when the ado-file ends. You can explicitly create these temporary variables with the `tempvar` command. When you need a variable within a program, use a `tempvar` to avoid possible name conflicts with the contents of the dataset. Since the variable is temporary, we refer to it as we would a local macro as `` `touse' ``, which is an alias to its internal (arbitrary) name. We must add `` if `touse' `` to each statement in the program that works with the input *varlist*. Here we need only modify the `summarize` statement.

(*Continued on next page*)

Let's try out the revised program using the abdata dataset:

```
. semean emp
Mean of emp = 7.891677 S.E. = .49627295
. return list
scalars:
                  r(N) =  1031
               r(mean) =  7.891677013539667
             r(semean) =  .4962729540865196
macros:
                r(var) : "emp"
. semean emp if year < 1982, noprint
. return list
scalars:
                  r(N) =  778
               r(mean) =  8.579679950573757
             r(semean) =  .6023535944792725
macros:
                r(var) : "emp"
```

The if *exp* qualifier works, and we can use noprint to suppress the printed output.

Two other features would be useful in this program. First, we would like it to be *byable*: to permit its use with a by *varlist*: prefix. Since we are creating no new variables with this program, we can make the program byable just by adding byable(recall) to the program statement (see [P] **byable** for details). Second, we might like to use time-series operators (L., D., F.) with our program. Adding the ts specifier to varlist will enable that. The improved program becomes

```
. capture program drop semean
. *! semean  v1.0.3  CFBaum  04aug2005
. program define semean, rclass byable(recall) sortpreserve
  1.          version 9.0
  2.          syntax varlist(max=1 ts numeric) [if] [in] [, noPRInt]
  3.          marksample touse
  4.          quietly summarize `varlist' if `touse'
  5.          scalar semean = r(sd)/sqrt(r(N))
  6.          if ("`print'" != "noprint") {
  7.                  display _n "Mean of `varlist' = " r(mean)
>              " S.E. = " semean
  8.          }
  9.          return scalar semean = semean
 10.          return scalar mean = r(mean)
 11.          return scalar N = r(N)
 12.          return local var `varlist'
 13. end
```

We can try out the new byable feature, first using an if *exp* to calculate for one year and then checking to see that the same result appears under the control of by *varlist*:

```
. semean D.emp
Mean of D.emp = -.30018408 S.E. = .0677383
. semean emp if year == 1982
Mean of emp = 6.9304857 S.E. = 1.2245105
. by year, sort: semean emp
```

```
-> year = 1976
Mean of emp = 9.8449251 S.E. = 2.1021706
```

```
-> year = 1977
Mean of emp = 8.5351159 S.E. = 1.393463
```

```
-> year = 1978
Mean of emp = 8.6443428 S.E. = 1.3930028
```

```
-> year = 1979
Mean of emp = 8.7162357 S.E. = 1.4311206
```

```
-> year = 1980
Mean of emp = 8.5576715 S.E. = 1.4611882
```

```
-> year = 1981
Mean of emp = 7.7214 S.E. = 1.3467025
```

```
-> year = 1982
Mean of emp = 6.9304857 S.E. = 1.2245105
```

```
-> year = 1983
Mean of emp = 5.2992564 S.E. = 1.3286027
```

```
-> year = 1984
Mean of emp = 2.2205143 S.E. = .48380791
```

Finally, for pedagogical purposes, I demonstrate how to add an interesting capability to the program: the ability to operate on a transformation of the *varlist* without first generating that variable.[17] We use the `tempvar` statement to allocate a temporary variable, `target`, which will be equated to the *varlist* in the absence of the `function()` argument or that function of *varlist* if `function()` is specified. The local macro `tgt` is used to store the *target* of the command and is used later to display the variable of interest and the returned local macro `r(var)`. We place the `if 'touse'` qualifier on the `generate` statement and `capture` the result of that statement to catch any errors. For instance, the user might specify an undefined function. The `_rc` (return code) is tested for a nonzero value, which will trap an error in the `generate` command. The revised program reads

17. This task mimics the behavior of Stata's time-series operators, which allow you to specify `D.emp` without explicitly creating that variable in your dataset.

```
. capture program drop semean
. *! semean  v1.1.0  CFBaum  04aug2005
. program define semean, rclass byable(recall) sortpreserve
  1.         version 9.0
  2.         syntax varlist(max=1 ts numeric) [if] [in]
>               [, noPRInt FUNCtion(string)]
  3.         marksample touse
  4.         tempvar target
  5.         if "`function'" == "" {
  6.                 local tgt "`varlist'"
  7.         }
  8.         else {
  9.                 local tgt "`function'(`varlist')"
 10.         }
 11.         capture tsset
 12.         capture generate double `target' = `tgt' if `touse'
 13.         if _rc > 0 {
 14.                 display as err "Error: bad function `tgt'"
 15.                 error 198
 16.                 }
 17.         quietly summarize `target'
 18.         scalar semean = r(sd)/sqrt(r(N))
 19.         if ("`print'" != "noprint") {
 20.                 display _n "Mean of `tgt' = " r(mean)
>               " S.E. = " semean
 21.         }
 22.         return scalar semean = semean
 23.         return scalar mean = r(mean)
 24.         return scalar N = r(N)
 25.         return local var `tgt'
 26. end
```

As the example below demonstrates, the program operates properly when applying a transformation that reduces the sample size. The log of `D.emp` is defined only for positive changes in employment, and most of the 140 firms in this sample suffered declines in employment in 1982.

```
. semean emp
Mean of emp = 7.891677 S.E. = .49627295
. semean emp, func(sqrt)
Mean of sqrt(emp) = 2.1652401 S.E. = .05576835
. semean emp if year==1982, func(log)
Mean of log(emp) = .92474464 S.E. = .11333991
. return list
scalars:
                  r(N) =  140
               r(mean) =  .9247446421128256
             r(semean) =  .1133399069800714
macros:
                r(var) : "log(emp)"
. semean D.emp if year==1982, func(log)
Mean of log(D.emp) = -2.7743942 S.E. = .39944652
```

```
. return list
scalars:
                  r(N) =  22
               r(mean) =  -2.774394169773632
             r(semean) =  .3994465211383764
macros:
                r(var) : "log(D.emp)"
```

The program can now emulate many of the features of an official Stata command while remaining brief. We have only scratched the surface of what you can do in your own ado-file. For instance, many user-written programs generate new variables or perform computations based on the values of options, which may have their own default values. User-written programs may also be used to define additional `egen` functions. Their names (and the file in which they reside) will start with _g: that is, `_gfoo.ado` will define the `foo()` function to `egen`.

Although many Stata users may become familiar with the program and its capabilities without ever writing an ado-file program, others will find that they are constantly rewriting quick-and-dirty code that gets the job done today, with minor variations, to perform a similar task tomorrow. With that epiphany, knowledgeable Stata users will recognize that it is a short leap to becoming more productive by learning how to write their own ado-files, whether or not those programs are of general use or meant to be shared with other Stata users. As suggested earlier, the would-be programmer should investigate StataCorp's NetCourses to formally learn these skills.

B.7 Using Mata functions in Stata programs

This last section briefly introduces the Mata matrix programming language added to Stata in version 9.[18] As Mata's online documentation indicates, you can use Mata in a purely interactive mode like other matrix languages, such as GAUSS, MATLAB, or Ox. However, the greatest benefit of Mata for applied economists is that it can speed up Stata programs that take advantage of its facilities by executing *compiled* code rather than the *interpreted* commands of an ado-file. There is no difference in computing speed between executing commands in a do-file and the same commands in an ado-file. But with Mata, functions can be compiled—a one-time task—and the resulting bytecode will execute many times faster than similar commands in the interpreted language. Two widely used SSC routines contributed by Stata users—Leuven and Sianesi's `psmatch2` for propensity score matching[19] and Roodman's `xtabond2` for extended Arellano–Bond estimation[20]—have been rewritten by their authors to take advantage of Mata's efficiency and speed.

Given that many common econometric procedures can be written concisely in matrix notation, Mata may make it easier to program a particular procedure for use in Stata.

18. If you are interested in using Mata, you should follow William Gould's "Mata Matters" columns in the *Stata Journal*, beginning in volume 5, number 3.
19. See http://ideas.repec.org/c/boc/bocode/s432001.html.
20. See http://ideas.repec.org/c/boc/bocode/s435901.html.

Code written in other languages—Fortran, C, or one of the matrix languages mentioned above—can readily be translated into Mata; for an excellent example, see Gould (2005). The suite of Mata functions contains the standard set of linear algebra functions (e.g., LAPACK and EISPACK) as well as all the matrix handling capabilities available in other programming environments.

Mata does not make a Stata program stored in an ado-file obsolete. Most new features are added to Stata as ado-file programs serving as *wrappers* for one or more Mata functions, so you can use each language—ado-file code and Mata—for what it does best. The high-level parsing functions available with the `syntax` statement provide important tools for creating Stata commands with an elegant user interface, error checking, and the like. Once the task is assembled in ado-file code, it can then be passed to Mata for rapid processing. As we shall see, Mata can both access Stata's variables, macros, and scalars and place results back into those objects.

We now construct a Mata function that solves a data transformation problem posed by a Statalist contributor.[21] This user had a set of variables, each containing N observations, and wanted to create a new set of variables in the same dataset. Each new variable's observations are the average of two consecutive observations. Thus the average of observations 1 and 2 of x becomes observation 1 of the new variable, the average of observations 3 and 4 becomes observation 2, and so on. If this transformation could be done, discarding the original data, the `collapse` statement could be used after defining an indicator variable that identifies the subgroups. Alternatively, the `egen group()` function would generate the averages but would align them with the even-numbered observations with missing values interspersed.

In contrast, the Mata function we construct will perform this task as originally specified. Since there may be limited use for such a specific tool, the function is designed to solve a more general problem. When working with time-series data, we often want to construct averages of p consecutive values of a variable as consecutive observations. We may want to juxtapose quarterly national income data with the average inflation rate during the quarter, with inflation reported monthly. Likewise, we may want to convert monthly data to annual format, quarterly data to annual format, or business-daily data to weekly data. Any of these tasks is generically similar to the more specific need expressed on Statalist.[22]

To meet this need, we write a Mata function named `averageper()`, which is called with three *arguments*: the name of a Stata variable, the number of consecutive periods (p) to be averaged, and the name of a `touse` variable. As discussed in section B.6, the `touse` construct allows us to specify which observations to include in a computation, as expressed by `if` *exp* and `in` *range* conditions. Here is the `averageper()` function:

---

21. See http://www.hsph.harvard.edu/cgi-bin/lwgate/STATALIST/archives/statalist.0507/Subject/article-296.html.

22. For the official Stata date frequencies (see section 2.2.5) this problem has already been solved by the author's `tscollap` routine (Baum 2000). However, that routine (like `collapse`) destroys the original higher-frequency data, requiring another `merge` step to emulate the Mata routine developed below.

```
. * define the Mata averageper function
. mata:
------------------------------------------------- mata (type end to exit) -----
:         void averageper(string scalar vname, real scalar per, string scalar t
> ouse)
>         {
> // define objects used in function
>         string scalar vnew
>         real scalar divisor
>         real scalar resindex
>         real matrix v1
>         real matrix v3
> // construct the new variable name from original name and per
>         vnew = vname + "A" + strofreal(per)
> // access the Stata variable, honoring any if or in conditions
>         v1=st_data(.,vname,touse)
> // verify that per is appropriate
>         if (per<=0 | per > rows(v1)) {
>                 _error("per must be > 0 and < nobs.")
>                 }
> // verify that nobs is a multiple of per
>         if (mod(rows(v1),per) != 0) {
>                 _error("nobs must be a multiple of per.")
>                 }
> // reshape the column vector into nobs/per rows and per columns
> // postmultiply by a per-element row vector with values 1/per
>         divisor = 1/per
>         v3 = colshape(v1',per) * J(per,1,divisor)
> // add the new variable to the current Stata data set
>         resindex = st_addvar("float",vnew)
> // store the calculated values in the new Stata variable
>         st_store((1,rows(v3)),resindex,v3)
>         }
: end
-------------------------------------------------------------------------------
```

The `mata:` command invokes Mata, allowing us to give Mata commands. Following the function definition, we declare several objects used within the function that are local to the function. These declarations are not required, but they are always a good idea. We first construct a new variable name from the `vname` passed to the routine with `A` and the value of `per` concatenated. Thus, if we specify variable `price` with a `per` of 3, the new variable name (`vnew`) will be `priceA3`.

We then use Mata's `st_data()` function to access that Stata variable and copy its values to Mata matrix `v1`. Mata also provides an `st_view()` function, which would allow us to create a view of that variable, or any subset of Stata's variables. In this function, we must use `st_data()`. This function will honor any `if` *exp* and `in` *range* conditions that have been stored in `touse`.

The following lines check for various errors. Is `per` less than one or greater than N? The function also verifies that N, the number of observations available for `v1`, is an even multiple of `per`. If any of these conditions is violated, the function will abort.

After these checks have been passed, we turn to the heart of the computation. The solution to the problem is a type of reshape. We can reshape the N-element column vector into which `vname` has been copied into a matrix `v3` with q rows and `per` columns. $q = N/\texttt{per}$ is the number of averaged observations that will result from the computation. If we postmultiplied the transpose of matrix `v3` by a `per`-element column vector ι, we would compute the sum over the `per` values for each new observation. The average would be $1/\texttt{per}$ times that vector. Thus we define the column vector's elements as `divisor = 1/per`. The resulting column vector, `v3`, is our averaged series of length q.

To illustrate, let x be the N elements of the Stata variable. Each element becomes a column of the reshaped matrix:

$$
\begin{pmatrix} x_1 \\ x_2 \\ \vdots \\ x_{\text{per}} \\ \\ x_{\text{per}+1} \\ x_{\text{per}+2} \\ \vdots \\ x_{2\text{per}} \\ \\ x_{2\text{per}+1} \\ x_{2\text{per}+2} \\ \vdots \\ \vdots \\ x_N \end{pmatrix} \Longrightarrow \begin{pmatrix} x_{1,1} & x_{1,2} & \cdots & x_{1,q} \\ x_{2,1} & x_{2,2} & \cdots & x_{2,q} \\ \vdots & & \ddots & \vdots \\ x_{\text{per},1} & x_{\text{per},2} & \cdots & x_{\text{per},q} \end{pmatrix}
$$

We then transpose the reshaped matrix and postmultiply by a `per`-element column vector to construct the `per`-period average:

$$
\begin{pmatrix} x_{1,1} & x_{2,1} & \cdots & x_{\text{per},1} \\ x_{1,2} & x_{2,2} & \cdots & x_{\text{per},2} \\ \vdots & & \ddots & \vdots \\ x_{1,q} & x_{2,q} & \cdots & x_{\text{per},q} \end{pmatrix} \begin{pmatrix} \frac{1}{\text{per}} \\ \frac{1}{\text{per}} \\ \vdots \\ \frac{1}{\text{per}} \end{pmatrix} = \begin{pmatrix} x_1^* \\ x_2^* \\ \vdots \\ x_q^* \end{pmatrix}
$$

The column vector x^*, labeled `v3` in the Mata function, contains the averages of each `per` element of the original Stata variable.

Finally, we attempt to add the variable `vnew`, declared as a `float`, to the Stata dataset with `st_addvar()`. This attempt will fail if this variable name already exists. If we are successful, we store the q elements of `v3` in the values of the new variable with the `st_store()` function, defining other elements of `v3` as missing. We then exit Mata back to Stata with the `end` command.

Stata will interpret this Mata code and flag any syntax errors. None having appeared, we are ready to *compile* this function and store it in object-code form. When you call a

Mata function, Stata looks on the ado-path for the object code of that function, here, `averageper.mo`. The `mata mosave` command creates (or revises) the object file:

```
. // save the compiled averageper function
. mata: mata mosave averageper(), replace
(file averageper.mo created)
```

Now that our Mata function has been completed, we can write the ado-file wrapper routine. Strictly speaking, this task is not necessary; we could call the Mata function directly from Stata as `mata: averageper(...)`. But we want to take advantage of `syntax` and other features of the ado-file language described in section B.6. Our program defining the `averageper` command to Stata is simple. We specify that one numeric `varname` must be provided, as well as the required `per()` option, necessarily an integer. We use `marksample` to handle `if` *exp* or `in` *range* conditions. The three arguments required by the Mata function are passed: the `varlist` and `touse` as strings, whereas the `per` is passed as a numeric value.

```
. * define the Stata averageper wrapper command
. *! averageper 1.0.0  05aug2005 CFBaum
. program averageper, rclass
  1.         version 9
  2.         syntax varlist(max=1 numeric) [if] [in], per(integer)
  3. // honor if and in conditions if provided
.         marksample touse
  4. // pass the variable name, per, and touse to the Mata function
.         mata: averageper("`varlist'",`per',"`touse'")
  5. end
```

Or we could have placed the Mata function definition in our ado-file rather than placing it in a separate `.mata` file and creating a `.mo` object file. If it is included in the ado-file, the Mata code will be compiled the first time that the `averageper` ado-file is called in your Stata session. Subsequent calls to the ado-file will not require compilation of the Mata function (or functions). An exchange on Statalist[23] suggests that if the Mata functions amount to fewer than 2,000 lines of code, incorporating them in ado-files will yield acceptable performance. However, if your Mata functions are to be called by several different ado-files, you will want to store their compiled code in `.mo` object files rather than duplicating the Mata code.

We now are ready to test the `averageper` command. We use the *Stata Time-Series Reference Manual* dataset `urates`, containing monthly unemployment rates for several U.S. states. We apply `averageper` to one variable—`tenn`, the Tennessee unemployment rate—using both `per(3)` and `per(12)` to calculate quarterly and annual averages, respectively.

23. See http://www.hsph.harvard.edu/cgi-bin/lwgate/STATALIST/archives/statalist.0508/date/article-358.html.

```
. use http://www.stata-press.com/data/imeus/urates, clear
. tsset
        time variable:  t, 1978m1 to 2003m12
. describe tenn
              storage  display     value
variable name   type   format      label      variable label
-------------------------------------------------------------------------
tenn            float  %9.0g
. averageper tenn, per(3)  // calculate quarterly averages
. averageper tenn, per(12) // calculate annual averages
. summarize tenn*
    Variable |       Obs        Mean    Std. Dev.       Min        Max
-------------+--------------------------------------------------------
        tenn |       312    6.339744    2.075308        3.7       12.8
      tennA3 |       104    6.339744    2.078555   3.766667   12.56667
     tennA12 |        26    6.339744    2.078075   3.908333   11.83333
```

The `summarize` command shows that the original series and two new series have identical means, which they must. To display how the new variables appear in the Stata data matrix, we construct two date variables with the `tsmktim` command (Baum and Wiggins 2000) and list the first 12 observations of Tennessee's data. You can verify that the routine is computing the correct quarterly and annual averages.

```
. tsmktim quarter, start(1978q1) // create quarterly calendar var
        time variable:  quarter, 1978q1 to 2055q4
. tsmktim year, start(1978)      // create annual calendar var
        time variable:  year, 1978 to 2289
. list t tenn quarter tennA3 year tennA12 in 1/12, sep(3)
```

	t	tenn	quarter	tennA3	year	tennA12
1.	1978m1	5.9	1978q1	5.966667	1978	5.8
2.	1978m2	5.9	1978q2	5.766667	1979	5.791667
3.	1978m3	6.1	1978q3	5.733333	1980	7.3
4.	1978m4	5.9	1978q4	5.733333	1981	9.083333
5.	1978m5	5.8	1979q1	5.733333	1982	11.83333
6.	1978m6	5.6	1979q2	5.7	1983	11.45833
7.	1978m7	5.7	1979q3	5.733333	1984	8.55
8.	1978m8	5.7	1979q4	6	1985	7.983334
9.	1978m9	5.8	1980q1	6.166667	1986	8.041667
10.	1978m10	5.9	1980q2	7.066667	1987	6.591667
11.	1978m11	5.7	1980q3	8	1988	5.775
12.	1978m12	5.6	1980q4	7.966667	1989	5.108333

Finally, we return to the original Statalist question: how can we perform this transformation for a whole set of variables? Rather than generalizing `averageper` to handle multiple variables, we just use a `foreach` loop over the variables:

```
. foreach v of varlist tenn-arkansas {
  2.         averageper `v', per(3)
  3. }

. summarize illinois*

    Variable |       Obs        Mean    Std. Dev.       Min        Max
-------------+--------------------------------------------------------
    illinois |       312    6.865064    1.965563        4.1       12.9
  illinoisA3 |       104    6.865064    1.964652        4.2   12.76667
```

Although we could do much to improve this routine, it gets the job done efficiently. This short excursion into the new world of Mata programming should give you some idea of the powerful capabilities added to Stata by this versatile matrix language. If you are going to write Mata functions, you should have a copy of the *Mata Reference Manual*.

References

Akaike, H. 1974. A new look at statistical model identification. *IEEE Transactions on Automatic Control* 19: 716–722.

Anderson, T. W. 1984. *Introduction to Multivariate Statistical Analysis.* New York: Wiley.

Arellano, M., and S. Bond. 1991. Some tests of specification in panel data: Monte Carlo evidence and an application to employment equations. *Review of Economic Studies* 58: 277–297.

Arellano, M., and O. Bover. 1995. Another look at the instrumental variables estimation of error components models. *Journal of Econometrics* 68: 29–52.

Bai, J., and P. Perron. 2003. Computation and analysis of multiple structural change models. *Journal of Applied Econometrics* 18: 1–22.

Baltagi, B. H. 2001. *Econometric Analysis of Panel Data.* 2nd ed. New York: Wiley.

Bartus, T. 2005. Estimation of marginal effects using margeff. *Stata Journal* 5: 309–329.

Basmann, R. 1960. On finite sample distributions of generalized classical linear identifiability test statistics. *Journal of the American Statistical Association* 55: 650–659.

Baum, C. F. 2000. sts17: Compacting time series data. *Stata Technical Bulletin* 57: 44–46. Reprinted in *Stata Technical Bulletin Reprints*, vol. 10, pp. 369–370. College Station, TX: Stata Press.

———. 2001. Residual diagnostics for cross-section time series regression models. *Stata Journal* 1: 101–104.

———. 2005. Stata: The language of choice for time series analysis? *Stata Journal* 5: 46–63.

Baum, C. F., N. J. Cox, and V. Wiggins. 2000. sg137: Tests for heteroskedasticity in regression error distribution. *Stata Technical Bulletin* 55: 15–17. Reprinted in *Stata Technical Bulletin Reprints*, vol. 10, pp. 147–149. College Station, TX: Stata Press.

Baum, C. F., M. E. Schaffer, and S. Stillman. 2003. Instrumental variables and GMM: Estimation and testing. *Stata Journal* 3: 1–31.

———. 2005. Software update: st0030_2. Instrumental variables and GMM: Estimation and testing. *Stata Journal* 5: 607.

Baum, C. F., and V. Wiggins. 2000. dm81: Utility for time series data. *Stata Technical Bulletin* 57: 2–4. Reprinted in *Stata Technical Bulletin Reprints*, vol. 10, pp. 29–30. College Station, TX: Stata Press.

Belsley, D. A. 1991. *Conditioning Diagnostics: Collinearity and Weak Data in Regression*. New York: Wiley.

Belsley, D. A., E. Kuh, and R. E. Welsch. 1980. *Regression Diagnostics: Identifying Influential Data and Sources of Collinearity*. New York: Wiley.

Blackburn, M., and D. Neumark. 1992. Unobserved ability, efficiency wages, and interindustry wage differentials. *Quarterly Journal of Economics* 107: 1421–1436.

Blundell, R., and S. Bond. 1998. Initial conditions and moment restrictions in dynamic panel data models. *Journal of Econometrics* 87: 115–143.

Bond, S. 2002. Dynamic panel data models: a guide to microdata methods and practice. Technical Report CWP09/02, Centre for Microdata Methods and Practice, Institute for Fiscal Studies.

Bound, J., D. A. Jaeger, and R. Baker. 1995. Problems with instrumental variables estimation when the correlation between the instruments and the endogenous explanatory variable is weak. *Journal of the American Statistical Association* 90: 443–450.

Box, G. E. P., and D. A. Pierce. 1970. Distribution of residual autocorrelations in autoregressive–integrated moving average time series models. *Journal of the American Statistical Association* 65: 1509–1526.

Breusch, T. S., and A. R. Pagan. 1979. A simple test for heteroskedasticity and random coefficient variation. *Econometrica* 47: 1287–1294.

———. 1980. The Lagrange Multiplier test and its applications to model specification in econometrics. *Review of Economic Studies* 47: 239–253.

Brown, M., and A. Forsythe. 1992. Robust test for the equality of variances. *Journal of the American Statistical Association* 69: 364–367.

Chao, J. C., and N. R. Swanson. 2005. Consistent estimation with a large number of weak instruments. *Econometrica* 73: 1673–1692.

Cochrane, D., and G. H. Orcutt. 1949. Application of least-squares regression to relationships containing autocorrelated error terms. *Journal of the American Statistical Association* 44: 32–61.

Cook, R. D., and S. Weisberg. 1983. Diagnostics for heteroscedasticity in regression. *Biometrika* 70: 1–10.

———. 1994. *An Introduction to Regression Graphics*. New York: Wiley.

Cox, D. R. 1961. Tests of separate families of hypotheses. In *Proceedings of the Fourth Berkeley Symposium on Mathematical Statistics and Probability*, vol. 1. Berkeley, CA: University of California Press.

———. 1962. Further results on tests of separate families of hypotheses. *Journal of the Royal Statistical Society, Series B* 24: 406–424.

Cox, N. J. 1999. dm70: Extensions to generate, extended. *Stata Technical Bulletin* 50: 9–17. Reprinted in *Stata Technical Bulletin Reprints*, vol. 9, pp. 34–45. College Station, TX: Stata Press.

———. 2000. dm70.1: Extensions to generate, extended: corrections. *Stata Technical Bulletin* 57: 2. Reprinted in *Stata Technical Bulletin Reprints*, vol. 10, p. 9. College Station, TX: Stata Press.

———. 2002a. Speaking Stata: How to face lists with fortitude. *Stata Journal* 2: 202–222.

———. 2002b. Speaking Stata: On numbers and strings. *Stata Journal* 2: 314–329.

———. 2003. Speaking Stata: Problems with lists. *Stata Journal* 3: 185–202.

Cox, N. J., and J. Weesie. 2001. dm88: Renaming variables, multiply and systematically. *Stata Technical Bulletin* 60: 4–6. Reprinted in *Stata Technical Bulletin Reprints*, vol. 10, pp. 41–44. College Station, TX: Stata Press.

———. 2005. Software update: dm88_1: Renaming variables, multiply and systematically. *Stata Journal* 5: 607.

Cragg, J. G., and S. G. Donald. 1993. Testing identifiability and specification in instrumental variables models. *Econometric Theory* 9: 222–240.

Cumby, R. E., J. Huizinga, and M. Obstfeld. 1983. Two-step two-stage least squares estimation in models with rational expectations. *Journal of Econometrics* 21: 333–355.

Davidson, R., and J. MacKinnon. 1981. Several tests for model specification in the presence of alternative hypotheses. *Econometrica* 49: 781–793.

Davidson, R., and J. G. MacKinnon. 1993. *Estimation and Inference in Econometrics*. 2nd ed. New York: Oxford University Press.

———. 2004. *Econometric Theory and Methods*. New York: Oxford University Press.

De Hoyos, R. E., and V. Sarafidis. 2006. XTCSD: Stata module to test for cross-sectional dependence in panel data models.
http://www.econ.cam.ac.uk/phd/red29/research.htm.

Delwiche, L. D., and S. J. Slaughter. 1998. *The Little SAS Book*. 2nd ed. Cary, NC: SAS Insitute.

Durbin, J. 1970. Testing for serial correlation in least squares regression when some of the regressors are lagged dependent variables. *Econometrica* 38: 410–421.

Durbin, J., and G. Watson. 1950. Testing for serial correlation in least squares regression I. *Biometrika* 37: 409–428.

Eichenbaum, M. S., L. P. Hansen, and K. J. Singleton. 1988. A time series analysis of representative agent models of consumption and leisure. *Quarterly Journal of Economics* 103: 51–78.

Godfrey, L. G. 1978. Testing for multiplicative heteroskedasticity. *Journal of Econometrics* 8: 227–236.

———. 1988. *Misspecification Tests in Econometrics: The Lagrange Multiplier Principle and Other Approaches.* Cambridge: Cambridge University Press.

———. 1999. Instrument relevance in multivariate linear models. *Review of Economics and Statistics* 81: 550–552.

Gould, W. 2005. Mata Matters: Translating Fortran. *Stata Journal* 5: 421–441.

Gould, W., J. Pitblado, and W. Sribney. 2006. *Maximum Likelihood Estimation with Stata.* 3rd ed. College Station, TX: Stata Press.

Greene, W. H. 2000. *Econometric Analysis.* 4th ed. Upper Saddle River, NJ: Prentice–Hall.

———. 2003. *Econometric Analysis.* 5th ed. Upper Saddle River, NJ: Prentice–Hall.

Griliches, Z. 1976. Wages of very young men. *Journal of Political Economy* 84: S69–S85.

Hahn, J., and J. Hausman. 2002a. A new specification test for the validity of instrumental variables. *Econometrica* 70: 163–189.

———. 2002b. Notes on bias in estimators for simultaneous equation models. *Economics Letters* 75: 237–241.

Hall, A. R., and F. P. M. Peixe. 2000. A consistent method for the selection of relevant instruments. In *Contributed Papers, Econometric Society World Congress 2000.* http://econpapers.repec.org/paper/ecmwc2000/0790.htm: EconPapers.

Hall, A. R., G. D. Rudebusch, and D. W. Wilcox. 1996. Judging instrument relevance in instrumental variables estimation. *International Economic Review* 37: 283–298.

Hansen, L. 1982. Large sample properties of generalized method of moments estimators. *Econometrica* 50: 1029–1054.

Hardle, W. 1990. *Applied Nonparametric Regression.* Cambridge: Cambridge University Press.

Hausman, J. 1978. Specification tests in econometrics. *Econometrica* 46: 1251–1271.

Hausman, J. A., and W. E. Taylor. 1981. Panel data and unobservable individual effects. *Econometrica* 49: 1377–1398.

Hayashi, F. 2000. *Econometrics.* Princeton, NJ: Princeton University Press.

Heckman, J. 1976. The common structure of statistical models of truncation, sample selection, and limited dependent variables and a simple estimator for such models. *Annals of Economic and Social Measurement* 5: 475–492.

———. 1979. Sample selection bias as a specification error. *Econometrica* 47: 153–161.

Hildreth, C., and J. Y. Lu. 1960. Demand relations with autocorrelated disturbances. Technical Report 276, Michigan State University Agricultural Experiment Station Technical Bulletin.

Hill, R. C., and L. C. Adkins. 2003. Collinearity. In *A Companion to Theoretical Econometrics*, ed. B. H. Baltagi. Malden, MA: Blackwell Publishing.

Hsiao, C. 1986. *Analysis of Panel Data.* New York: Cambridge University Press.

Huber, P. J. 1967. The behavior of maximum likelihood estimates under non-standard conditions. In *Proceedings of the Fifth Berkeley Symposium in Mathematical Statistics and Probability*, vol. 1, 221–233. Berkeley, CA: University of California Press.

Jann, B. 2005. Making regression tables from stored estimates. *Stata Journal* 5: 288–308.

Johnston, J., and J. DiNardo. 1997. *Econometric Methods.* 4th ed. New York: McGraw–Hill.

Judge, G. G., R. C. Hill, W. E. Griffiths, H. Lütkepohl, and T. C. Lee. 1985. *The Theory and Practice of Econometrics.* 2nd ed. New York: Wiley.

Koenker, R. 1981. A note on Studentizing a test for heteroskedasticity. *Journal of Econometrics* 17: 107–112.

Levene, H. 1960. Robust tests for equality of variances. In *Contributions to Probability and Statistics*, ed. I. Olkin, 278–292. Palo Alto, CA: Stanford University Press.

Ljung, G. M., and G. E. P. Box. 1979. On a measure of lack of fit in time series models. *Biometrika* 65: 297–303.

Long, J. S., and J. Freese. 2006. *Regression Models for Categorical and Limited Dependent Variables using Stata.* 2nd ed. College Station, TX: Stata Press.

McFadden, D. 1974. The measurement of urban travel demand. *Journal of Public Economics* 3: 303–328.

Mitchell, M. 2004. *A Visual Guide to Stata Graphics.* College Station, TX: Stata Press.

Munnell, A. H., G. Tootell, L. Browne, and J. McEneaney. 1996. Mortgage lending in Boston: Interpreting HMDA Data. *American Economic Review* 86: 25–53.

Newey, W. K., and K. D. West. 1987. A simple, positive semi-definite, heteroskedasticity and autocorrelation consistent covariance matrix. *Econometrica* 55: 703–708.

Nickell, S. 1981. Biases in dynamic models with fixed effects. *Econometrica* 49: 1417–1426.

Pagan, A. R., and D. Hall. 1983. Diagnostic tests as residual analysis. *Econometric Reviews* 2: 159–218.

Pesaran, M. 1974. On the general problem of model selection. *Review of Economic Studies* 41: 153–171.

Pesaran, M., and A. Deaton. 1978. Testing non-nested nonlinear regression models. *Econometrica* 46: 677–694.

Poi, B. P. 2002. From the help desk: Demand system estimation. *Stata Journal* 2: 403–410.

Poirier, D. 1981. Partial observability in bivariate probit models. *Journal of Econometrics* 12: 209–217.

Prais, S. J., and C. B. Winsten. 1954. Trend estimators and serial correlation. Technical Report 383, Cowles Commission Discussion Paper Series.

Ruud, P. A. 2000. *An Introduction to Classical Econometric Theory*. Oxford: Oxford University Press.

Sargan, J. 1958. The estimation of economic relationships using instrumental variables. *Econometrica* 26: 393–415.

Schwarz, G. 1978. Estimating the dimension of a model. *Annals of Statistics* 6: 461–464.

Shea, J. 1997. Instrument relevance in multivariate linear models: A simple measure. *Review of Economics and Statistics* 79: 348–352.

Staiger, D., and J. H. Stock. 1997. Instrumental variables regression with weak instruments. *Econometrica* 65: 557–586.

Stock, J., and M. Watson. 2006. *Introduction to Econometrics*. 2nd ed. Reading, MA: Addison–Wesley.

Stock, J. H., J. H. Wright, and M. Yogo. 2002. A survey of weak instruments and weak identification in generalized method of moments. *Journal of Business and Economic Statistics* 20: 518–529.

Tobin, J. 1958. Estimation of relationships for limited dependent variables. *Econometrica* 26: 24–36.

Van de Ven, W., and B. M. S. Van Pragg. 1981. The demand for deductibles in private health insurance: A probit model with sample selection. *Journal of Econometrics* 17: 229–252.

Welsch, R., and E. Kuh. 1977. Linear regression diagnostics. Technical Report 923–977, Sloan School of Management, MIT.

White, H. 1980. A heteroskedasticity-consistent covariance matrix estimator and a direct test for heteroskedasticity. *Econometrica* 48: 817–838.

———. 1982. Instrumental variables regression with independent observations. *Econometrica* 50: 483–499.

Windmeijer, F. 2005. A finite sample correction for the variance of linear efficient two-step GMM estimators. *Journal of Econometrics* 126: 25–51.

Wooldridge, J. M. 2002. *Econometric Analysis of Cross Section and Panel Data*. Cambridge, MA: MIT Press.

———. 2006. *Introductory Econometrics: A Modern Approach*. 3rd ed. New York: Thomson.

Zellner, A. 1962. An efficient method of estimating seemingly unrelated regressions and tests of aggregation bias. *Journal of the American Statistical Association* 57: 500–509.

Author index

Subject index